# PREFACE

Milič Čapek has devoted his scholarship to the history and philosophy of modern physics. With impeccable care, he has mastered the epistemological and scientific developments by working through the papers, treatises, correspondence of physicists since Kant, and likewise he has put his learning and critical skill into the related philosophical literature. Coming from his original scientific career with a philosophy doctorate from the Charles University in Prague, Čapek has ranged beyond a narrowly defined philosophy of physics into general epistemology of the natural sciences and to the full historical evolution of these matters. He has expounded his views on these matters in a number of articles and, systematically, in his book *The Philosophical Impact of Contemporary Physics*, published in 1961 and reprinted with two new appendices in 1969. His particular gift for many of his readers and students lies in the great period from the mid-nineteenth century through the foundations of the physics and philosophy of the twentieth, and within this spectacular time, Professor Čapek has become a principal expositor and sympathetic critic of the philosophy of Henri Bergson. He joins a distinguished group of scholars – physicists and philosophers – who have been stimulated to some of their most profound and imaginative thought by Bergson's metaphysical and psychological work: Cassirer, Meyerson, de Broglie, Metz, Jankélevitch, Zawirski, and in recent years, Costa de Beauregard, Watanabe, Blanché, and others. Now, with his monograph on *Bergson and Modern Physics*, Professor Čapek has set forth the little-attended and interesting physical doctrine of Bergson in its entirety, and with its systematic connections to actual physics of the time and later. This monograph completes Čapek's earlier papers on these questions; it was anticipated by his noted essay on 'Bergson's Theory of Matter and Modern Physics', *Revue philosophique* 77 (1953) which has recently appeared in English in *Bergson and the Evolution of Physics* (ed. by P. A. Y. Gunter), University of Tennessee Press, Knoxville, 1969, pp. 297–330.

The young Čapek, working as a country schoolteacher, was happily startled by Bergson's cordial letter about Čapek's dissertation, complimenting him for his insight into the significance of Bergson's views on matter and their relations to modern physics. It is our own pleasure to publish the mature reflections and new insights of our dear colleague on the same topic to which he brought a clear perception in his doctoral thesis.

*           *           *

*Boston Studies in the Philosophy of Science* are devoted to symposia, congresses, colloquia, monographs and collected papers in the philosophical foundations of the sciences. Professor Čapek has joined energetically and incisively with the scientists and philosophers who gather together for the discussions at our Colloquia. Some of his contributions appear in *Boston Studies*: 'The Myth of Frozen Passage: the Status of Becoming in the Physical World', vol. II, pp. 441–463; 'Ernst Mach's Biological Theory of Knowledge', vol. V, pp. 400–420; 'Two Types of Continuity' [forthcoming]. His Bergsonian studies not only reflect his own commitment and philosophical labor but also suggest the interest, sensitivity, and stimulation which may be brought to the central tasks of philosophical analysis of the sciences by research into metaphysics, phenomenology, and philosophies of nature.

*Boston Center for the Philosophy*       ROBERT S. COHEN
*of Science, Boston University*       MARX W. WARTOFSKY

BOSTON STUDIES IN THE PHILOSOPHY OF SCIENCE

VOLUME VII

EDITED BY ROBERT S. COHEN AND MARX W. WARTOFSKY

MILIČ ČAPEK

# BERGSON
# AND MODERN PHYSICS

## A REINTERPRETATION AND RE-EVALUATION

D. REIDEL PUBLISHING COMPANY / DORDRECHT-HOLLAND

Library of Congress Catalog Card Number 79–146967

ISBN 90 277 0186 5

Printed in The Netherlands by D. Reidel, Dordrecht

# AUTHOR'S PREFACE

An old proverb says that 'the books have their destinies' (*habent sua fata libelli*) and this is certainly true of this one. The first idea of writing it occurred to me in 1940 when as a student at the Sorbonne I was encouraged by Professor Emile Bréhier to write an article about Bergson and modern physics for *Revue philosophique*. This must have taken place in the late spring since the German offensive was already in a full swing; yet, neither I nor Mr Bréhier expected that Paris would be in Nazi hands within a month. This is why I did not even begin to write the article; the whole project was forgotten for years. In 1950, long after the war was over, I hardly hoped that Professor Bréhier would remember his kind invitation; but he did – and thus the article "La théorie bergsonienne de la matière et la physique moderne" was finally written and appeared in *Revue philosophique* in January 1953. This article together with another one which appeared nearly at the same time in *Revue de métaphysique et de morale* represented a sort of outline for the book to be written. But various circumstances – and various interests – prevented me from working on it. All that I was able to do in this respect was several articles in English and French and my participation in the discussion section "Bergson et la physique" at the centenary *Congrès Bergson* at Paris in 1959. Thus I did not begin to write this book systematically before the fall of 1965 and, while its major part had been finished in the middle of 1968, some parts were added and some re-written as late as the fall of 1970. But this delay proved to be salutary; for although the most decisive and revolutionary changes in physics took place in the first four decades of this century, important discussions about their epistemological and metaphysical significance continued and I was thus able to take them into account. Thus the book in its final form is different from what was originally envisioned as. None of its parts has been previously published except Chapter 2 of Part I which appeared with some insignificant modifications under the title "Ernst Mach's Biological Theory of Knowledge" in *Boston Studies for the Philosophy of Science* vol. V (1969).

In referring to Bergson's writings I use abbreviations, that is, the initial letters in the titles of the English translations; for instance, *C.E.* for *Creative Evolution*, *C.M.* for *The Creative Mind* etc. For the convenience of the reader references are made to the paperback editions of Bergson's works, for all hard cover editions, except the two books just mentioned, are out of print.

Finally, I wish to express my thanks to all those who in various ways made the publication of this book possible: in the first place, to the editors of this series, Professors Robert Cohen and Marx Wartofsky; to all those, besides the editors, who read the parts of the manuscript and made some suggestions; to Boston University for several generous grants; to those departmental secretaries and graduate students, in particular to Mr and Mrs B. Hallen, who helped me to type the manuscript; finally to Professors Robert Cohen and Paul Sagal as well as to my wife who read the proofs. It is needless to say that for the book in its final form the responsibility is entirely my own.

*Boston, May 14, 1971*                                            MILIČ ČAPEK

# INTRODUCTION

When Bergson's views about the nature of matter were formulated in *Matter and Memory*, especially in its fourth chapter, in 1896, they appeared, in contrast to the prevailing classical picture of the physical world, so grotesquely improbable, that they were largely ignored. Even later, after the publication of *Creative Evolution*, when Bergson's philosophy became both fashionable and controversial, his views of matter were rarely analyzed; if they were, they were compared to the irresponsible speculations of German Romantic *Naturphilosophen*. This was at least the view of René Berthelot, who concluded his criticism of Bergson's 'philosophy of physics' on the following harsh note: "Thus the way in which Bergson's views on matter developed show the uncertainty of his thought and the extravagance of his results. Do not mistake for the rise of a star this unsteady light of the will-o'-the wisp floating over the swamps of Romanticism." He explicitly compared Bergson's views with the speculations of Goethe and Schelling on nature. This was written in 1913, one year after Russell's scathing attack on Bergson's philosophy in general.[1] But Russell, unlike Berthelot who was one of a few opponents who read attentively all Bergson's works, did not even mention Bergson's views about the physical world; he thus missed an opportunity for further invective and ridicule.

It is interesting to observe what the attitude of Bergson's admirers was in this respect. By 'admirers' I do not mean the uncritical public, which at the time when Bergson's philosophy was fashionable, was crowding the halls of the Collège de France and applauding its master. This public was mainly responsible for the appearance of pseudo-Bergsonism, which was nothing but a mere literary fashion, comparable to existentialism today. It is hardly any exaggeration to say that the content of this pseudo-Bergsonism consisted in the enthusiastic response to the emotional color of certain words, like 'intuition', 'création', 'élan vital', without the slightest effort at critical analysis. In this sense what Julien Benda called "le succès du bergsonisme"[2] was in truth the greatest damage done to the authentic Bergson's thought; authentic Bergsonism was misunderstood because it was wrongly identified with its fashionable and literary counterfeit. But even critical and responsible disciples of Bergson and those who showed a disciplined, positive attitude to his thought were mostly embarrassed by Bergson's treatment of the physical world; either they passed it over completely or they politely expressed their embarrassment.

There is one explanation for this attitude of Bergson's opponents and disciples: the year 1896 – the date of the publication of *Matter and Memory*. Although at that time the first rumblings under the foundations of classical physics were discernible, hardly anybody could then guess even remotely the extent of the coming scientific revolution. The classical corpuscular-kinetic view of nature still remained unchallenged. The evidence for the corpuscular structure of matter and electricity was steadily increasing and the epistemological doubts of Stallo and Mach about the fruitfulness of mechanistic explanations were both isolated and premature. Even when the complexity of the atom was discovered, its constituent parts – the electrons and the nuclei – still retained a number of the classical features of the ancient Lucretian atom: its permanence through time (the materialization and dematerialization of the electrons was discovered only in the fourth decade of this century), its definite shape and definite spatio-temporal location (which only de Broglie's discovery of the undulatory nature of matter in 1924 made questionable). Similarly, despite the increasing difficulties in constructing a satisfactory mechanical model of the aether, nobody doubted its analogies with an elastic medium in which transverse vibrations take place. Certainly nobody anticipated that the aether would eventually lose – under the impact of Michelson's experiment – even the most basic kinematic properties. The view that matter and its spatio-temporal framework eventually would be stripped of their classical, mechanistic features, which yielded so easily to pictorial models, was at that time looming on a very distant horizon, indeed – and only in a few and heretically daring minds.

Bergson was one of these, and he was fully aware why his theory of matter was either ignored or misunderstood: "This particular point [i.e. his philosophy of physics] has been hardly noticed for one very simple reason: since my views about this question were formulated at the time when it was regarded as self-evident that the ultimate material elements should be conceived in the image of the [macroscopic] whole, they confused the readers and were most frequently set aside as an unintelligible part of my work. It was probably assumed that this was an accessory part. Nobody, with a possible exception of the profound mathematician and philosopher Whitehead, noticed... that this was for me something essential which was closely related to my theory of duration and which lay in the *direction* in which physics would move sooner or later."[3] In other words, Bergson was fully aware that it was the non-pictorial character of his 'model' of matter, conceived as consisting of imageless events, which was the main stumbling block for his readers. Writing these words in 1938 he was also aware how far physics had moved since the years 1912–1913 – the time of Russell's caricature and Berthelot's indignation. It has moved still farther since 1938.

An interesting question arises in this context: to what extent can a thinker imaginatively anticipate the future development of science? Bergson, in stressing the word 'direction' ('in which physics would move') rightly disclaimed the anticipation of any specific discovery. For only the general line of the future development can be foreseen – and that only by a thinker who has an unusual insight into the inadequacies of the accepted conceptual scheme. Yet even the anticipation of the general trend sometimes does imply important specific features. Democritus did not anticipate the law of multiple proportions nor the size of the atoms; but he *did* anticipate the atomistic structure of matter and the limitlessness of space – not a small achievement at his time! To use a less well known example: when Nicolas d'Autrecourt was forced in 1348 to recant his view that light moves with a finite velocity, and that this velocity is too great to be perceived, he was forced to deny what Olaf Roemer experimentally confirmed more than three centuries later. It was Bessel who first determined the parallax of the 'fixed' stars in 1837, thus confirming Giordano Bruno's view, expressed two and a half centuries earlier, that "the stars beyond Saturn" are only apparently "fixed", their motion being imperceptible because of their enormous distance. Bacon and Descartes did not anticipate the kinetic theory of heat in its details: but they *did* anticipate one of its basic ideas – that the difference between the solid and liquid state consists in the degree of internal molecular motion. Now it would be absurd to say that Bergson anticipated Heisenberg's principle; but he *did* anticipate the elementary indeterminacy of microphysical processes which restricts the applicability of Laplacean determinism to the physical world. (In this respect, he was not alone; before him Boutroux and Peirce were bold enough to affirm it.) He certainly did not anticipate wave mechanics; yet – like Whitehead later – he did assert that matter consists of imageless vibrations devoid of any intuitive, material substratum. But more about this in the text of this book.

The book consists of three parts. The first deals with the biological theory of knowledge in the way it was amended by Bergson; comparisons are made with the older views of Spencer, Helmholtz, Mach, and Poincaré, and with more recent views of Reichenbach and Piaget, which Bergson's view resembles, but from which it also differs. This aspect of Bergson's thought was largely ignored; René Berthelot, who was one of a few who was aware of it, rejected it because of his traditional rationalism. Yet without an understanding of this aspect of Bergson's philosophy, no genuine understanding of his other views is possible. The second part deals with the meaning of the controversial term 'intuition'; it explores in great detail the structure of the Bergsonian *durée réelle*, shows its implications, and takes into account some significant criticisms. The third part deals with the Bergsonian theory of matter and its relation to contemporary physics. It shows that his central ideas are closely related both to his general theory of duration and to his biological theory of knowledge; it traces Bergson's relations to other thinkers, especially to Leibniz, Boutroux, Whitehead, Bohm and de Broglie; it analyzes what is valid and what is wrong in his later comments on the theory of relativity. Two appendices deal with Russell's complex relations to Bergson and with the much discussed relation of microphysical indeterminacy to freedom. Appendix III deals with Bergson's views on entropy and their relations to modern cosmogony.

## NOTES

[1] René Berthelot, *Un romantisme utilitaire*, Paris, 1913, II, p. 213, Bertrand Russell, 'The Philosophy of Bergson', *The Monist* **22** (1915) 321–347.
[2] Julien Benda, *Sur le succès du bergsonisme*, Paris, 1914.
[3] Bergson's letter to the author, July 3, 1938.

# TABLE OF CONTENTS

# PART I

# BERGSON'S BIOLOGICAL THEORY OF KNOWLEDGE

CHAPTER 1

# THE CLASSICAL BIOLOGICAL THEORY OF
# KNOWLEDGE: HERBERT SPENCER

'Anti-intellectualism' is a label frequently applied to Bergson's philosophical thought, but only rarely is the meaning of this term properly
understood and fully clarified. It is generally not realized, or at least not
sufficiently stressed, that his alleged anti-intellectualism is opposed only
to a certain, though historically the most important and still the most
dominant, form of human intellect; that is, to what he himself called
the 'logic of solid bodies' and what, perhaps more appropriately, may
be called the Newton-Euclidian form of intellect. This form, as Bergson
stressed, is a result of a long evolutionary process the duration of which
certainly transcends the duration of the human species. This duration
itself is the ground for its practical justification and its approximate
applicability to our ordinary daily experience; but when we are con-
fronted with the new types of experience which do not fit the traditional
patterns of thought, a new and fresh effort is required to create more
flexible and more adequate forms of understanding. Although Bergson's
language is not always as specific as we would wish to have it, the meaning
of his anti-intellectualism is unmistakably clear for an attentive and
unprejudiced reader, especially as he becomes aware that it is an organic
part of Bergson's *biological theory of knowledge*. In my opinion no fruitful
study of Bergson's thought, and, in particular, no true understanding of
his sometimes paradoxical philosophy of nature is possible without
taking his biologically oriented epistemology into consideration.
   The view that the cognitive functions of the human mind are not static
and immutable entities but, like all other physical and psychological
features of man, are subject to gradual growth and development, began
to gain ground in the second half of the nineteenth century after the
advent of the theory of evolution. It seemed natural and consistent to
apply the theory of evolution not only to the physical organism of man,
but also to his psychological features, including that which is usually

designated by the words 'reason' or 'thought'. It was Darwin himself who, in his study of the *Expression of Emotions in Man and Animals*, applied the evolutionary method to one specific problem of human and animal psychology (1872). But there is no question that Darwin was in this respect not the first one; as early as 1855, that is, even prior to the publication of Darwin's *Origin of Species*, Herbert Spencer applied the evolutionary and genetic method in the first edition of his *Principles of Psychology*. This book, especially in its later editions, was, and probably still remains, the most systematic exposition of biologically oriented epistemology.

That Spencer's philosophy reflects the intellectual atmosphere of the second half of the nineteenth century is well known. His influence was very extensive and culminated around 1900; in many respects he was one of the most representative and influential thinkers of that period. The fact that this philosophy is now regarded as outdated merely shows how quickly philosophical fashions change. Unfortunately, changing fashions frequently affect terminology rather than the ideas themselves; and it is amusing to observe how those who would be offended by being called Spencer's disciples still retain his basic philosophical outlook. Charles Singer pointed out how, despite the prevailing neglect and even contempt for Spencer's philosophy, his ideas and even their verbal expressions, which he coined, continue to form a part of the twentieth century language, especially in what is today called the 'behavioral sciences'.[1] Certainly, the present psychological and sociological determinism and cultural relativism are not significantly different from the view which dominates both Spencer's *Principles of Psychology* and his *Principles of Sociology*. We do not share the view of Professor Aiken[2] that Spencer's philosophy is bound for revival; unfortunately his influence, notwithstanding verbal declarations to the contrary, is still very much alive and does not need to be revived. We say 'unfortunately' because this unacknowledged influence is in many respects retarding. But this does not mean that future generations will not show a greater fairness to his thought than is outwardly being shown now. At present, Spencer is too outdated to be regarded as important, and not ancient enough to be included among the classical thinkers of the past. Together with a more just evaluation of his persistent, though unacknowledged, influence, posterity will probably more clearly differentiate between the truly outdated parts of his thought and those whose fruitfulness is far from being exhausted. The present situation is curiously reversed; the most influential parts of Spencer's thought are those which are most clearly marked by the nineteenth century atmosphere while its most interesting aspects with the greatest potential bearing on some contemporary problems are now almost universally overlooked. To these potentially fruitful aspects belongs his biologically oriented theory of knowledge, especially when it is freed from its nineteenth century coloring.

It was precisely Bergson who made an attempt to amend Spencer's biological theory of knowledge in order to free it from its outdated and unsatisfactory features. We know from Bergson himself that at the beginning of his academic career he was a fervent disciple of Spencer, and that his main ambition at that time was to improve Spencer's system in some details without altering its essential features.[3] It is true that very early, as early as the time of writing his doctoral thesis, he radically departed from Spencer's mechanistic monism, and that his main work, *Creative Evolution*, was written in a conscious opposition to what he termed the "false evolutionism" of the English philosopher. But this fact tends to obscure another equally important truth, that traces of Spencer's influence never completely disappeared from Bergson's thought. This is only apparently paradoxical; we have only to remember that we are often as strongly influenced by that against which we react as by what we like, and that underlying different and even opposite solutions of one and the same problem there is often a tacit agreement about the way in which the problem is formulated. The thesis which remained common to both Bergson and Spencer even after their ways parted was the view that *the theory of life and the theory of knowledge are inseparable*[4]; more specifically, that no theory of knowledge can be adequate without relating the genesis of the cognitive forms to the whole evolutionary process of life. For this reason an outline of Spencer's evolutionary epistemology will provide us with the necessary contrasting backdrop for Bergson's own version of the biological theory of knowledge.

The general character as well as the specific features of Spencer's epistemology are determined by his views concerning the nature of life and its development. The following passage contains the essential features of his theory of knowledge which follows naturally from his whole evolutionary and mechanistic outlook:

Whence it becomes manifest, that while Life in its simplest form is the correspondence of certain inner physico-chemical actions with certain outer physico-chemical actions, each advance to a higher form of Life consists in a better preservation of this primary correspondence by the establishment of other correspondences.

Divesting this conception of all superfluities and reducing it to its most abstract shape, we see that Life is definable as the continuous adjustment of internal relations to external relations. And when we so define it, we discover that physical and psychical life are equally comprehended by that definition. We perceive that this which we call Intelligence, shows itself when the external relations to which the internal ones are adjusted, begin to be numerous, complex and remote in time or space; that every advance in Intelligence essentially consists in the establishment of more varied, more complete, and more involved adjustments; and that even the highest achievements of science are resolvable into mental relations of co-existence and sequence, so co-ordinated as exactly to tally with certain relations of co-existence and sequence that occur externally. A caterpillar, wandering at random and at length finding its way on to a plant having a certain odor, begins to eat – has inside of it an organic relation between a particular impression and a particular set of actions, answering to the relation outside of it, between scent and nutriment. The sparrow, guided by the more complex correlation of impressions which the color, form, and movements of the caterpillar gave it; and guided also by other correlations which measure the position and the distance of the caterpillar; adjusts certain correlated muscular movements in such way as to seize the caterpillar. Through a much greater distance in space is the hawk, hovering above, affected by the relations of shape and motion which the sparrow presents; and the much more complicated and prolonged series of related nervous and muscular changes, gone through in correspondence with the sparrow's changing relations of position, finally succeed when they are precisely adjusted to these changing relations. In the fowler, experience has established a relation between the appearance and flight of a hawk and the destruction of other birds, including game; there is also in him an established relation between those visual impressions answering to a certain distance in space, and the range of his gun; and he has learned, too, by frequent observation, what relations of position the sight must bear to a point somewhat in advance of the flying bird, before he can fire with success. Similarly if we go back to the manufacture of the gun, by relations of co-existence between colour, density, and place in the earth, a particular mineral is known as one which yields iron; and the obtainment of iron from it, results when certain correlated acts of ours are adjusted to certain correlated affinities displayed by ironstone, coal, and lime, at a high temperature. If we descend yet a step further, and ask a chemist to explain the explosion of gunpowder, or apply to a mathematician for a theory of projectiles, we still find that special or general relations of co-existence and sequence between properties, motions, spaces, etc., are all they can teach us. And lastly, let it be noted that what we call *truth*, guiding us to successful action and the consequent maintenance of life, is simply the accurate correspondence of subjective to objective relations; while *error*, leading to failure and therefore towards death, is the absence of such accurate correspondence.[5]

I quote this lengthy passage purposely; it shows, with a Victorian over-explicitness, that what we call *truth* and *error* on the human scale can be defined in behavioral terms as the presence or absence of adjustment to the environment and that in this respect the human situation is merely the last and culminating stage in the general evolutionary process. The terms 'knowledge' and 'intelligence' are used by Spencer in a very broad sense in order to be applied to what may be called the 'knowledge' and 'intelligence' of animals which resemble human knowledge and intelligence at least in important behavioral aspects. (Unlike Descartes and unlike behaviorists, Spencer does not deny the existence of mental qualities in animals; for him the introspective data and the cerebral processes are two aspects of the same reality in the sense of the double-aspect or identity theory which is still popular in some philosophical circles.) Defined in this broader sense, knowledge in both its human and subhuman forms has primarily a practical and biological function; it is a weapon in the struggle for life. A better knowledge means a better chance for survival. There is no question that in stressing the biological and practical character of knowledge Spencer anticipated later important trends in epistemology. The empirico-criticism of Mach and Avenarius, various forms of pragmatism, including the 'partial pragmatism' of Bergson and the instrumentalism of Dewey tread, whether consciously or unwittingly, in the footsteps of Spencer.

It is true that present human thought, especially in its more complex and abstract forms, seems to be purely speculative and disinterested. For this reason the emphasis on the biological and practical character of human thought is resented by rationalistically minded thinkers, who are committed by their metaphysics to regard reason as a faculty *sui generis*, and who consequently resent any naturalistic attempt to regard it as a continuation and culmination of subhuman forms of intelligence. It is resented also by abstract logicians who, though generally free of similar metaphysical inhibitions, remain by the very nature of their limited interests either hostile or indifferent to any kind of genetic approach. There is hardly any question that the hostility towards pragmatism stemmed largely from these two main sources. But it must not be overlooked that this apparently disinterested and purely speculative character of human thought is of relatively recent origin; it would have been impossible without the economic and technological conditions which freed the human mind from its subservience to immediate material needs and released its energy into new channels not directly related to practical activity. Who would seriously question that even the most abstract sciences can be traced to humble, purely utilitarian origins? Has not geometry developed from the art of surveying, astronomy from astrology and chemistry from alchemy? Does not this practical character of human thinking become quite obvious when we consider the less developed forms of human societies?

It becomes even more obvious when we consider rudimentary forms of intelligence in various subhuman species, although only in the higher vertebrates is the affinity to human intelligence clearly discernible. It is true that cognitive functions in animals, with the exception of the higher species, consist mainly of the faculty of perception; but it is precisely the character of perception which exhibits most convincingly the close connection between knowledge and biological needs. In the ascending hierarchy of organic beings, the field of animal perception is gradually widened as the corresponding sense organs grow in complexity and in adaptation to various external stimuli. Thus the surroundings, to which the animal becomes adapted, is gradually wider and wider; at the same time, with the appearance of memory and imagination, that is with the increasing capacity to retain traces of past impressions and rearrange them into new patterns, the animal ceases to be confined within the narrow limits of the present moment and is able to become adjusted even to conditions not yet existing. Thus by the appearance of memory and anticipation the field to which an organism is adjusted is widened in time as well. Needless to say the greater number and greater precision of external stimuli registered by the increasingly more complex and more accurate sensory organs is reflected in the growing complexity and accuracy of the concomitant motor reactions; and the quoted passage of Spencer shows how the anticipatory character of both perception and motor reaction becomes more prominent in higher species. This widening in space and in time of the milieu to which the organism reacts is one of the most conspicuous features of organic evolution; we may say that the place of a living being in the hierarchy of organisms is generally determined by the spatial extent of its sensory field as well as by the temporal span of its memory and anticipation.

The evolution of human intelligence is ruled by the same law. In Spencer's own words:

From early races acquainted only with neighboring localities, up to modern geographers who specify the latitude and longitude of every place on the globe – from ancient builders and metallurgists, knowing but surface deposits up to the geologists of our day whose data in some cases enable them to describe the material existing at a depth never yet reached by the miner – from the savage barely able to know in how many days a full moon will return, up to the astronomer who ascertains the period of revolution of a double star – there has been a gradual widening of the surrounding region through which the adjustment of inner to outer relations extends.[6]

In other words, the evolution characterized above as a steady process of adjustment to the increasingly wider spatio-temporal surrounding culminates, in the species *homo sapiens*, and especially in its civilized specimens. This was made possible by the fact that the field of sensory perception in man was enormously increased by means of artificial instruments like the telescope, microscope, photographic plate, radio telescope, seismograph, Wilson cloud chamber, Geiger counter, electron microscope and others; thus the objects normally imperceptible, like distant stars and galaxies, or the microscopic world of molecular and atomic dimensions are open to human senses. Secondly, by a capacity of reasoning never reached by his animal ancestors man is able to transcend the narrow interval of time to which he is confined by his own 'specious present' and to reconstruct past situations as well as to anticipate future events which are both enormously remote in time.

It was only natural at the end of the last century to believe that the surroundings to which human intellect is adjusted is not only incomparably wider than the narrow *milieu* of animals and even of primitive man, but is practically co-extensive with the whole of spatio-temporal reality. In other words, "the adjustment" – to use Spencer's and Mach's term – of the cognitive faculties to the objective order of things is *complete* in the human species in which the development of these faculties not only culminates, but *ends*. This was a conclusion naturally inspired by the spectacular triumphs of science in the previous three centuries. For the unity and coherence of the scientific world-view in the last decades of the last century was such that it was unreasonable at that time to expect any essential modifications in any future time. In 1875 the French philosopher and historian Hyppolite Taine, writing of classical science, concluded: "Sauf des corrections partielles, nous n'avons rien à effacer."[7] Ten years later Marcellin Berthelot wrote even more strongly: "The world is now without mysteries."[8] It is true that some changes in detail were regarded as possible and some gaps expected to be filled; but almost nobody, except those who completely ignored everything that happened between the time of Galileo and that of Darwin, dared to challenge the basic features of classical science. "We supposed that nearly everything of importance about physics was known", Whitehead conceded in recalling the belief of his young days. "Yes, there were a few obscure spots, strange anomalies having to do with radiation which physicists expected to be cleared by 1900."[9] We know today that these hopes were not fulfilled; on the contrary, in 1900 a new theory of radiation came into being which represented the first radical departure from the traditional scientific view. But this does not mean that at the beginning of this century the faith in the final character of the classical picture of reality was seriously shaken. As late as 1903 Bertrand Russell, in surveying the main features of the mechanistic view of the universe, concluded confidently: "All these things, if not quite beyond dispute, are yet so nearly certain, that no philosophy which rejects them can hope to stand."[10]

Note the date: the year 1903 was the year of Spencer's death. But his belief in the final character of classical science obviously lived on and is far from being dead even now. Spencer simply made explicit the assumption which was silently accepted by the overwhelming majority of his contemporaries and which, as we shall see, is still tacitly being made even now. It is the assumption that the evolution of human reason is already completed and that the mechanistic science of the last century represents the last stage of the adaptive process by which human mind gradually adjusted itself to the structure of reality. In other words, the structure of human reason as exhibited in classical physics, in particular in the Newtonian mechanics and Euclidian geometry, is an adequate replica of the objective order of nature in the human mind or, if we prefer a more naturalistic term, in the human brain. This structure of human reason is essentially nothing but a system of mental habits established and strengthened by the continuous pressure of the objective order of things by which, during long geological periods, the human mind has been moulded into its present forms. From this point of view Kant was not mistaken when he believed that Euclidian geometry and Newtonian mechanics will remain forever valid as they will never be contradicted by any future experience. But this is not because of the a priori character of Euclidian geometry and Newtonian mechanics; they both will remain valid not because they *precede* experience, as Kant erroneously believed, but because they were both implanted in our mind by experience. But this experience, according to Spencer, should not be understood in an individual ontogenetic sense, but as an *experience of the whole species*. In this sense even the concept of a priori may be regarded as admissible provided it is carefully redefined. For there is no question that we are born with certain intellectual dispositions which merely wait for external stimuli to be fully unfolded; but what is a priori for an individual is a posteriori for the whole species.[11] Thus the term 'individual a priori' is merely an abbreviation designating the condensed experience of countless preceding generations. Although Spencer believed that he offered a solution intermediary between traditional rationalism and traditional empiricism, it is clear that he basically remained loyal to the empiricist trend in British philosophy; he merely widened the Lockean and Humean concept of *individual* experience by substituting for it the concept of the experience of the whole species.

It is clear that in the naturalism and positivism of the last century there was a strong underlying note of *epistemological optimism*. The confidence that, as Ernest Häckel put it, all "the riddles of the universe" are solved or, as Marcellin Berthelot declared, that "the world is now without mysteries", was inspired by the same idea of evolution conceived as a gradual process of adjustment which culminated in the human species. For natural selection automatically eliminated all biological maladjustments, including wrong beliefs and wrong theories. An animal whose behavior would implicitly ignore Euclid's theorem that among all lines joining two points a straight line is the shortest one, would be definitely handicapped in its effort to reach its target, or to run before its enemies; it would be hopelessly outrun by its Euclidian competitors. Similarly, an organism deprived of the capacity of associating effects with their causes would not be able to profit from its past experience and would not be able to learn anything. Thus the allegedly ultimate triumph of Euclidian geometry and of the principle of causality was regarded by the positivistic evolutionists as a special instance of the principle of survival of the fittest. In the words of Hyppolite Taine, the universe itself formed its own accurate image within the human brain[12]; as the universe itself consists of the masses moving according to Newton's laws in the infinite Euclidian space, there is nothing more natural than that the final picture of the universe within the human mind is Newtonian and Euclidian in its character. Thus what was called by Kant the "transcendental structure of human mind" is merely a final product of the gradual process of adjustment to the objective order of nature. Thus Kant, in whose system evolution is *absent*, was led to the same conclusion as Spencer and positivism in general, in whose systems evolution is already *completed*.

Evolutionary epistemology only at first glance seemed to imply a sheer Protagorean relativism and agnosticism; on the contrary, in its classical Spencerian form it was as dogmatic as the rationalism of the eighteenth century. It is true that the positivistic evolutionists of the last century rarely claimed explicitly that their view gives the ultimate answers to all philosophical questions. On the contrary, they were outwardly very modest and stressed quite emphatically their complete lack of interest in ontological or, as they said, 'metaphysical' questions. But we should not be deceived by their agnostic and phenomenalistic language which merely veiled their hidden metaphysical assumptions. Behind their verbal agnosticism and anti-metaphysical effusions there was a definite, though often unconscious, commitment to a very precise view of reality. A closer analysis shows that, for instance, Spencer's "Unknowable Reality" is merely a verbal disguise for the substantialized concept of energy borrowed from nineteenth century physics. In this respect Spencer again is fairly representative of the whole classical positivistic trend which, in spite of loud verbal denials and misleading phenomenalistic terminology, accepted consciously or semi-consciously a definite type of metaphysics, usually in the form of mechanistic monism. There is no place here to show it in detail; but even a cursory glance at the leading positivists of the last century will convince us how much it is there.

While in the founders of classical positivism, Auguste Comte and John Stuart Mill, the agnostic and antimetaphysical note was unquestionably more genuine, the situation became very different in the second half of the nineteenth century, after the discovery of the law of conservation of energy and the coming of the idea of evolution. The law of conservation of energy anticipated by Leibniz and Huygens and formulated and verified by Mayer and Helmholtz, was regarded as the logical basis of the law of causality; the equality of cause and effect, expressed in the classical saying 'causa aequat effectum', found its expression in the quantitative equality of successive energetic equivalents. The majority of physicists and philosophers of the nineteenth century sincerely believed they had found the mysterious link joining each cause to its effect for which Hume was looking in vain. This link was found in the relation of equivalence

between the successive transformations of energy. Thus what we call
'cause' and 'effect' is in the language of classical physics nothing but two
successive forms of the constant quantity of energy. No cause can be
without its effect because no particular quantity of energy can disappear
without being transformed into its equivalent; and no effect can be
without its cause because no quantity of energy in the world can arise
out of nothing, but only from the transformation of a previous energetic
equivalent. Thus the basic identity and quantitative constancy of the
unitary physical stuff underlies the successive causal series in nature.[13]
There is no need to stress how much this view was consonant with
classical philosophical prejudices about the unity and constancy of substance underneath its changing modifications. By natural extrapolation
of the validity of the law of conservation of energy in both directions
of time, the uniformity of nature in time was secured; the evolutionary
views of Lyell and Darwin were nothing but a concrete application of the
doctrine of uniformity of nature to the past.

Nothing shows better the close alliance of positivism and mechanistic
monism than the sympathetic attitude which the leading positivistic
thinkers of that period showed to Spinozism. There is no question that
Spinoza was the first thinker who freed the Cartesian mechanism of all
inconsistencies in extending physical laws to human behavior and re-
legating consciousness to the limbo of causal inefficacy. For this reason
the outspoken sympathy of the leading positivists of the second half
of the last century for Spinoza is hardly accidental; in truth the major part
of the positivistically oriented physiological psychology of that period
was inspired by the double aspect or identity theory which clearly has its
roots in Spinoza. In this respect even some contemporary positivists
follow Spinoza whether they are aware of it or not. It is then hardly
surprising that the terms 'positivistic' and 'mechanistic' were at that time
– in truth they often are even now – synonymous, and that mechanistic
metaphysics, disguised by phenomenalistic terminology, was regarded
as the last stage in the cultural development of mankind whether it was
called the "third positive stage" by Auguste Comte or "the complete
adjustment of inner to outer relations" by Herbert Spencer.

The situation was even less ambiguous in the materialistic branch of
last century naturalism; its blunt language, completely free of terminological subtleties and obscurities, did not leave the slightest doubt about

its real stand. For this reason, materialists sometimes found the positivistic language irritatingly obscure or at least not sufficiently explicit. But in spite of the differences in language and in spite of the generally greater epistemological caution of positivists there was no significant disagreement between both groups as far as their acceptance of the mechanistic view was concerned. From this point of view Lenin's anger against Ernst Mach was completely groundless because the latter, in spite of his critical attitude toward some Newtonian assumptions, never basically departed from classical physics and from the philosophy which grew out of it; he even at the end of his life showed toward the revolutionary ideas of the special and the general theory of relativity a reluctant attitude similar to that shown by some Marxists. Thus Lenin's angry attack against Mach and Avenarius in his book *Materialism and Empiriocriticism* is an almost ludicrous instance of the sectarian intra-mechanistic and intra-deterministic disputes.

## NOTES

[1] Ch. Singer, *A Short History of Scientific Ideas to 1900*, Clarendon Press, Oxford, 1959, p. 512–513.

[2] Henry D. Aiken, *The Age of Ideology. The Nineteenth Century Philosophers*, New York 1956, p. 170.

[3] *C.M.*, p. 10–13; 109. Spencer's influence was rightly stressed by Albert Thibaudet, *Le Bergsonisme*, Paris 1923, II, p. 218.

[4] *C.E.*, p. XXIII.

[5] H. Spencer, *First Principles*, 4th Ed., Appleton, New York, 1896, pp. 86–87.

[6] *The Principles of Psychology*, 3rd Ed., Appleton, New York, 1897, I, pp. 318–319.

[7] H. Taine, *Les origines de la France contemporaine*, 34th Ed., Paris, 1947, I, p. 270.

[8] M. Berthelot, *Les origines de l'alchemie*, Paris 1885, p. 151.

[9] *Dialogues of Alfred North Whitehead as Recorded by Lucien Price*, New American Library, New York, 1960, p. 12.

[10] B. Russell, 'A Free Man's Worship', in: *Mysticism and Logic and Other Essays*, Pelican Books, 1953, p. 51.

[11] *The Principles of Psychology*, II, 1, p. 195.

[12] H. Taine, *De l'intelligence*, 16th Ed., Paris 1927, I, p. 236.

[13] H. Spencer, *First Principles*, Appleton, New York, 1896, Chapters VII and VIII, in particular pp. 203–204; 229–231; W. Ostwald, *Vorlesungen über Naturphilosophie*, Leipzig 1902, p. 296; *Grundriss der Naturphilosophie*, Leipzig 1919, p. 144.

# CHAPTER 2

# THE INTERMEDIATE STAGE:
# HELMHOLTZ, MACH AND POINCARÉ

The example of Mach is especially instructive because by his criticism of some classical concepts he prepared the way for the future far-reaching revision of the New.o.iian physics in the twentieth century. Although in this respect he was not alone, there were only a very few thinkers who expressed equally prophetic views at that time. For our purposes it will be sufficient to describe briefly besides the views of Mach also those of Helmholtz and Poincaré, to show how all three in spite of their bold anticipations of certain aspects of contemporary physics remained essentially loyal to the most important postulates of the classical science. This loyalty was logically correlated with their biologically oriented theory of knowledge which either explicitly or implicitly was identical with the theory of Spencer described above.

Chronologically, we have to consider Helmholtz first. By the extent of his interests and the importance of his achievements he is one of the most outstanding representatives of the culminating period of classical science. The extent of his interests was truly impressive; medicine, physiology of sense organs, psychology, physics, the logical foundations of geometry were the fields to which he made important and lasting contributions. His formulation of the law of conservation of energy five years posterior to, but independent of, Robert Mayer may be regarded as a culmination of classical physical science; as pointed out earlier, the transformation of the subsequent energetic equivalents was regarded as the answer to Hume's search for the causal link between the phenomena, and Helmholtz spoke of this law with the same nearly religious emotion as other nineteenth century scientists and philosophers. His *Physiological Optics* as well as his work on the *Sensations of Tone* may be regarded as the last stage in the formulation of the doctrine of the primary and secondary qualities on which classical physical science was based. He was the first to understand the significance of Faraday's laws of electrolysis in seeing that they imply the atomistic structure of electricity; and his mathematical investigations in hydrodynamics inspired Lord Kelvin's famous theory of vortex-atoms. But in other respects Helmholtz's own thought clearly pointed beyond the classical period. He explicitly challenged Kant's view that the axioms of Euclid are a priori necessities of thought, although he still regarded spatiality in general as an a priori form of intuition. Using even the Kantian vocabulary, he claimed that the specific axioms determining this general form are subject to empirical verification or disconfirmation.[1] Long before Einstein, he introduced the term 'physical geometry'. He expressed a similar view about arithmetic, although in a more obscure form.[2] In spite of his natural preference for intuitive models in physics, his doctrine that *all* sensations are mere 'signs' (*Zeichen*) of objective reality clearly points *beyond* the classical distinction between primary and secondary qualities.[3] It is hardly surprising that Helmholtz's significance for contemporary science and epistemology was recognized; Moritz Schlick listed his prophetic insights in his centenary speech of 1921, while a few years later he re-edited a number of Helmholtz's essays, carefully and extensively annotated.[4]

Although Helmholtz was a contemporary of Spencer (who was only one year older), there is hardly any evidence of mutual influence between the two, certainly no evidence of the influence of the latter's evolutionary thought on the former. Moreover, explicit references by Helmholtz to the biological evolution are surprisingly few; although he accepted Darwin, so far as I know no application of the evolutionary theory to epistemology can be found in his books. In this respect Helmholtz's empiricism was of the traditional type in confining his attention to *individual* experience, like the pre-evolutionary empiricism of Locke and Hume. Yet, Helmholtz's psycho-physiological reinterpretation of Kant's a priori[5] almost required an evolutionary explanation. If the a priori forms are merely parts of our psycho-physiological organization, and if this organization is our phylogenetic heritage – and both of these theses were accepted by Helmholtz – can we escape Spencer's conclusion that "what is a priori for an individual is a posteriori for a species"? Although this conclusion never was formulated by Helmholtz, he could hardly have rejected it.

In his particular views Helmholtz was a little more open-minded than Spencer. The latter retained belief in the absolute validity of the propositions of Euclidian geometry, and justified this by our subjective incapacity to conceive their denial.[6] This criterion of truth based on 'the inconceivability of the opposite' followed *apparently* from his biological theory of knowledge. For, if our generalizations are based on experience, or more specifically, on repeated perceptual patterns, then the solidity and unbreakability of the associative links between two neural patterns measures, so to speak, their phylogenetic age, the duration of the objective influences which created these corresponding links. 'The feeling of evidence' then is merely an introspective counterpart of these associative links, and their unbreakability underlies our feeling of absurdity when we try to frame mentally the denial of evident propositions. In other words, according to Spencer, intense feelings of evidence and of absurdity cannot be deceptive. To illustrate this by one example: our imaginative resistance to any attempt at drawing *two* parallel lines through a point in respect to a given line is based, according to Spencer, on the resistance of the whole network of associations which constitute our Euclidian subconscious and which were shaped during enormously long periods by the continued pressure of objective stimuli. In other words, our belief in Euclidian geometry is, so to speak, a mirror which the objective Euclidian space gradually created in our mind. As stated above, Helmholtz, in this respect, was more cautious; he certainly would have agreed with Spencer in principle about the ultimately empirical origin of our belief in Euclidian space, but he was aware that the structure of the real space may be *slightly* different from the Euclidian one and thus admitted the possibility of the disconfirmation of the Euclidian propositions on a larger scale, by the observations of the parallaxes of very distant stars.[7] He claimed that non-Euclidian geometries are not only free of contradiction, but are even *imaginable in principle*; this was a controversial and perhaps too optimistic belief in the plasticity of our imagination which understandably gave rise to protracted discussions.[8] But what is significant is that Helmholtz apparently did not realize that the law of causality in its classical Laplacean form can have the same merely approximate validity as the Euclidian geometry. He regarded the law of causality as a priori in his sense, i.e. as a part of our psycho-physiological organization; and of the law of conservation of energy which for him, as for Spencer, was merely a concrete and quantitative expression of the principle of causality, he spoke with the same religious

emotion as did other nineteenth century thinkers.[9] In accepting the
mechanistic and monistic view of reality he did not differ in any sense
from the prevailing view of his period of which Spencer gave the most
systematic exposition.

Similar features are present in the thought of Ernst Mach. His criticism
of Newton's concepts of absolute space, time, and motion, in particular
his criticism of Newton's interpretation of the rotating bucket experiment,
are sufficiently known; equally known is the influence of his thought
on Einstein and on the genesis of modern relativistic ideas.[10] Mach's
distrust of the corpuscular-kinetic models in physics clearly heralded a
new era, although it led him to reject these models even on the molecular
and atomic level where their fruitfulness was later established beyond
doubt. His empiricism and his rejection of Kantian apriorism were even
more unambiguous than that of Helmholtz, probably because of Mach's
greater emphasis on the evolutionary approach to epistemology. In this
respect he was closer to Spencer whom he frequently quotes, especially
in one of his last books, *Erkenntniss und Irrtum*.[11] Mach's biological
theory of knowledge is most concisely stated in one essay of his *Popular-
Wissenschaftliche Vorlesungen* under the title 'On Transformation and
Adaptation in Scientific Thought', but its central ideas are scattered
through all of his main works. Like Peirce and, before, Dewey he defines
any problem which man faces as a result of conflict between an established
mental habit and a widened field of observation; when our thought
becomes adjusted to this widened experience, the problem disappears or,
using the words of C. S. Peirce, "doubt is appeased."[12] The adjustment of
our thoughts to experience is, according to Mach, quite analogous to the
adjustment of physical organisms to their environment; in fact this change
in thoughts (*Gedankenumwandlung*) which we observe in ourselves is
merely a part of the wider evolutionary process. All our concepts, even
those which are the most abstract, are results of adaptive experience.
Even the concepts of space, time and causality have basically a biological,
that is, utilitarian and economic, character; none of them is inborn in the
Kantian sense. We shall later have the opportunity to meet some of Mach's
anticipations of some more recent trends in philosophy of science. In this
mere sketch it will be sufficient to say that Mach's view of geometry as an
empirical science strongly influenced Einstein; Mach's book *Space and
Geometry*[13] is a direct predecessor of Einstein's *Geometry and Experience*.

His view that underlying the concept of unidirectional time is the elementary sensation of time, *Zeitempfindung*, which is an introspective counterpart of the ceaseless consumption of nervous energy, was one of the first speculations about the nature of the 'physiological entropy-clock' which play such a prominent role in the problem of the direction of time.[14] Even the concept of causality is a result of experience, although experience in this case should not be understood in the ontogenetic sense, but as the experience of the whole species:

Much of the authority of the ideas of cause and effect is due to the fact that they are developed *instinctively* and involuntarily, and that we are distinctly sensible of having nothing contributed to their formation. We may, indeed, say that our sense of causality is not acquired by the individual, but has been perfected in the development of the race.[15]

This is clearly as close as possible to Spencer's claim that 'what is a priori for the individual is a posteriori for the species' or to Helmholtz' physiological reinterpretation of the Kantian a priori.

Yet, in spite of his distrust of 'mechanistic mythology'[16], his critical attitude toward Newton's absolutism, and his openminded approach to non-Euclidian geometry, Ernst Mach clearly belonged to the nineteenth century as much as did Spencer and Helmholtz. While emphasizing the economic and utilitarian character of the concept of causality, it apparently never occurred to him that this concept itself may be a generalization based on a limited experience or, to use his own terminology, a result of adaptation to only a certain sector of reality. His determinism is as absolute as that of Spencer, Helmholtz, Laplace, or Spinoza; the behavior of man is as strictly determined 'as the movements of a fly' and the only difference between man and fly is that one is a much more complex automaton than the other.[17] In explaining his principle of economy Mach stressed explicitly that despite its superficial similarity to the minimum principles in physics, this principle really *does not apply to nature*, for there is no *choice* in physical processes. It is therefore, strictly speaking, meaningless to speak, for instance, of the shortest of all *possible* trajectories along which a physical body actually moves, for no trajectory but that along which a body really moves is possible.[18] For a rigorous determinist it is only consistent to deny *any genuine possibility in nature*; in doing this Mach follows the same logic which guided Spinoza and long before him, the father of modern determinism, Democritus. It is true that Mach

occasionally adopted a more open-minded view of the problem of
causality, as when he says at the end of his *Science of Mechanics*:

Again, I have repeatedly remarked that all forms of the law of causality spring from
subjective impulses, *which nature is by no means compelled to satisfy*. (Italics mine.)[19]

This, however, was merely a passing prophetic glimpse; already in the
next paragraph he defends and commends Robert Mayer for being guided
in his discovery of the law of conservation of energy by the principle of
equivalence of cause and effect:

But in Mayer's case these maxims are, in my judgment, not weaknesses. On the contrary, they are with him the expression of a powerful *yearning*, as yet unsettled and
unclarified, for a sound, substantial conception of what is now called energy. This
desire I should not exactly call metaphysical.

If nature is not bound to satisfy our causal impulse, why should it be
compelled to satisfy Mayer's yearning for quantitative constancy in
nature? This is especially difficult to see because the principle of causality
and the law of constancy of energy are very closely related and, as
mentioned above, the latter was in the last century regarded as the basis
of the former. Thus 'causal instinct' does not seem to be essentially
different from Mayer's 'yearning for the constancy in nature' and it is
plainly illegitimate to divide our cognitive 'impulses' into harmful (or
'metaphysical') and harmless (or non-metaphysical) as Mach does. This
apparent discrepancy in Mach is really nothing but the conflict between
a passing anticipatory insight and his firm commitment to last century's
mechanism. Having, like Spencer, defined truth and error in the terms of
success and failure, Mach simply could not believe that the belief which
survived a practically limitless number of tests can ever in the future be
revised or modified. The belief in causality in its precise form of the law
of constancy of energy was for him that uniformity of thought which
reflected the most general and most pervasive feature of nature – its
constancy and uniformity.

  This was the reason why Mach, again like Spencer, regarded the
inconceivability of opposites as the most reliable epistemological criterion.
Because the uniformities of nature generate the uniformities of thought,
the solidarity of the associative links between two ideas constitute in
Mach's empiricist and associationist psychology the basis of *logical
necessity* which is a subjective counterpart of the objective order of nature.

Like Spencer, Mach illustrates the inconceivability of the opposite by an example borrowed from geometry; but while Spencer uses as an illustration an example of an Euclidian theorem, Mach is more cautious in using theorems whose validity is not confined to the Euclidian geometry only:

> But I notice that in imagining two couples of points I already imagine four points at the same time. In the same way I can mentally conceive that the angle of the triangle is growing and in a separate act of attention notice that its opposite side grows at the same time. Then I notice that the increase of the opposite side is contained in the mental representation of the growing angle.[20]

It is needless to say that the mere subjective inconceivability of the opposite, is a very questionable epistemological criterion. The feeling of inconceivability, which is often a mere difficulty to imagine, is relative and may change considerably through centuries; the apparent absurdity of the existence of antipodes is a good example. Moreover, Mach was clearly inconsistent in applying this criterion; unlike Spencer he did not use it to justify Euclidian geometry, but apparently he did use it in asserting the classical concept of causality. This clearly reflects Mach's intermediate position between classical and modern physics.

Of all three thinkers, Helmholtz, Mach and Poincaré, the last was certainly closest not only to the spirit of modern physics, but also to its concrete content. His anticipation of the principle of relativity was such that E. T. Whittaker did not hesitate to call this principle by his name.[21] He showed an equally open-minded receptiveness toward the emerging quantum theory whose revolutionary significance he fully appreciated. He was one of a few thinkers who envisioned the startling possibility of giving up the traditional concept of continuity of time in physics, and he was probably the first to use the term 'atom of time'.[22] Yet, despite this positive attitude toward the incipient revision of the classical conceptual framework, Poincaré's thought in two important aspects remained close to the thought of Mach, Helmholtz and Spencer. In the first place his epistemology was clearly biologically oriented. This may sound surprising because Poincaré's theory of knowledge is known under the name "conventionalism", and it can hardly be doubted that this term describes correctly an important aspect of his thought.[23] It is true that the essence of Poincaré's view was seemingly contained in the following passage of his article in *The Monist* in 1898:

Our choice is therefore not imposed by experience. It is simply guided by experience. But it remains free; we choose this geometry rather than that geometry, not because it is more *true*, but because it is more *convenient*. To ask whether the geometry of Euclid is true and that of Lobatchevski is false, is as absurd as to ask whether the metric system is true and that of the yard, foot, and inch is false.[24]

But, unfortunately, as René Berthelot pointed out in his monograph, the term 'convenient' is not always used by Poincaré in the same sense.[25] There are at least two senses of this word in his writings: (a) convenience in the sense of *logical simplicity* and (b) convenience in the sense of *biological usefulness*. It is clear that these two meanings are different; logical simplicity does not always mean a practical advantage in a biological sense. In choosing a geometry which is logically simpler, we are, according to Poincaré, making a *free choice*, a *free convention* which is not imposed by experience, even though, as Poincaré concedes, we are 'guided' by it. But what is the exact meaning of the word 'guided'? Poincaré never succeeded in making this term clear as the following two quotations from his writings plainly show:

We choose the geometry of Euclid because it is the simplest. If our experiences should be considerably different, the geometry of Euclid would no longer suffice to represent them conveniently, and we should choose a different geometry.[26]

This was written in 1898 and sounded open-minded enough; in truth, in retrospect, it sounds almost prophetic. But let us compare it with what Poincaré wrote only four years later:

What is important is the conclusion: experiment cannot decide between Euclid and Lobatchevski. To sum up, whichever way we look at it, it is impossible to discover in geometric empiricism a rational meaning.[27]

And even more emphatically:

The Euclidean geometry has, therefore, nothing to fear from fresh experiments.[28]

The contradiction is obvious; Poincaré first considered the possibility that we may in the future, 'guided' by experience, 'prefer' the geometry of Lobatchevski to that of Euclid; then, only a few years later, he declared, in what H. P. Robertson characterized as the 'pontifical pronouncement'[29], that such possibility will never occur. The explanation for this oscillation of Poincaré's attitude should be sought, in my view, in those features of his epistemology which attracted less attention, especially since the classification of him as a conventionalist definitely obscured them. As already mentioned, René Berthelot recognized the importance of these features which show the affinity of Poincaré's thought with the biological theory of knowledge.

There is no question that Poincaré's epistemology was basically empiricist. But his empiricism was different from that of Hume and Mill in that it stressed the importance not only of individual, but of ancestral experience as well. In truth, in the formation of our representation of space, Poincaré regarded the ancestral experience as far more decisive:

It has often been said that if individual experience could not create geometry the same is not true of ancestral experience. But what does it mean? Is it meant that we could not experimentally demonstrate Euclid's postulate, but that our ancestors have been able to do it? Not in the least. It is meant that by natural selection our mind had *adapted* itself to the conditions of the external world, that it has adopted the geometry most *advantageous* to the species: or in other words the *most convenient*. This is entirely in conformity with our conclusions; geometry is not true, it is advantageous.[30]

Similarly, in considering the possibility of imagining a four-dimensional space, he says:

That is possible, but that is difficult because we have to overcome a multitude of associations of ideas, which are the fruit of a long personal experience and of still longer experience of the race. It is these associations (or at least those of them which we have inherited from our ancestors), which constitute this *a priori* form of which it is said that we have pure intuition. Then I do not see why one should declare it refractory to analysis and should deny me the right of investigating its origin.[31]

This genetic approach is even more explicit in the following passage in *Science and Method*:

It is this complex system of associations, it is this table of distributions, so to speak, which is all our geometry or, if you wish, all in our geometry that is instinctive. What we call our intuition of the straight line or of a distance is the consciousness of these associations and of their imperious character itself. An association will seem to us by so much the more indestructible as it is more ancient. But these associations are not, for the most part, conquests of the individual, since their trace is seen in the newborn babe: they are conquests of the race. Natural selection had to bring about these conquests by so much the more quickly as they were the more necessary...

We see to what depths of the unconscious we must descend to find the first traces of these spatial associations, since only the inferior parts of the nervous system are involved. Why be astonished then at the resistance we oppose to every attempt made to dissociate what so long has been associated? Now, it is *just this persistence that we call the evidence for the geometric truths*; this evidence is nothing but the repugnance we feel toward breaking with very old habits which have always proved good.[32]

Finally, there is his concluding statement which sheds light on the nature of his so-called conventionalism:

We see that if geometry is not an experimental science, it is science born apropos of experience; that we have created the space it studies, but adapting it to the world wherein we live. We have selected the most convenient space, but experience has guided our choice; as this choice has been unconscious we think it has been imposed on us; some say experience imposes it, others that we are born with our space ready made; we see from these preceding considerations, what in these two opinions is the part of truth, what of error.[33]

The last passage indicates how close Poincaré's thought was to that of Herbert Spencer, even though no explicit reference was made to the latter. True, there were also important differences in language. Contrary to the view of Spencer and Taine, Euclidian space was, according to Poincaré, not imposed on us by experience, but 'chosen' by us. But Poincaré's own language makes this difference far less significant than it originally appeared; in truth, we are tempted to ask whether it is not merely a difference in words. By admitting that our choice is 'guided' by experience and that it is 'unconscious', Poincaré moved far closer to Spencer's view than is generally realized. Is it still meaningful to use the word 'choice' in its usual conventionalistic sense if it is 'unconscious', i.e. an impersonal result of the accumulated ancestral experience? The term 'adaptation' is far more accurate to describe it; this is also the term which Poincaré occasionally used. But if this alleged choice is a result of natural selection or of the 'adaptation' to the objective features of reality, it ceases to be an arbitrary choice; it is, whether Poincaré liked the word or not, really *imposed* by experience, and we cannot do anything about it. Does this implied consequence explain Poincaré's statement that no future experience can ever be a threat to our 'choice' of the geometry of Euclid? I am inclined to believe so, even though Poincaré never explicitly conceded it. He always insisted that we are free to choose a different geometry; moreover like Helmholtz he believed that we are not only able to abstractly conceive, but even to *imagine* the relations in a non-Euclidian space; he even went further than Helmholtz in claiming that we are capable of psychological representation of a four-dimensional space.[34] There is no doubt that this belief in the plasticity of the human imaginative capacities represents the most important difference from Spencer's view, which was the traditional view in general, on the inconceivability of any non-Euclidian theorem. But Poincaré also always insisted that the adoption of any geometry different from that of Euclid as well as any effort to imagine it, though possible in principle, would be highly cumbersome and uneconomical. Even if the discovery of the negative parallax would suggest that cosmic space has a slightly Riemannian curvature, it would be more *convenient*, more *economical*, logically *simpler* to assume a slight modification of the laws of optics rather than to adopt the geometry of Riemann. It is clear that in this particular case Poincaré used the two basic terms in considerably different senses. In this context, 'choice' means a *conscious* choice in the true conventionalistic sense while 'convenient' means 'logically simpler'. But apparently he never anticipated the possibility that new empirical discoveries may be of such kind that the adoption of non-Euclidian geometry *together with the modification of the laws of dynamics* may represent a logically simpler alternative. As Moritz Schlick pointed out, Poincaré overlooked that what should be taken into consideration is the logical simplicity of the *whole system of explanation*, not merely of one single field like geometry.[35] But why did he overlook it? Was it not because he believed that our acceptance of Euclidian geometry is a result of the ancestral experience, of a *complete* adaptation to the *objective* features of reality? And consequently, that while some future empirical discovery may impose a modification of some specific laws of physics, it will *never* suggest that the space in which we live is only *approximately* Euclidian? In this respect the implications of Poincaré's view were as conservative as the confident beliefs of the nineteenth century positivism and neo-Kantianism.

This view is substantiated by Poincaré's attitude toward the classical principle of causality. Nothing shows this more convincingly than two essays contained in his posthumously published *Dernières pensées*. In the essay 'L'evolution des lois' he considers the question raised by Émile Boutroux, whether the laws of nature themselves can vary with time. He answers that any observed variability of any law would merely induce scientists to search for *an explanation* of such variability; and explanation would finally be found in another more comprehensive law from which the observed variation would follow:

We would therefore conclude that the laws are variable, but we must note carefully that this would be by virtue of the principle of their immutability.[36]

Thus certain regularities of the macroscopic world may be found variable; but their variations will be explained by the kinetic theory of matter as resulting from different arrangements of the molecules. In other words, the laws of molecular motion would retain the immutability in virtue of which some apparent irregularities would be explained. It is clear that at the time of his death Poincaré firmly believed in the rigorous determinism of the microphysical world; similarly, he did not have the slightest doubt about the absolute validity of Newton's law:

Leverrier, knowing the present orbits of the planets, can calculate by means of Newton's law what these orbits will be 10,000 years hence. Whatever method he uses in his calculations, he will never be able to find Newton's law to be false in a few millennia. He could have calculated, simply by changing the sign for the time factor in his formulae, what these orbits were 10,000 years ago. *But he is certain beforehand* not to discover that Newton's law has not always been true. (Italics mine.)[37]

Poincaré expressed himself even more strongly in the last essay of the same book:

Science is deterministic; it is so a priori; it postulates determinism because, without it, it could not exist. It is deterministic a posteriori also; if it began by postulating it, as an indispensable condition for its existence, it proves it later by the very fact of existing, and each of its conquests is a victory of determinism.[38]

It is interesting to note that *Dernières Pensées* appeared in 1913; only a few years before the astronomical observation verified one specific prediction of Einstein's new theory of gravitation which explained the facts unexplainable by Newton's law. Hardly more than a decade later the classical principle of causality began to be questioned, in direct opposition to Poincaré's belief in the a priori, deterministic character of science. In other words, Poincaré's prediction that no future experience will ever contradict the validity of Newton's law or the immutability of the molecular laws of motion was as little borne out by the subsequent development of physics as was his overconfident claim that "no future experience will ever constitute a threat to the Euclidian geometry". It would, of course, be unfair to blame Poincaré for this lack of foresight; the developments of the twentieth century in physics were so revolutionary that they could have hardly been anticipated. But René Berthelot was correct when he pointed out that Poincaré never considered seriously Boutroux's bold and prophetic claim about the genuine contingencies in nature.[39] It is true that he briefly touched on this question only a few lines after the passage quoted above when he asked:

Can we admit that this forward march of determinism will continue without halting and without backward movement, without encountering insurmountable obstacles, and that nevertheless we do not have the right to pass to the limit, as we mathematicians say, and to infer an absolute determinism because at the limit determinism *would vanish in a tautology or in a contradiction?* (Italics mine.) [40]

We shall understand later the profound significance of the italicized words; but Poincaré merely raised this question and then passed it by. Although less conservative than Mach and although witnessing the first decade of the twentieth century revolution in physics, Poincaré remained loyal to three basic dogmas of the Newtonian age: Euclidean geometry, Newton's gravitation law, and microphysical determinism.

## NOTES

[1] H. von Helmholtz, 'Über die Ursprung und Bedeutung der geometrischen Axiome', in: *Vorträge und Reden*, Braunschweig 1884, II, pp. 1–31; 'Die Tatsachen in der Wahrnehmung', *ibid.*, II, pp. 217–271, in particular, Beilage II: 'Der Raum kann transcendental sein, ohne dass es die Axiome sind'. The first article was translated with insignificant modifications under the title 'The Origin and Meaning of Geometrical Axioms' in *Mind* III (July 1876), 302–321; a large part of the second article was incorporated into #26 of the second edition of Helmholtz's *Treatise on Physiological Optics* (the English translation by James P. C. Southall, Dover, New York, 1962), III, pp. 1–37.

[2] 'Zahlen und Messen', in *Heinrich von Helmholtz Schriften zur Erkenntnistheorie* (edited and annotated by Moritz Schlick), Berlin 1921; hereafter referred to as Helmholtz-Schlick. The English translation *Counting and Measuring*, (translated by C. L. Bryan, with an introduction and notes by H. T. Davis), Van Nostrand, Princeton, 1930. Helmholtz points out that even the applicability of the simple commutation law presupposes that the objects remain permanent, distinct and that none of them either disappear or merge with another – the properties which can be known only by experience. (Cf. Helmholtz-Schlick, *op. cit.*, p. 82.)

[3] Cf. 'Die Tatsachen in der Wahrnehmung', p. 226; Helmholtz-Schlick, *op. cit.*, p. 116; *Treatise on Physiological Optics*, II, p. 4: "The quality of the sensation is thus, so to speak, merely a *symbol* to our imagination, a sort of earmark of objective reality."

[4] Cf. M. Schlick, 'Helmholtz als Erkenntnistheoretiker', in *Helmholtz als Physiker, Physiologe und Philosopher*, Karslruhe 1922, pp. 29–39. Cf. also Note 2. A comprehensive evaluation of Helmholtz' significance can be found in Victor F. Lenzen's essay 'Helmholtz's Theory of Knowledge', in *Studies and Essays in the History of Science and Learning in Honor of George Sarton* (1946), pp. 301–319.

[5] This is what Schlick calls "psychological a priori" in contrast to the Kantian "transcendental a priori". Helmholtz-Schlick, *op. cit.*, pp. 117–118; Note 23, p. 158.

[6] H. Spencer, *The Principles of Psychology*, 3rd Ed., Appleton, New York, 1897, II, 2, § 428, p. 412.

[7] *Vorträge und Reden*, II, p. 22; *Mind* III, *loc. cit.*, p. 314.

[8] Cf. *Vorträge und Reden*, II, pp. 26–28 (*Mind* III, 316–318) where Helmholtz describes the hypothetical experiences in the spherical and pseudospherical spaces. On the other hand, against the later view of Poincaré he insists on the absolute impossibility to imagine the fourth dimension (p. 28). Helmholtz's view about the imaginability of

non-Euclidian geometries was challenged by Alois Riehl, 'Helmholtz in seinem Ver-
hältnis zum Kant', *Kantstudien* IX (1904) 278–279; by J. P. N. Land, 'Kant's Space and
Modern Mathematics', *Mind* IV (1877) 38–46. On the other hand, Helmholtz's
criterion of imaginability i.e. visualization ("the power of imagining the whole series of
sensory impressions that would be had in such a case") was accepted by Reichenbach
in *The Philosophy of Space and Time*, Dover, New York, 1956, ≠11 ('Visualization of
non-Euclidian Geometry').

[9] *Vorträge und Reden*, II, p. 247: "Das Kausalgesetz ist ein a priori gegebenes, trans-
cendentales Gesetz." See his quotation of Schiller on p. 244.

[10] Mach's influence on Einstein discussed extensively and many times, in particular by
Philip Frank, was admitted by Einstein himself, though not without qualifications. Cf.
*Albert Einstein: Philosopher-Scientist* (ed. by Paul Schilpp), Evanston, Ill., 1949, p. 21;
also the essays of H. Margenau, V. Lenzen and P. Frank in the same volume; Gerald
Holton, 'Mach, Einstein and the Search for Reality', *Daedalus* 19 (1968).

[11] *Erkenntnis und Irrtum*, Leipzig 1906, p. 145–146; 426.

[12] Cf. *Popular Scientific Lectures* (transl. by Thomas J. McCormack), The Open Court,
Chicago, 1910, p. 223: "with the new adaptation to the enlarged field of observation
the problem disappears, or, in other words, is solved."

[13] *Space and Geometry*, The Open Court, Chicago, 1906.

[14] E. Mach, *The Analysis of Sensations*, Dover, New York, 1959, p. 250.

[15] E. Mach, *The Science of Mechanics* (transl. by Thomas J. McCormack), The Open
Court, Lasalle, Ill., 1942, p. 582.

[16] *Ibid.*, p. 560.

[17] Quoted by R. Bouvier, *La pensée d'Ernst Mach*, Paris, 1923, p. 170. Bouvier
repeatedly stressed the biological orientation of Mach's theory of knowledge and its
affinity with that of Spencer. The quotation is from *Erkenntnis und Irrtum*, 2nd Ed.,
Leipzig 1906, p. 27.

[18] *Die Principien der Wärmelehre*, 4th Ed., Leipzig 1923, pp. 392–393. Mach answers
here Petzoldt's criticism. Cf. also *Erkenntnis und Irrtum*, p. 64: there is as little choice in
will as in the geotropism and heliotropism of plants or in the fall of a stone.

[19] Both passages are on the same page of *The Science of Mechanics* (p. 609).

[20] *Principien der Wärmelehre*, pp. 455–456.

[21] E. T. Whittaker, *A History of the Theories of Aether and Electricity*, II, London 1953,
p. 27.

[22] *Dernières Pensées*, Paris 1923, p. 188.

[23] On this aspect of Poincaré's thought cf. A. Grünbaum, *Philosophical Problems of
Space and Time*, Alfred Knopf, New York, 1963, pp. 115–131.

[24] 'On the Foundations of Geometry', *The Monist* IX (1898) 42.

[25] René Berthelot, *Un romantisme utilitaire*, Vol. I, *Le pragmatisme chez Nietzsche et
Poincaré*, Paris 1912. On different meanings of the term "commodité" in Poincaré see
Ch. III where the author makes an interesting comparison of Poincaré with Mach; also
pp. 286–308.

[26] Cf. Note 24.

[27] 'Science and Hypothesis', in *The Foundations of Science* (transl. by G. B. Halsted),
The Science Press, Lancaster, 1946, p. 86. To be referred to as *F.S.*

[28] *Ibid.*, p. 81.

[29] H. P. Robertson, 'Geometry as a Branch of Physics', in *Albert Einstein: Philosopher
Scientist*, p. 325.

[30] *F.S.*, p. 91.

[31] *The Value of Science* (in *F.S.*), p. 274.
[32] *F.S.*, pp. 420–421.
[33] *F.S.*, p. 428.
[34] *F.S.*, pp. 78–79.
[35] Helmholtz-Schlick, *op. cit.*, Note 38.
[36] H. Poincaré, *Last Essays* (transl. by John W. Dolbuc), Dover, New York, 1963, p. 12.
[37] *Ibid.*, p. 3.
[38] *Ibid.*, p. 112.
[39] *Un romantisme utilitaire*, I. p. 256.
[40] *Op. cit.*, p. 112.

# CHAPTER 3

# BERGSON'S AMENDMENT OF THE CLASSICAL
# BIOLOGICAL THEORY OF KNOWLEDGE

The biological orientation of Bergson's epistemology has been already mentioned; so was his own admission of Spencer's influence on his thought. It remains to be shown in what sense and to what degree Bergson's epistemology went beyond the classical biological theory of knowledge sketched in the preceding paragraphs. As is the case with Spencer, Helmholtz, Mach and Poincaré, Bergson accepts the view that the present form of human intellect is a result of the gradual evolutionary adaptation of the human psychophysical organism to the order of nature. But is this adaptation complete? Is the objective order of nature represented adequately, completely and without distortion in the present Newtonian-Euclidian form of human intellect? This is the question which Bergson raised, the question which Spencer answered affirmatively or rather did not even raise, so obvious did the answer appear to him. Although Helmholtz, Mach and Poincaré sometimes asked to what extent our cognitive forms correspond to the objective features of nature, they did it only occasionally or hesitatingly, without challenging in any radical way the basic dogmas of the nineteenth century mechanistic science.

If Bergson's answer to the question above differs radically from that of Spencer, Helmholtz, Mach and Poincaré, it is because he made a more consistent use of the biological theory of knowledge from which he drew consequences which his predecessors failed to draw. According to Bergson, the Newtonian-Euclidian, or rather the Laplacean-Euclidian form of intellect *does not adequately represent nature in its entirety, but merely the sector of it which is of vital importance for the human organism.* In other words, the basic error of most theories of knowledge is, according to him, "that nature is one and that the function of intellect is to embrace it in its entirety."[1] But this is not so; while nature is *ultimately* one – and Bergson, whose thought is, despite frequent misrepresentations, much closer to Leibniz' panpsychism than to Cartesian dualism, is certainly far from denying it – it is diversified into strata which, while not being separated by sharp boundaries, are still sufficiently heterogeneous. It is precisely their heterogeneity which prevents them from being all "intelligible" in the same sense, at least as long as we define "intelligibility" in the narrow sense of classical physical science. It seems that in truth, what is "intelligible" in the classical sense is what Reichenbach appropriately called "the world of middle dimensions"[2] located between microcosmos and "megacosmos", i.e. between the zone of atomic events and the universe as a whole: *if* there is a whole! It is clear that the zone of applicability of the classical form of intellect, no matter how large, is still limited; it does not extend either below or above the limits of the middle dimension.

This arresting fact is largely ignored by most epistemological theories today. Even biologically oriented John Dewey hardly paid any attention to it; and when it is recognized, its significance is usually not fully grasped and not all the consequences are drawn from it. In any case, hardly ever was any attempt made to explain this fact. Why do certain types of explanation which were so beautifully and triumphantly successful for more than two centuries within such a large field of experience, cease to be applicable beyond the limits of this field? Aristotle observed that philosophy begins with wonder; the reason why the capacity for wonder is so conspicuously lacking in this particular case is probably due to the fact that a wide gulf still separates epistemological speculations (exclusively preoccupied with the abstractions of methodology or with irrelevant linguistic games) from concrete problems which emerge in concrete sciences. I am not aware of any other plausible explanation of the fact mentioned above except that given by the amended version of the biological theory of knowledge. This theory may be summarized as follows: *the limited applicability of the classical (i.e. Newtonian-Euclidian) modes of thought is due to the fact that they themselves are products of evolutionary adjustment to a limited segment of reality; consequently, when by the process of extrapolation we try to apply them outside of the zone to which they are adjusted, their inadequacy becomes obvious – the more so, the further beyond the limits that they are applied.*

It is now time to consider this explanation in a more detailed way. Using Spencer's terminology; although there is a general correlation "between the internal and external relations", this correlation is far from being complete even when human perception is considered; no "objective relations" whose practical significance is negligible, produce any consciously registered repercussion in the organism. As early as 1879 William James in his penetrating criticism of Spencer's definition of mind showed that human knowledge is essentially *selective* even in its apparently most passive form, that of the level of sensory perception.[3] What we call "external perception" is never a photographic replica of the surrounding reality; our sensory apparatus register only some physical influences, and by their very structure eliminate others. The number of examples which James gives can be greatly increased; to the common knowledge that we cannot hear with our eyes or see with our ears, we have to add the limitations of each sense organ in its responsiveness to its specific stimuli. Our sight, for instance, not only ignores the minute amount of light coming from bodies which are either too distant or too small, but by its own structure does not respond to radio-waves, infrared, ultraviolet and X-rays; the visible colors registered by our sight are merely a very narrow portion of the immense electromagnetic spectrum. This is true of other senses as well; our hearing does not register ultrasonic waves; the skin of a normal person is insensitive to tiny variations in the pressure of the air to which the skin of a blind person reacts; our senses of taste and smell are very limited, at least in comparison with the senses of other animal species. As we shall see, the sensory organs of animals are as much selective as human organs, though in a startlingly different way. What is the cause of this selective character of human as well as animal perception?

Bergson's answer to this question is contained in the first chapter of *Matter and Memory*, completed later by a number of passages of *Creative Evolution*. In the latter book the analysis of sensory perception, originally confined to the human organism only, was set into a broader evolutionary setting by pointing out briefly the selective nature of animal perception in general. Bergson's answer can hardly be doubted by any competent biologist; in truth, it is now regarded as a truism. It is neither original – it was fully anticipated by Spencer, Helmholtz, Mach and by the physiology of sense organs in general – nor startling; what is original and startling is, as we shall see, the epistemological consequences which he drew from it. Briefly, the answer is as follows; the selective nature of human and animal perception is determined by the *general teleology of organism*. It is vitally important for every living being that those features of its surroundings which have a bearing on its survival and well-being are signaled; while it is *economical* that other features of the same surrounding which have no or negligible significance for the organism are not registered. (We use here the term "teleological" in an entirely neutral and noncommittal way; it is nothing but an adjective designating the observed features of the sensory receptors in man and animals.) It is irrelevant for our present purpose to ask whether a mechanistic explanation of these features is adequate or not. Although the answer to this question sharply separated Bergson from Spencer, Helmholtz, Mach and mechanism in general, it may temporarily be set aside, since the existence of the teleological features cannot be doubted whatever genetic explanation of them is accepted.

In reading the key passages in *Matter and Memory* we have to bear in mind that biological, physical and psychological considerations are couched by Bergson in the language of methodological solipsism or phenomenalism, which at first glance appears to be in strange contrast to the terminology of critical realism present especially in the last chapter. A phenomenalistic terminology, however, was purposely chosen to forestall possible accusations of naive realism. Thus the body and the brain are called 'images' and their privileged character in an individual perception is shown by their comparison with other 'images', that is the physical bodies which constitute 'the external world';

In fact, I note that the size, shape, even the color of external objects is modified according as my body approaches or recedes from them; that the strength of an odour, the intensity of the sound, increases or diminishes with distance; finally that this very distance represents, above all, the measure in which surrounding bodies are insured, in some sort against the immediate action of my body. In the degree that my horizon widens, the images which surround me, seem to be painted upon a more uniform background and become to me more indifferent. The more I narrow this horizon, the more the objects which it circumscribes space themselves out distinctly according to the greater or less ease with which my body can touch and move them. They send back, to my body, as would a mirror, its eventual influence; they take rank in an order corresponding to the growing or decreasing powers of my body. *The objects which surround my body reflect its possible action upon them.*[4]

Bergson completes this statement by stressing the reciprocal character of the relation between organism and environment; the decreasing distance means not only a greater probability of the action of our body on its environment, but also an intensification of the action of the surrounding objects on our organism, or at least on the special sensitive areas of its surface which we call 'sensory organs'. The action of a distant object on our body is conveyed either by luminous or sound waves or by the molecules escaping from odorous substances. When the distance is zero, the action takes a form of direct contact whether an object touches various areas of our skin or is dissolved on the surface of the tongue. It is obvious that perception generally depends on the distance of an object from our sensory organ; the smaller its distance is, i.e. the more imminent its virtual promise or threat to our body becomes, the more distinct it appears; while, conversely, the greater its distance, the smaller is its action on our body. Consequently, the more distant the objects are, the more "both promises and threats defer the date of their fulfillment"[5]; at the same time the details of the perceived object, whether it is a color and shape for the eyes or the structure of sound for the ears, grow gradually dimmer and dimmer until they eventually shrink into single faintly perceived qualities. Thus the perception of a huge celestial body is reduced to the sensation of a single luminous point, the perception of distant gun fire shrinks to a hardly perceived noise. When the distance is further increased, a corresponding sensation disappears; thus the gunfire from the Dunkirk battlefield which was audible on the southern coast of England, could not be heard in Scotland; the planets outside of Uranus's orbit cannot be seen by a naked eye.

But the last example shows that the influence of an object and its corresponding perception does not exclusively depend on its distance; there are hundreds of asteroids inside of the orbit of Jupiter which nevertheless do not affect our naked eye; furthermore, there are tiny objects in an immediate neighborhood of our body or even on its surface, like microorganisms, which remain unperceived and can be observed only by microscope or some other special means of detection. Obviously, the things which are too small have as little effect on our body or its sense organs as the things which are too far. This follows from the well known psychophysical law: all physical influences must attain a certain degree of intensity, the so called 'threshold of excitation', to be registered physiologically and psychologically. All influences which lie below this threshold are by definition ineffective and therefore unregistered by our senses. It matters little whether the inefficacy of the stimulus is due to the smallness of an object or the effect of large distance; the inverse square law produces the same effect as a small size – a visual angle of Jupiter is the same as that of a grain of sand. What is epistemologically most important is the general conclusion; not only the distant regions of the universe, but also *its whole microphysical layer* are beyond the limits of spontaneous human perception. This is in conformity with the biological character of perception; generally those physical objects which are either too small or too far away to produce a significant physiological effect do not have any practical significance for the organism; their presence in our perception would be a sheer luxury while their absence does not generally constitute a serious biological disadvantage.

The apparent exceptions merely confirm the general rule. It is known that certain sensations of odor and taste are astonishingly precise; the amount of substance which they register is so minute that it escaped the detection of the nineteenth century chemical and spectroscopic analysis. This fact inspired Nietzsche to his hymnical praise and defense of the senses against "the slanders of idealists".[6] Thus the sweetness of sugar is perceived by man in the solution of the ratio 1:100; but the presence of sulphuric acid is noticed when the ratio is 1:10000; of strychnine when it is 1:2000000! To produce the sensation of scent only five thousandths of a milligram of camphor are necessary; in the case of ether one thousandth, for citral one half a thousandth, for vanillin five millionths of a milligram. Ethyl mercaptan is perceived when only one hundredth of a milligram is present in 230 $m^3$.[7] But we must not overlook the fact that the substances which are perceptible in such minute quantities are generally those which have a *vital importance* for the organisms whether in a positive or negative sense; whether they are poisons like strychnine or alimentary substances like sugar it is important that the organism should register them as quickly as possible. Thus it is not only the size and the distance if an object which indicate whether it is a potential promise or a threat for the organism, but also its *chemical composition*; when this composition has either a favorable or a dangerous effect on the human body, it is vitally important that our perception would detect it even if it is present in minute quantities. Thus an apparent exception merely confirms the teleological character of human perception asserted by the biological theory of knowledge. It is true that, like in organic nature in general, this teleology is far from perfect since it is complicated by certain accidents of evolution or by sheer inertia of organic tissues. There is hardly any question that in some instances the sensitivity to certain minute stimuli survived its biological usefulness in a way similar to that by which the vestigial organs survived their original function. This is especially true of man whose sensory organs are more and more extended and even replaced by artificial instruments of observation.

A comparison of animal sense organs with those of men will bring even more into focus the utilitarian and biologically conditioned character of sensory perception in general. Out of the immense network of physical actions constituting 'the external world', the sensory apparatus, which are prodigiously different in different species, eliminate, select, or magnify different parts. Hence the enormous variety of 'the animal milieu' or *Umwelt*, stressed by Bergson, and elucidated more systematically by Jacob von Uexküll two years after the publication of *Creative Evolution*.[8] Every such milieu perceived by different sensory organs represents a *useful selection* out of the total physical surrounding – needless to say, useful in the evolutionary sense, that is for the species in the question rather than for an individual.

Every animal is a subject, which, in virtue of the structures peculiar to it, selects stimuli from the general influences of the outer world, and to these it responds in a certain way. These responses, in their turn, consist of certain effects on the outer world, and these again influence the stimuli. In this way there arises a self-contained cycle, which we may call the *function-circle* of the animal[9].

Illustrations abound. We have already mentioned the blindness of man to the ultraviolet rays which are perceived by the bees; this undoubtedly has a great biological importance for them since some flowers or their parts like, for instance, the tips of the petals of the common yellow daisy reflect the ultraviolet part of the spectrum. Both the male and female of the luna moth which appear pastel green to our eyes are differently colored in ultraviolet light. On the other hand, bees are red-blind; and if they still visit red poppies it is because they perceive the ultraviolet rays reflected by these flowers.[10] The rattlesnake has close to its eyes a pair of dimples which are sensitive to infrared rays by which it can detect its prey inconspicuous in the normal range of vision.[11] The visible range of the electromagnetic spectrum varies widely from species to species and the selective sensibility to various wave-lengths always has a biological significance.

Nor is it different when we turn our attention to other senses. The vital significance of the perception of ultrasonic vibrations in bats, moths and porpoises was only recently discovered; it plays the essential role in 'echolocation' by which food, obstacles and enemies are detected.[12] Even more obvious is the biological character of perception in the remaining senses. The action of condensed moisture on the nostrils and of the dissolved substance on our tongue was called by Spencer "evanescent form of nutrition"[13]; but this is only partly true. For the sensations of taste and scent, besides signalling the presence of food in very minute quantities, register perhaps with even a greater precision tiny traces of destructive and poisonous substances as the above mentioned example of human sensitivity to sugar and strychnine shows. The acuity of taste and scent in certain species transcends widely that of human senses. Sugar which is tasted by man in the solution 1:100 is perceived by flies in the ratio 1:1800. On the other hand, out of 34 sugars and sugar-like substances which taste sweet for human beings only nine are sweet for bees; and trisaccharide rafinose, which is tasteless for bees is the sweetest to the ants tested by von Frisch.[14] As far as scent is concerned there are some insects which perceive sexual odors at the distance of several miles.[15] The enormous significance of odor in the sexual life of animals shows most clearly how Spencer's characterization of scent as an "evanescent nutrition" is far too narrow.

Besides the fact that the selection of *Umwelt* out of the total physical surroundings is determined by general biological needs, two other facts should be borne in mind:

(1) Despite the fact that the human senses are often much blunter than the corresponding sense of certain animals, the total *Umwelt* of the human species is incomparably wider than any world perceived by any animal. And (2) No perception, whether human or animal, no matter how astonishingly precise this may be, can ever register the inner structure of the microcosmos.

Re (1), the superiority of human perception to that of other animals is due not so much to the superiority of his sense organs, but to what Herbert Spencer appropriately called "artificial senses"[16] by means of which the sensory field of perception was enormously widened in space. We shall return to this point soon. But this spatial widening of our sensory perception went concomitantly with the increasing imaginative and conceptual grasp of the more and more distant events in time. Both processes were actually inseparable; in Spencer's jargon, "the correspondence extending in space" went together with "the correspondence extending in time"; in Bergson's words, "*la perception dispose de l'espace dans l'exacte proportion où l'action dispose du temps*".[17] By its very definition, our imaginative and conceptual grasp of the wider sequences of events depends on the development of memory and anticipatory imagining and reasoning without which no planned action, no matter how rudimentary, is possible. Thus the zone of reality which man perceived and to which he reacted both in imagination and action increased continually both in space and time. The passages in Spencer's *Principles of Psychology* illustrating by numerous examples this "concomitant extension of correspondence in both space and time" are still relevant and convincing despite their Victorian style.

Re (2), in spite of the prodigious diversity of the sense organs and of the perceptions which are conditioned by them, there is one feature common to the whole animal realm, including *homo sapiens*; all perceptions, even those which are the most subtle and most discerning, are too crude to register the ultimate elements of the physical world, and they barely reach the upper boundaries of the microcosmos. No matter how astonishingly minute is the minimum quantity of ethyl mercaptan registered by our sensation of odor – no more than 0.000 000 000 000 071 ounces must be inhaled – it still contains no less than 19 400 000 000 molecules![18] And we must not forget that molecules, especially the organic ones, are very complex aggregates, dimensionally still far removed from the world of electrons and quanta. Yet, it is this molecular level only which our vision can reach when improved by the electron microscope. Hence a conclusion of the foremost importance for the theory of knowledge: the external world such as it appears to the perception of man and of higher vertebrates does not represent the totality of the physical world, but only its part or rather its *parts*. For the perception of each species registers out of the total surrounding only those segments which are biologically significant, i.e., those which are related to its vital needs; and since these needs vary from species to species, the character and the extent of the corresponding perceptions vary accordingly. In Uexkull's words, "there are as many surrounding worlds (*Umwelten*) as there are animals."[19] But even the world of man, though incomparably wider and more accurate than the *Umwelt* of animals, still remains, to use Henri Piéron's concise expression, merely *l'univers partiel*, a mere part of the total picture.[20] It represents fairly accurately a wide segment of the physical reality located between the microcosmos and megacosmos; but the minute world of electrons, nucleons and quanta as well as the huge spaces of the extragalactic dimensions remain forever beyond its reach.

## NOTES

[1] *Creative Evolution* (transl. by Arthur E. Mitchell), The Modern Student Library, New York, 1944, p. 209. Hereafter referred to as *C.E.*

[2] H. Reichenbach, *Atom and Cosmos* (transl. by Edward S. Allen), G. Braziller, New York, 1957, p. 38, 237, 288. Cf. also my article 'The Development of Reichenbach's Epistemology', *The Review of Metaphysics* **XI** (1957) 42–67.

[3] W. James, 'Remarks on Spencer's Definition of Mind as Correspondence', in *Collected Essays and Reviews*, Longmans, Green, London, 1920, pp. 43–68; *The Principles of Psychology*, Dover, New York, 1950, I, pp. 284–289.

[4] *Matter and Memory*, Doubleday Anchor Books, 1955, p. 5. (Hereafter referred to as *M.M.*)

[5] *M.M.*, p. 17.

[6] F. Nietzsche, 'Twilight of Idols', in *Portable Nietzsche* (ed. by W. Kaufmann), Viking Press, New York, 1963, p. 481.

[7] J. Parry, 'Sur quelques minimes perceptibles d'odeur', *Compt. Rend.* **CXIV** (1892) 786–788. H. Piéron, 'Les univers des animaux et l'univers de l'homme', *J. de Psychologie* **34** (1941), 1–28.

[8] J. von Uexküll, *Die Umwelt und die Innenwelt der Tiere*, Berlin, 1909; *C.E.*, pp. 193, 207.

[9] J. von Uexküll, *Theoretical Biology*, Harcourt & Brace, New York, 1926, p. 126.

[10] Karl von Frisch, *Bees, their Vision, Chemical Senses and Language*, Cornell University Press, Ithaca, 1950, pp. 6–11.

[11] Vitus S. Dröscher, *The Mysterious Sense of Animals*, Dutton, New York, 1965, pp. 23–26.

[12] Dröscher, *op. cit.*, pp. 9–23.

[13] H. Spencer, *The Principles of Psychology*, 3rd Ed., I, p. 309.

[14] H. Piéron, *loc. cit.*, p. 21; K. von Frisch, *op. cit.*, p. 40.

[15] For instance, the marked males of the silkworm moth have been known to fly upwind seven miles to a fragrant female of their own kind. Cf. Lorus and Margery Milne, *The Senses of Animals and Men* (Atheneum, New York, 1962), p. 131. Thus Piéron's estimation of the odor range of the insects – one kilometer – was far too conservative. (Piéron, *loc. cit.*, p. 23).

[16] H. Spencer, *op. cit.*, I, p. 365.

[17] *M.M.* p. 17.

[18] Milne, *op. cit.*, p. 121. The tiniest quantity of sexual odor detected by the male silkworm moths still contains 200000 molecules.

[19] Uexküll, *Theoretical Biology*, p. 176.

[20] Piéron, *loc. cit.*, p. 1.

# CHAPTER 4

# WHY MECHANICAL-PICTORIAL MODELS FAILED

It is true that classical science was fully aware of the limitations of human perception; but it firmly believed that by the artificial lowering of the threshold of perception this limitation could be overcome. Objects too far away as well as objects too small still can be perceived when their physical action on our sensory organs, which is too small to be effective, is concentrated by means of specially constructed devices like the telescope, microscope, seismograph, acoustic concave mirror and other means of detection; and also when conditions can be so arranged that the individual physical action, too weak in itself, can be repeated long enough to produce finally a perceptible trace. The long-time exposure of the photographic plate joined to the telescope is illustrative of both methods combined into one. The general pattern of all such instances was the same: to construct artificially a causal chain leading from the unperceived physical agency to the physical effect perceptible by human senses. The last term in the chain was always either a retinal image or an audible vibration. The very continuity of this chain seemingly guaranteed the belief that its first link cannot be *essentially* different from the last one; in other words that the microscosmic or megacosmic object differs merely in its dimensions from its retinal image. It is true that both telescope and microscope discovered things never seen before and in many respects strange and surprising. It was surprising to see the mountains and the valleys on the moon, yet these objects were *essentially similar* to the valleys and the mountains of the earth; Leeuwenhoek's "strange little animals" were unquestionably weird, yet not weirder than the dragons of our fairy tales or the reconstructed reptiles of the Cretaceous era. Thus the artificial widening of our sensory perception merely strengthened the central inspiring idea of the cosmological revolution of the sixteenth and seventeenth centuries – the unity of nature in space.

This belief was not shaken even when, with the advent of physical optics, it was found that the wave character of light imposes a definite limit on the resolving power of both telescope and microscope. This was regarded as a mere technical limitation which in no way affected the magnification power of what may be called the 'mental microscope', that is of our ability to construct similar figures *at any scale of magnitude*, no matter how large or small. This was the root of the prevailing belief that the microcosmos is a mere miniature of our world of visual and tactile sensations from which it differs only by its dimensions. From the time of Democritus to that of Lorentz the atom was *imagined* as an extremely minute replica of the solid body of our experience; the clashes of atoms and molecules were *imagined* to be similar to the clashes of billiard balls. Even when the solidity of the atomic particles was dissolved into the continuity of the aether, its various hydrodynamical models were borrowed from the same perceptual world of the middle dimensions; the condensations, vortices and waves in aether were believed to be of *essentially* the same kind as the condensations, vortices and waves in the familiar elastic substances of our daily experience, whether air, water or jelly. Similarly, the picture of the universe at large as a huge island of stars and galaxies did not differ except by its dimensions from the image of a cloud of dust particles floating in the air. All explanatory models of classical science were thus mere extrapolations of our sensory experience beyond its upper and lower limit. In John Tyndall's words "the subsensible world"[1] was not essentially different from the world of sensory experience.

As hinted above, the artificial widening of the field of our sensory perception in both directions of the microcosmos and megacosmos seemed at first to confirm the legitimacy of such extrapolation. At the end of the last century Mach, Ostwald, Stallo, Duhem and others, guided by epistemological considerations, believed that the fruitfulness of mechanistic explanations had been already exhausted and that in any case molecules and atoms are mere conceptual constructs which, though methodologically useful, ought not to be naively reified. But the subsequent development of physics quickly discredited all such claims. In 1914 Jean Perrin in his now classical book *Les atomes* summarized all the evidence then available, and convincingly supported the kinetic and corpuscular character of matter. It was impossible then as it is impossible now to deny the objective existence of atoms as Ostwald himself eventually conceded.[2] It is true that the atom ceased to be indivisible; but its very complexity strongly suggested the existence of its corpuscular components which were first called 'corpuscles' before they acquired the present name 'electrons'.[3] Being the atoms of both matter and electricity, electrons provided the unifying substrate for the mechanical and electromagnetic phenomena. Their orbital arrangements around the positive nucleus provided a satisfactory explanation for the periodical regularities of the Mendelejev system and reduced the allegedly qualitative differences between chemical elements to differences of configuration only. The separation from and addition of an electron to a complex atom accounted for the phenomena of positive and negative ionization; chemical affinity itself ceased to be an occult irreducible quality since it seemed but a special case of electrostatic attraction. Finally, the empirical evidence for the correlation between the atomic spectra and the displacements of orbital electrons apparently strengthened the hope of finally finding the clue to the origin of electromagnetic waves; it is the electrons, it was believed, whose vibratory displacements inside the atom generate in the aether the undulatory disturbances of which luminous waves are only a very small part. In the Wilson cloud chamber and in the spinthariscope it was possible directly to perceive, if not the electrons themselves, at least their tracks marked by droplets of condensed vapour or the traces of their impact on the fluorescent screen; in Millikan's oil drop experiment it was even possible to manipulate a single electron. For all practical purposes their existence seemed to be directly verified. This allowed Abel Rey and Léon Brunschwicg to write in 1918 about "the renaissance of kinetism" in physics.[4]

But this was merely the first more or less visualizable phase of the electron theory which represented the last success of the intuitive and mechanistic models in microphysics. There is no place here to list and analyze all the failures of the visual-tactile models in microphysics which are not only the most spectacular, but also the most significant features of contemporary physics. Let us merely recall the most important trends. The increasing difficulties in constructing a satisfactory model of the aether culminated in the negative result of Michelson's experiment which showed that the hypothetical medium does not possess even the most elementary classical kinematic properties; if we continue to insist on using the word 'aether', then, as Einstein observed, we should say that it is "neither at rest nor in motion".[5] It is sufficiently known how first the special and later the general theory of relativity grew out of Michelson's revolutionary discovery and Einstein's parallel fundamental analysis. At about the same time the quantum theory came into being (1900) whose later development into the wave mechanics theory established the undulatory nature of the electron and this utterly ruined the intuitive idea of the ultimate units of matter conceived as nothing-but-corpuscles. On the other hand the quantum theory in its first phase with an equal definiteness ruined the attractive intuitive idea of light as 'nothing but waves'. Even the most fundamental classical concepts like space, time and causality which to the rationalists of Kant's type appeared the very essence of the unchangeable intellectual structure of man, and to the evolutionary empiricists as the *final* result of the evolutionary adjustment to the objective reality, are now being superseded by concepts which defy any appealing intuitive interpretation.

The concept of the universe at large underwent perhaps a less conspicuous, but ultimately no less revolutionary modification. With the coming of the Einsteinian cosmology for the first time the concept of finiteness was logically separated from that of limitedness; yet, our Euclidian imagination will forever resist this abstractly *conceivable*, but *not imaginable* separation. To Immanuel Kant, who in his first antinomy asked the question whether the universe is finite or infinite *in* space and *in* time, it never occurred even to raise the question whether space and time themselves may not be finite; he would be equally shocked by the Riemannian geometry as by Lemaître's and Gamow's cosmogonies assuming the finiteness of the cosmic past bounded by the initial 'zero-time'. (Hoyle's alternative cosmogony of 'continuous creation of matter' would hardly make him happier since this would show that his 'First Analogy of Experience' – the principle of the constancy of substance – does not apply to our experience any longer.) But it is not only the introduction of non-Euclidian models which represents a radical departure from the classical modes of thought. One revolutionary, and often overlooked implication of the relativization of simultaneity is that it definitely excludes any model of the universe consisting of co-existing (i.e. *simultaneously* existing) parts. For if there is no objective cosmic *Now* spreading instantaneously through the universe, then no intuitive model whose parts in virtue of their *visual* character are simultaneous with each other, can adequately represent the cosmic space-time except in small regions where the dislocations of simultaneity are negligible. This is merely another aspect of the inseparability of space from time, about which more later. We may then safely conclude that the widest regions of space-time show themselves equally recalcitrant to the application of our intellectual habits which were so successfully applied to the realm of the middle dimensions.

From the standpoint of Bergson's modified biological theory of knowledge the failure of intuitive models of the microcosmos as well as of the megacosmos had to be expected. How could our modes of thought be adjusted to the areas which were until recently altogether beyond the reach of our sensor-motor responses? The opposite would be, if not a miracle, then at least the most astonishing lack of economy in the general economy of the organism; it would be an unexplainable exception in the general rule according to which the sensory perception registers from the total *milieu* only those parts which have a biological significance for the species. The fact that our sensory perception ignores the microscopic quantum discontinuities as well as the vanishing non-Euclidian curvature of the space on our scale clearly did not in the past represent any bio-logical disadvantage to the human species. Not only this; the technical and theoretical triumphs of classical science, by which man gradually became master of the planet, showed the enormous practical usefulness of thought which was exclusively fashioned by the macroscopic surrounding. Only today can we realize how limited this surrounding was; and only today do we begin to realize how limited was our imagination and thought shaped by this surrounding. And this realization was brought about by the spectacular failure of the visual-tactile models outside the domain of the middle dimensions. Bergson as early as in 1889, deep in the classical era, anticipated the growing tendency of physicists to turn away from concrete intuitive models of matter, when he wrote that "the concrete existence of the natural phenomena tends to vanish in the dust of algebraic formulae".[6] More specific and more systematic comments on the in-adequacy of such models can be found in Bergson's later works as his biological theory of knowledge became more complete and more definite and as the trends in contemporary physics became more and more discernible.

## NOTES

[1] John Tyndall, *Lectures on Light*, Appleton, New York, 1873, p. 34.

[2] Cf. E. Cassirer, *Determinism and Indeterminism in Modern Physics*, Yale University Press, 1956, p. 141.

[3] This was the name J. J. Thomson originally gave to the cathode ray particles. Cf. E. T. Whittaker, *A History of the Theories of Aether and Electricity*, Philosophical Library, New York, 1951, I, p. 361.

[4] Abel Rey, 'La renaissance du cinétisme' *Scientia* **XXIII** (1918) 249–258; 329–340; Léon Brunschwicg, 'Le renouvellement des theories atomistiques', *Revue philosophique* **93** (1922) 325–380.

[5] A. Einstein, 'Relativity and the Ether', in *Essays in Science*, Philosophical Library, New York.

[6] *Time and Free Will*, Harper Torchbook, 1960, p. 207.

# THE CONTRAST BETWEEN TECHNICAL CONTROL AND INTELLECTUAL INSIGHT: THE PERSISTENT INFLUENCE OF MACROSCOPIC IMAGERY

Two objections can be raised against the theory just described. In the first place, the claim about 'biological unimportance' of the microcosmos seems to be clearly contradicted by the technology of the twentieth century. Who can reasonably deny the practical importance of the electrons now harnessed to perform useful services for man in vacuum tubes, photocells, electronic brains, electron microscopes and other such devices? To use an even more spectacular example; who could deny the vital importance of the use of atomic energy for the future and the very survival of the human species? Yet, this objection, even when only superficially analyzed, turns out to be a confirmation of the theory against which it is raised. For there is no question that the widening of the field of our sensory perception at the same time widened our biological surrounding; more especially, *it increased the region biologically important for our organism.* Of this, the founder of classical biological epistemology was clearly aware when he stressed that the creation of 'supplementary senses' went together with the creation of 'supplementary limbs':

All observing instruments, all weights, measures, scales, micrometers, verniers, microscopes, thermometers, etc., are artificial extensions of the senses; and all screws, hammers, wedges, wheels, lathes, etc. are artificial extensions of the limbs. The magnifying glass adds but another lens to the lenses existing in the eye. The crowbar is but one more lever attached to the series of levers forming the arm and hand. And the relationship which is so obvious in these first steps, holds throughout. This being perceived, a meaning becomes manifest in the fact that the development of these supplementary senses is dependent on the development of these supplementary limbs and *vice versa.* Accurate measuring instruments imply accurate instruments for turning and planing; and these cannot be made without the aid of previous measuring instruments of some accuracy. A first-rate astronomical quadrant can be produced only by a first-rate dividing engine; a first-rate dividing engine can be produced only by first-rate lathes and cutting tools; and so, tracing the requirements backwards, it becomes obvious that only by repeated actions and reactions on each other, can directive and executive mplements be brought to perfection. Only by means of artificial limbs can artificial senses be developed; and only through artificial senses does it become possible to improve artificial limbs.[1]

The development of microphysics and astronomy would provide us a practically limitless number of examples illustrating the continual and complex interaction between the development of technology and that of theoretical science. Whitehead observed that Michelson's experiment, on which the relativistic reconstruction of physics was ultimately based, "could not have been made earlier than it was" since the instrumental design of Michelson's interferometer would not have been possible prior to the development of metallurgy and of the German optical industry.[2] He might well have added that the development of the optical industry was possible only after the previous development of geometrical and physical optics, that is, only *after* the discoveries of Descartes, Snell, Newton, Huygens, Young and Fresnel. The electron microscope, which so significantly lowered the threshold of our visual perception, clearly was possible only after de Broglie's theoretical discovery of the wave associated with the particle; and experimental verification of this discovery by Davisson and Germer was likewise possible only at the highly developed stage of the experimental technics which in its turn depended on the whole previous development of both microphysics and microtechnology. The whole science of astrophysics was impossible before the invention of the telescope, spectroscope and the photographic plate; we hardly need to say that the design of the spectroscope in the last analysis would have been impossible without Newton's early discovery of the dispersion of light while no construction of the photographic plate was possible prior to Scheele's observation of the chemical effect of the ultra-violet part of the solar spectrum.[3] These three examples must suffice. They are obvious enough; if we stress the correlation between science and technology at all, it is because the historians of ideas sometimes tend to overlook it when they focus their attention exclusively on the logical and genetic relations between theories and their constitutive ideas. For our purpose it is important to bear the conclusion in mind: in virtue of the perpetual intercourse between the development of science and technology, of the "artificial sense organs" and "artificial limbs", the limited opening of the microcosmos to our perception implied its opening to our technical control. The situation is not the same at the opposite scale of magnitudes; our technological control is still confined to our planet, even though the development of guided interplanetary rockets certainly represents an enormous step toward the widening of our control beyond the upper limits of the world of our daily experience.

The second objection is apparently far more serious. If our present form of understanding is adjusted only to the world of the middle dimensions, can there be any hope that we shall ever intellectually assimilate the fresh empirical material furnished by the recent discoveries in microphysics? Would not our intellectual horizon be hopelessly confined to the zone of the middle dimensions? Would we not be thus bogged down in an epistemological *impasse*? But this is apparently contradicted by the very existence of contemporary physics which, while giving up the visual-tactile models of the classical era, creates mathematical formalisms sufficiently flexible and coherent to logically organize the empirical material coming from outside of the zone of the middle dimensions. In other words, the way out of the apparent epistemological *impasse* is by means of abstract formalism; all intuitive models borrowed from our macroscopic experience are eliminated to make place for differential equations and tensors. The abstract theories of mathematical physics which are apparently divested of any concrete sensory content are spectacularly successful outside the limits of the middle dimensions; does not this fact invalidate the basic claim of biological epistemology according to which "our logic is a logic of solid bodies" whose applicability is restricted to our macroscopic environment only?

Several points should be stressed in facing this objection. In the first place, it is highly questionable whether even the most abstract theories of mathematical physics are *completely* free of any sensory macroscopic associations. In my previous book I tried to show that the classical (that is, macroscopically conditioned) habits of thought die hard and that, "notwithstanding our declarations to the contrary, they persist and the fact that they are driven into subconsciousness by being consciously rejected makes their influence only less easily detectible and far more insidious."[4] I analyzed a number of instances in which physicists or, even more frequently, philosophical interpreters of modern physics failed to draw all the consequences from some new revolutionary discovery mainly because their thinking remained tinged by the influence of some hidden classical habit. I tried to show that the task of an epistemologist in contemporary physics is comparable to that of the psychoanalyst; to detect the remnants of classical thought beneath verbal denials and conscious rejections; to eliminate the parasitic associations by which our Newtonian-Euclidian subconscious leaks into even most abstract operations. The opposite would be miraculous; how could the human mind, which was shaped during the whole geological periods by contact with the limited macroscopic surroundings, suddenly, that is, within a few decades, become completely emancipated from such perennial influence? It is true, as we shall see, that we have the right to doubt the truth of Locke's claim that "there is nothing in intellect which has not been previously in the senses"; yet it would be equally hazardous to hold the very opposite view that "there is nothing in the intellect (including the intellect of Einstein, Planck, de Broglie, Heisenberg, Schrodinger, Dirac etc.) which *has been* in the senses". Or less aphoristically: that the abstract symbolism of the present physical theories is completely untinged by unconscious or semi-conscious macroscopic imagery, however subtle and disguised its influence may be. We shall see, for instance, that the persistent belief of the majority of mathematical physicists in spatio-temporal continuity is grounded upon the sneaking influence of intuitive geometrical imagery which is clearly inadequate in dealing with the sub-microscopic processes.

It is this fact, I think, which is responsible for the deep feeling of contrast between the technical mastery of the microcosmos and our intellectual insight into its real structure. This feeling is present not only in the educated layman, but in a number of philosophically minded physicists as well. We can manipulate the electron without knowing what the electron itself is; we know it possesses both corpuscular and undulatory properties, but no effort has been successful in reconciling in an intuitive model the image of corpuscle and that of wave. But why not then forget about the pictorial models altogether? Why not retain the mathematical formula purged of all parasitical macroscopic associations? Why, for instance, in the case of the electron should we not retain de Broglie's equation $\lambda = h/mv$ while forgetting the intuitive connotation of both $l$ and $mv$? After all, it was long ago that Pierre Duhem observed that theoretical physicists fall generally into two main groups – one which insists on the construction of pictorial and mechanical models of either the Cartesian or of the atomistic type, the other which maintains that the construction of a consistent mathematical scheme, representing physical reality symbolically, is all we can ever hope for; there is no need to insist on a concrete pictorial interpretation of the formal structure of the physical theory. There is no question that the second group definitely prevailed; and if the word "model" is still being used by theoretical physicists, it has almost completely lost its original concrete pictorial connotation. To return to our example: from this standpoint it is meaningless to ask what is the intuitive meaning of $m$, $v$ and $h$; all we should be concerned with is the way in which these symbols are connected in the equation and the way – a very indirect one – by which they are correlated with the empirical data.

This seems to be a reasonable attitude; for from the point of view of biologically oriented epistemology it is futile to search for any intuitive pictorial interpretation for one simple reason – *no such interpretation is possible.* Any search for such an interpretation is nothing but an extrapolation of our imaginative habits beyond the limits of their applicability. Nothing should be more welcomed than the effort to dissociate the formulae of theoretical microphysics from parasitic pictorial associations as the formalists in physics, though not always consistently, recommend. Yet the situation is not as simple as that. Not speaking of the enormous difficulties which any serious effort of de-pictorialization faces, difficulties which are vastly underestimated by rank-and-file theoretical physicists – there is yet one questionable assumption which underlies the attitude of the formalists. It is the confident and dogmatic belief that logico-mathematical formalism represents a magical clue to our understanding of physical reality outside of the zone of the middle dimensions; while it is uncertain *which* specific formal pattern will be the required magical key, no doubt is entertained that *some one will be found* or *can at least in principle be found.* The question whether the limitations of our macroscopically conditioned perception, imagination and thought are not also embodied in the very structure of mathematical formalism is practically never even raised.

### NOTES

[1] H. Spencer, *The Principles of Psychology*, 3rd Ed., I, p. 365–366.
[2] A. N. Whitehead, *Science and the Modern World*, Macmillan, New York, 1926, p. 167.
[3] Ernst von Meyer, *A History of Chemistry* (transl. by George McGowan), Macmillan, London, 1891, p. 458.
[4] M. Čapek, *The Philosophical Impact of Contemporary Physics*, Van Nostrand, Princeton, 1964, p. XV.

CHAPTER 6

# LIMITATIONS OF PANMATHEMATISM

The confident belief in the unrestricted applicability of logico-mathematical formalism to experience has its historical roots in metaphysical panmathematism, according to which the world itself consists of mathematical entities and their relations. This was the belief of the Pythagoreans when they claimed that things are made of numbers; also of Plato when, in the *Timaeus*, he constructed matter from elementary plane triangles. In such an extreme form panmathematism, or rather pangeometrism, hardly occurred in the modern era, even though Descartes in his ambitious attempt at reducing matter to geometrical space came very near to it. But once the distinction between homogeneous causally inert space and its material content had been made, pangeometrism in its radical form was all but impossible. Even the homogeneous Cartesian matter was not, as Leibniz pointed out, completely reducible to space since its impenetrability, not to speak of its dynamical manifestations, was *not* a geometrical property. Similarly, the atoms of Democritus, Gassendi and Dalton are *not* equivalent to the places they occupy; the former are indivisible and moveable, the latter are mathematically continuous (infinitely divisible) and motionless, being the eternally unmoveable portions of static Euclidian space. The situation did not become different when the atoms of finite size were replaced by the extensionless material points of Boscovich, Ampère, Cauchy and Cantor; by being moveable and by possessing dynamical properties the material points still differ from the geometrical points they occupy; they are *in* space while their underlying positions constitute space itself. In this way classical physics, in adopting the dualism of space and matter, has *not* adopted radical panmathematism.

Yet, there was always a strong tendency in this direction. It was this tendency to reduce to the barest minimum the properties of matter irreducible to geometry, the process which started already with ancient atomists when they banished the secondary qualities from the physical world. Furthermore, with the foundation of modern dynamics there began a gradual mathematization of the dynamic properties of matter which Descartes had still failed to reduce to geometry. When at the beginning of the twentieth century Poincaré wrote that 'masses are nothing but the coefficients which enter the equations'[1] he was closer to Plato than he realized; his mass-coefficients were merely more abstract entities than Plato's triangles. The substitution of the relativistic and quantum-mechanical formulae for the equations of Newton as well as the replacement of the geometry of Euclid by that of Riemann or Lobachevski showed an enormous fruitfulness of mathematical formalism as well as the futility of pictorial models. Furthermore, since the publication of *Principia Mathematica* in 1912 it was believed that mathematics is a mere branch of logic; it was natural then to regard the triumphs of mathematical physics as a decisive vindication of panlogism, even if of a panlogism less naive and more coherent than that of Hegel. Yet, it is hardly accidental that Einstein, who had a deep admiration for Spinoza, in 1928 enthusiastically welcomed Meyerson's comparison of the relativistic explanation of physical reality with Hegel's all-embracing deductive synthesis.[2]

In any case the development of mathematical physics seemingly substantiated Kant's famous words that "in every area of knowledge there is as much science as there is mathematics." On this point neo-positivists fully agree with neo-Kantians. The attitude of this latter group was consistent from their philosophical standpoint for they shared with Kant, some modifications apart, his view about the a priori character of the logico-mathematical framework. Since this framework is a priori, that is, independent of experience, and at the same time "the transcendental condition of any experience", it must be applicable to *any* empirical material, whether of the past, present or future. Thus Kant believed that *any* sensory experience will always appear in the Euclidian form, and that *all* empirically observable changes will appear to us, according to his "Anticipations of Perception" as mathematically continuous. Since these dogmatic claims are today, to put it mildly, far less certain than at the time of Kant, neo-Kantians now are more careful; but although their apriorism is far more liberal than that of Kant, they still remain adamantly opposed even to a hint that the logico-mathematical framework could be conditioned by limited experience.

The position of neo-positivists is more complex. Since their epistemology is basically sensualistic, they naturally reject the Kantian apriorism. But they do not in general reinterpret the Kantian a priori in the evolutionary sense as the psycho-physiological structure, inborn in the individual, though acquired by the species. They returned to the empiricism of the pre-evolutionary, i.e. Lockean-Humean kind; and show little interest in psychological and genetic questions. Undoubtedly the positivists of the Vienna Circle were lastingly impressed by the severe criticism which Frege, Husserl and phenomenology made of psychologism in general and in particular of psychologism in epistemology; this criticism apparently convinced them that the genetic and psychological approach to epistemological questions is both irrelevant and misleading. They believe that to judge the value of any conceptual framework by analyzing its psychological genesis is to commit the 'genetic fallacy', that is, to confuse *questio facti* and *questio iuris*; logical and epistemological analysis should be free of any psycho-genetic considerations. (This only shows how profoundly a certain philosophy can be influenced by its opponents.) Thus the only a priori which neo-positivists retain is the *a priori of the analytical propositions*, i.e. of those propositions which are true in virtue of their meaning, independently of any experience. Since logical and arithmetical propositions are in their view (and contrary to Kant's view) of this kind, they do apply to our world. This is certainly quite different from the empiricism of John Stuart Mill who regarded even the law of contradiction as being ultimately dependent on human experience.[3] Thus the neo-positivists, in spite of their return to pre-evolutionary empiricism, arrived by a curious detour at basically the same conclusion as the neo-Kantian panmathematicists; in truth they even went further than Kant. For while Kant restricted the applicability of even such basic categories as the categories of quality and quantity to the realm of human experience, the neo-positivists insist on the applicability of the logico-mathematical framework to *any possible world*. As A. J. Ayer put it:

We saw that the reason why they [i.e. analytical propositions] cannot be confuted by experience is that they do not make any assertion about the empirical world. They simply record our determination to use words in a certain fashion. We cannot deny them without infringing the conventions which are presupposed by our very denial, and so falling into self-contradiction. And this is the sole ground of their necessity. As Wittgenstein put it, our justification for holding that the world could not conceivably disobey the laws of logic is simply that *we could not say of an unlogical world how it would look*. And just as the validity of an analytic proposition is independent of the nature of the external world; so it is independent of the nature of our minds. (Italics mine.)[4]

This passage graphically illustrates the detour mentioned above; it starts with linguistic conventionalism and ends with a sort of epistemological and even metaphysical Absolute. This Absolute is represented by analytical propositions, that is by logico-mathematical tautologies, applicable to any kind of world! The exclusion of genetic considerations is clearly implied in the statement that the validity of these tautologies is "independent of the nature of our minds". This sentence shows how far we are from the epistemological modesty of the psychologically and biologically oriented epistemology.

Wittgenstein's claim that "the world could not conceivably disobey the laws of logic because we could not say of an unlogical world how it would look" raises a number of questions. Is the two-valued logic to which Wittgenstein clearly refers the only possible one? Furthermore, is his claim that "we could not say how a different world would look" relevant at all? Could Aristotle have said how the world of Newton would look? Could Newton have said how the universe of Einstein and de Broglie would look? Do these two examples not show plainly that man's failures to anticipate any radically new experience are due to temporary limitations of human mind and to its pretentious claim that the forms of understanding adequate for certain strata of experience must be adequate for *any* possible experience?

A lurking suspicion that mathematical formalism does not exhaustively describe the nature of physical reality was summarized by Bertrand Russell in 1959 in the following manner:

It [i.e. theoretical physics] lays down fundamental equations which enable it to deal with the logical structures of events, *while leaving it completely unknown what is the intrinsic character of the events that have the structure*. We know only the intrinsic character of events when they happened to us. *Nothing whatever in theoretical physics enables us to say anything about the intrinsic character of events elsewhere*. They may be just as the events that happen to us, or they may be totally different in strictly unimaginable ways. All that physicists give us is certain equations giving abstract properties of their changes. But as it is what is that changes, and what it changes from and to – as to this theoretical physics has nothing to say.[5] (Italics mine.)

Briefly, the intrinsic qualities of physical events, according to Russell, do exist, even though they are not describable mathematically. On this point Russell was, without knowing it, very close to Bergson. Bergson would certainly agree with Russell's characterization of theoretical physics; he would welcome Russell's reference to the transcendent qualities of matter. In conceding their existence Russell certainly departed from the radical methodological panmathematism which claims that to speak of any properties of matter undescribable by mathematical language is altogether meaningless. Bergson like Russell insisted on the event-like character of matter. Like Russell he held that the transcendent qualities of physical events forever elude human *perception*, though not, as we shall see, human knowledge in general. One of the central ideas of *Matter and Memory* is that "pure perception" (*la perception pure*), un-tinged by our subjective sensory qualities, is a limit concept, never realized in a concrete process of perception; in other words, that our perception of reality is never direct. Neo-realists and John Dewey claimed the possibility of direct immediate perception of physical reality at the price of reinstating the secondary qualities into nature.[6] Bergson's view was very opposite; the secondary qualities remain subjective for him since they are due to the greater temporal extension of the human specious present; and in this respect the status of the primary qualities is not essentially different. This will be discussed later in detail. But the point which Bergson would certainly dispute is Russell's claim about the absolute unknowability of physical events. On what grounds?

In answering this question, we are returning after a long detour to the second objection against the biological theory of knowledge; if our mind is conditioned by our macroscopic environment, does it not remain forever incapable of overcoming the resulting limitations? Such an objection, indeed, *does* apply, but only to the classical form of this theory, *not* to its form as amended by Bergson. For according to the classical theory, the evolution of our forms of understanding is already completed, and consequently it is not possible to make them more adequate or to widen them. But this claim was made precisely in the belief that, in Berg-son's words, "our intellect embraces the whole of nature"; in other words, that within nature there are no zones impervious to our Euclidian-Newtonian insight. But precisely this is not true; the adjustment of our intellect to reality is only *partial* and not definitive. In insisting on this point Bergson's biological theory of knowledge opens the way to a further development of our forms of understanding beyond its classical

Newtonian-Euclidian form and probably even beyond the concomitant two-valued logic.

This, however, implies that the human mind is not completely and exclusively conditioned by sensory, that is, macroscopic experience as the sensualists of all ages, including Herbert Spencer, the author of the biological epistemology, claimed. Rationalists of all ages made an opposite claim, and on this particular point Bergson sided with them. But while rationalists from Plato to Cassirer believed that the only area within our mind uncontaminated by sensory experience is abstract mathematical and logical thought, Bergson was on this point far more cautious. One of the central, doubtless most controversial, but also most interesting, theses of his epistemology is that the sensory elements of our macroscopic experience are subtly and insidiously present even on the highest level of logical and mathematical abstraction. This is the meaning of his claim that our logic is the "logic of solid bodies" and that the operations of conceptual thought, at least in its classical form, betray the influence of our macroscopic perception of solid bodies as well as of our technique by which these bodies are manipulated. Since the discussion of this thesis is inseparable from its applications, it is beyond the scope of this chapter. Suffice it to say that "the area uncontaminated by our macroscopic perception" is located by Bergson in the imageless introspective thought, intuited in its ceaseless becoming and purged of all parasitic spatial metaphors. This will be discussed extensively in Part II. What will be given in a subsequent chapter is a mere anticipatory sketch of the *negative* aspect of Bergson's epistemology, that is, of the method by which, Bergson hopes, the human mind can be freed from the exclusive domination by macroscopic imagery.

## NOTES

[1] H. Poincaré, 'Science and Hypothesis', in *F.S.*, p. 102.

[2] A. Einstein, 'A propos de la *Déduction rélativiste* de M. Emile Meyerson', *Revue philosophique* CV (1928) 164.

[3] John Stuart Mill, *A System of Logic, Ratiocinative and Inductive*, Book II, Chapter VII, in particular #5. The whole chapter VII is a polemic against Spencer's view that "the inconceivability of the opposite" is the ultimate epistemological criterion.

[4] A. J. Ayer, *Language, Truth and Logic*, Denver 1946, p. 84; *Tractatus logicophilosophicus*, 3.031.

[5] B. Russell, *My Philosophical Development*, Simon & Schuster, New York, 1959, pp. 17–18. A very similar passage can be found in another of Russell's books written more

than three decades prior to that quoted above: *Philosophy*, Norton, New York, 1927, p. 157.

[6] Cf. in particular John Dewey, 'The Postulate of Immediate Empiricism', *J. Phil.* **II**, No. 15 (1905). According to this postulate "things are what they are experienced as". This view was upheld by Dewey through all his life; see his defense of it against Reichenbach in *The Philosophy of John Dewey* (ed. by Paul Schilpp), Evanston, Ill., 1939, pp. 534–543.

# NEGATIVE ASPECTS OF BERGSON'S EPISTEMOLOGY – ITS RELATIONS TO BACHELARD, BRIDGMAN AND EMPIRIO-CRITICISM

The negative aspect of Bergson's epistemological method consists in a difficult and perpetually renewed effort to break certain intellectual habits, that is, certain associations of ideas which, by the effect of familiarity and repetition, appear solid and unbreakable. They appear so solid that the rationalists endowed them with the status of *a priori* principles constituting the immutable or, as Kant called it, "transcendental" structure of the human mind which no future experience can ever challenge. In truth this alleged transcendental structure is nothing but what Gaston Bachelard called "geometrical subconscious" or, in the terms strongly reminiscent of Bergson's language, "the Euclidian infrastructure which is formed in a mind subject to experience of solid bodies both natural and manufactured".[1] It is extremely difficult to keep in check our Euclidian kinetic-corpuscular subconscious which is the depositary of our daily individual, as well as ancestral, experience. This is why, as the late Professor Bridgman observed not without humour, longing for mechanical explanation has "all the tenacity of original sin" and "just as the old monks struggled to subdue the flesh, so must the physicist struggle to subdue this sometimes nearly irresistible, but perfectly unjustifiable desire".[2] The effort which Bergson's epistemology requires is of the same kind; it tries to dissociate certain associations of ideas which under the persistent pressure of the microscopic environment became ingrained mental habits by which we try stubbornly to interpret or rather misinterpret new recalcitrant empirical data. In this sense Bergson's philosophy is a true "negative philosophy" – *la philosophie du non* – in Bachelard's sense.[3]

In one well written chapter of his classic work Albert Thibaudet stressed this particular aspect of Bergson's thought which the diluted literary Bergsonism consistently overlooked.[4] "I am going to ask you", Bergson wrote himself, "to make a strenuous effort to put aside some of the artificial schema we interpose unknowingly between reality and us. What is required is that we should break with certain habits of thinking and perceiving that have become natural to us." [5] What is required is "tearing oneself away from deeply rooted habits, veritable extensions of nature". [6] Or in the less fortunate words of *Creative Evolution*, "one must thrust intelligence outside itself by an act of will". [7] Needless to say that by the word 'intelligence' in this context Bergson means its mechanistic Euclidian-Newtonian form; while the effort to which he refers is clearly 'intellectual effort' of which he himself gave a fine analysis in one of his essays. [8] Only by such intellectual effort can the required widening of our intellectual frames ("l'élargissement des cadres de l'intelligence") be achieved. "What a careless attitude of intellect had done, an effort on the part of intellect can undo." [9]

The result of this epistemological effort is called by Bergson – after some hesitation – "intuition", the full discussion of which will be given in Part II. But the passages just noted suffice to show the *active* and *voluntaristic* character of the Bergsonian intuition. This was explicitly stressed by Bergson as early as in 1903 in the very first article in which the term 'intuition' was used. To intuit, he wrote, does not mean simply to live, to contemplate oneself passively in a similar way "as a sleepy shepherd watches the water flow" [10]; such a view would completely ignore the essentially active character of intuition. In his last book, undoubtedly frustrated by frequent misinterpretations of his thought, he emphasized it again in unmistakeable terms:

Thus I repudiate facility. I recommend a certain manner of thinking which courts difficulty; I value effort above everything. How certain people have mistaken my meaning? To say nothing of the kind of person who would insist that my 'intuition' was instinct or feeling. Not one line of what I have written could lend itself to such an interpretation. And in everything I have written there is assurance to the contrary: my intuition is reflection. [11]

This is explicit enough. But the confusion was facilitated by Bergson's ill chosen terminological opposition of 'intellect' and 'intuition'; also by the fact that Bergson did not always stress that by 'intellect' he meant only its particular Newtonian-Euclidian or Spinozist-Laplacean form; finally, and most importantly, by the general tendency, only natural at the beginning of this century, to confuse the reflecting activity of the human mind in general with its particular, although most persistent, form. André Lalande's distinction between 'la raison constituante' and 'la raison constituée' certainly applies here. Or if we prefer to use Plato's language, we may say that Bergson sees the essence of reason in its dynamic aspect, in what Plato called *Eros*, in the perpetual search for truth, in the persistently renewed effort of invention and reinterpretation rather than in the static framework of immutable Platonic ideas. Whether Bergson's dynamic conception of mind and of reality in general does not also imply, as Roman Ingarden claimed, the existence of certain constants, though undoubtledly more elusive than those upheld by the processless view, is another question which will certainly not be neglected in our subsequent discussion.

The activistic and voluntaristic aspect of Bergson's epistemology is one of the main features which differentiates it from the empirio-criticism of Mach and Avenarius with which it shares their biological orientation. The main error of this school was that they confused the *actual psychological* processes of human thought with the *epistemological criterion of the correctness of thought*; *questio facti* was confused with *questio juris*. Mach, and in particular, Avenarius, greatly stressed the principle of economy; Avenarius even defined philosophy as "the thinking of the world according to the principle of the smallest expenditure of effort".[12] Now it is true that in human thinking as well as in any other human activity there is an inherent tendency to reduce the amount of effort required at the beginning. Every idea, every intellectual discovery, no matter how original, that is, how great an effort was required to acquire it, tends to automatize itself in a ready-made formula; this is clearly a part of the general economy of the psychophysiological human organism. It would be a mistake to underestimate its positive significance; but it would be an equally serious mistake to elevate the principle of the least expenditure of effort to the dignity of an epistemological criterion. Théodule Ribot, Avenarius's French contemporary, was undoubtedly right when he characterized general ideas as "intellectual habits".[13] The sum total of these habits constitutes what we call a *conviction, belief, or view* which displays a certain inertia, especially when it faces a new experience not fitting into the preconceived intellectual framework.

When the cohesiveness of the accepted belief is threatened, our automatized association network resists every fresh fact which tends to dissociate it. The subjective aspect of this resistance may even acquire an emotional intensity: new experience appears as strange, grotesque, absurd, impossible. Of this, an historian of science could give a number of examples, some of them amusing, some of them tragic. This is what Vladimir Jankélevitch, a French interpreter of Bergson, called "constitutional misonéisme"[14] of the human mind; the new is instinctively rejected, and only when the pressure of recalcitrant facts is irresistible are new assumptions reluctantly introduced in such a way that the imposed modification of our intellectual habits would be as small as possible. This is the meaning of the 'heterotic minimum' of Avenarius, of the principle of economy of Mach and – at least in certain contexts – of the 'maximum convenience' of Poincaré. Now experience requires a certain reorganization of our mental structure which cannot take place without effort; the natural inertia of our thought tends to reduce this effort to a minimum; we try to reinterpret new facts in the terms of old experience, only reluctantly do we introduce new assumption and even more reluctantly do we change the total intellectual perspective. In Quine's recent words, "our tendency is to disturb the total system as little as possible".[15]

But such psychological facts should not be confused with epistemological criteria. Applied to modern physics the empirio-criticist criterion clearly showed its helplessness. Physicists of the last century certainly did their best to apply 'the principle of the least expenditure of effort' in interpreting new discoveries. Let us only recall a great number of mechanical models of the aether and the even more numerous, though less known, models of gravitation. All these models had one feature in common; they appealed to the habits of our imagination fashioned by the ages of our macroscopic, especially visual and tactile, experience. For this reason they required a relatively small effort of comprehension, as all familiar patterns of thought do. Yet, they failed, and when Michelson's experiment showed that aether cannot possess even the most elementary kinematic properties, it became painfully clear that their failure was definitive. The fate of the intuitive models of the atom in this century was the same. First there was an understandable tendency to interpret it in the terms of familiar macroscopic experience; the atom was first a minute billiard ball or a minute vortex in the sea of aether; then it was regarded as a system of still tinier balls – electrons, a sort of solar system on a minute scale with the atomic nucleus instead of the sun. The quantification of the electronic orbits by Bohr showed the fallacious character of this alleged analogy, and when the undulatory nature of the particles was established, the physicists generally realized that the period of pictorial atomic models was over, as much as that of mechanistic models of aether. Long before that, in 1896, Bergson anticipated the crisis of the mechanical models in the following words:

But the materiality of the atom dissolves more and more under the eyes of the physicist. We have no more reason, for instance, to represent the atom to ourselves as solid rather than as liquid or gaseous, nor for picturing the reciprocal actions of atoms by shocks rather than in any other way. Why do we think of solid atoms, and why of shocks? Because solids, being the bodies on which we clearly have the most hold, are those which interest us most in our relations with the external world; and because contact is the only means which appears to be at our disposal in order to make our body act upon other bodies. But very simple experiments show that there is never true contact between two neighboring bodies; and, besides, solidity is far from being an absolutely defined state of matter. Solidity and shock borrow then, their apparent clearness from the habits and necessities of practical life; *images of this kind throw no light on the nature of things.* (Italics mine.)[16]

The comprehension of physical reality, or more accurately, the creation of more adequate imageless models of the ultimate elements of matter will be possible only with an enormous and utterly uneconomical expenditure of intellectual effort. The resulting reorganization of the traditional patterns of thought will be so radical that Gustave Juvet did not hesitate to compare it with mutations in the biological sphere.[17] Whatever our final judgment of Bergson's imageless model of matter may be, there is no question that it is of the same kind. Bergson was fully aware of it when he wrote that it will appear "strange and difficult to our mind" and "fatiguing for our imagination"[18], to wit, to our Newtonian-Euclidian imagination. Albert Thibaudet justly compared Bergson's intellectual effort to free the human mind from the obsession of pictorial and mechanistic thought to the effort of Lobachevski and Riemann to emancipate our thought from being exclusively dominated by the geometry of Euclid.[19] And in a similar way, just as the end of the privileged status of the geometry of Euclid did not mean the end of geometry in general, Bergson's strictures upon mechanistic thinking, instead of ruining human thought in general, merely liberates its wider potentialities from biological and local limitations.

This liberating significance of the biological theory is understood by the idealistically oriented philosophers as little as by their positivistic and analytical opponents. Thus N. Berdyaev, one of the leading Christian existentialists, made the following pontifical pronouncement: "Bergson's biologism is a scandal in philosophy. Metaphysics falls into dependence upon a special science." [20] When Ludwig Wittgenstein says that "Darwin's theory has no more to do with philosophy than any other hypothesis in natural science" [21], he displays essentially the same indifference to genetic investigations as Berdyaev; according to both, philosophy (Wittgenstein would probably substitute 'logic' for 'metaphysics') should not fall into dependence upon special sciences like biology or psychology. That philosophy thus conceived cuts itself off from a very important class of facts; that it is based on an arbitrary narrowed experience and that it thus becomes itself narrow, is apparently not clear to those who uphold it. Nor is it realized that the dogmatic and unwarranted assumption about the static and unchanging character of the cognitive functions of man is due to the arbitrary exclusion of the facts of biology and genetic psychology.

## NOTES

[1] G. Bachelard, *Le nouvel esprit scientifique*, Paris 1946, p. 37.

[2] P. W. Bridgman, *The Logic of Modern Physics*, Macmillan, New York, 1927, p. 47.

[3] G. Bachelard, *La philosophie du non. Essai d'une philosophie du nouvel esprit scientifique*, Paris 1949.

[4] A. Thibaudet, *Le bergsonisme*, I, pp. 165–169. (Chapter 'Les dissociations des idées'.)

[5] Cf. 'The Perception of Change', in *The Creative Mind* (transl. by Mabelle L. Andisom), The Philosophical Library, New York, 1946, p. 167. Hereafter referred to as *C.M.*

[6] *C.M.*, p. 80.

[7] *C.E.*, p. 212.

[8] Cf. the essay 'L'effort intellectuel', in *L'énergie spirituelle*. The English translation by H. Wildon Carr appeared under the title *The Mind Energy*, Macmillan, London, 1920. Hereafter referred to as *M.E.*

[9] *C.M.*, p. 83.

[10] *C.M.*, p. 217.

[11] *C.M.*, p. 103.

[12] Richard Avenarius, *Philosophie als Denken der Welt gemäss dem Princip des kleinsten Kraftmasses*, 3rd Ed., Berlin 1917.

[13] Théodule Ribot, *L'évolution des idées générales*, 4th Ed., Paris 1915, p. 149: "Les idées générales sont les habitudes dans l'ordre intellectuel." Ribot, despite his completely different philosophical view stressed the significance of the *dissociation* of ideas in epistemology as unambiguously as Thibaudet and Bergson: "Bien des vérités (par exemple l'existence des antipodes) se sont imposées difficilement, parce qu'il fallait briser des associations indissolubles." (Cf. *Essai sur l'imagination creative*, 6th Ed., Paris 1921, p. 18).

[14] Vladimir Jankélevitch, 'La signification spirituelle du principe d'économie', *Revue*

*philosophique* 53e année, **105**, 88–126, in particular p. 92. This essay was later incorporated in his book *L'alternative*, Paris 1938, Chapter II.

15 Willard Van Orman Quine, *From a Logical Point of View*, Harper Torchbook, 1961, p. 44.

16 *M.M.*, p. 195.

17 Gustave Juvet, *La structure des nouvelles théories physiques*, Paris, 1933, p. 135.

18 *M.M.*, p. 203, 205.

19 A. Thibaudet, *op. cit.*, I, p. 46.

20 N. Berdyaev, *The Nature of the Free Act* (transl. by D. C. Lowrie), Harper, 1954, p. 40.

21 *Tractatus logico-philosophicus*, 4.1122.

CHAPTER 8


# BERGSON, REICHENBACH AND PIAGET


An interesting affinity with Bergson's biologically oriented epistemology can be found in certain aspects of Hans Reichenbach's theory of knowledge and in "the genetic epistemology" of Jean Piaget. Reichenbach's biological orientation, only one ingredient in his epistemology, is not always compatible with other ingredients, especially with his conventionalism.[1] In this respect there is a similarity between him and Poincaré. Reichenbach's emphasis on the steady adaptation (*ständige Anpassung*)[2] of experience shows his affinity with Mach's similar view from which it differs, however, in one important aspect: Reichenbach is far more explicit than Mach in stressing the *incomplete* character of this adaptation. Already in his early book *Relativitätstheorie und Erkenntnis apriori* Reichenbach upheld the *mutability of reason*.[3] In *Atom und Cosmos*, published a decade later, he coined the term "the world of the middle dimensions" designating the zone intermediate between the microphysical world of quanta and the world of the fleeting galaxies; only to this zone is the conceptual framework of classical physics applicable.[4] This today sounds as a mere truism; but this truism acquires a full significance only in the evolutionary perspective. If the traditional frame of thought developed by the interaction between the human psycho-physiological organism and the "world of the middle dimensions", then its limited applicability is not only understandable, but inevitable; as we said before, its unlimited applicability would be nothing short of a miracle. Although this explanation is still far from being generally accepted, it begins to find its way even into the textbooks of atomic physics:

The first major case where theoretical results seem irreconcilable with our normal concepts of physics, or even with apparently self evident philosophical concepts, was encountered by the theory of relativity. Velocities of the order of the speed of light were just as inaccessible to man's experience before the age of physics as distances of millions of light years, where the apparently so self-evident Euclidian geometry seems to fail. The same applies, however, to the atomic dimensions of microphysics which are, in the other direction, beyond our direct reach. The phenomena in this region are, for all we know, described correctly by quantum mechanics, but this description leads to the conceptual and philosophical difficulties outlined above. Is it really incidental that we run into such difficulties whenever we go beyond the range of the world "given" to man? Could it not be that these conceptual and philosophical difficulties stem from the fact that all our elements ("tools") of thinking have been developed and derived from man's interaction (i.e. experience) with "his" region of the world? Our tools of thinking, then, are not suited for the immensely expanded world of our modern science, and our attempt to "understand" this wider world in a pictorial way necessarily leads to conceptual difficulties.[5]

Compare this passage to the words which may be found on the very first page of Bergson's *Creative Evolution*:

The history of the evolution of life, incomplete as it yet is, already reveals to us how the intellect has been formed by an uninterrupted progress, along a line which ascends through the vertebrate series up to man. It shows us in the faculty of understanding an appendage to the faculty of action, a more and more precise, more and more complex and subtle adaptation of the consciousness of living beings to the conditions of existence which are made for them. Hence should result this consequence that our intellect, *in the narrow sense of the word*, is intended to secure the perfect fitting of our body to its environment, to represent the relations of external things among themselves – in short, to think matter. Such will indeed be one of the conclusions of the present essay. We shall see that the human intellect feels at home among inanimate objects, more especially among solids, where our action finds its fulcrum and our industry its tools; that our concepts have been formed on the model of solids; that our logic is pre-eminently the logic of solids...[6] (Italics mine.)

Obviously, the evolutionary setting is more explicitly stressed by Bergson; but in other respects, if we substitute the words "macroscopic bodies of our daily experience" for Bergson's "solids" the meaning of both passages is the same. How close Reichenbach himself was to Bergson is clear from the following paragraph:

The conception of a corporeal substance, similar to the palpable substance shown by the bodies of our daily environment, has been recognized as an extrapolation from our sensual experience. What appeared to the philosophy of rationalism as a requirement of reason – Kant called the concept of substance a synthetic a priori – has been revealed as being the product of a conditioning through environment. The experience offered by atomic phenomena makes it necessary to abandon the idea of a corporeal substance and requires a revision of the forms of the description by means of which we portray physical reality. *With the corporeal substance goes the two-valued character of our language, and even the fundamentals of logic are shown to be the product of an adaptation to the simple environment into which human beings were born.*[7] (Italics mine.)

Reichenbach would probably have been deeply shocked had he realized the affinity of his epistemology with that of Bergson; there are only a few other thinkers who are more disliked – and more ignored – by the majority of positivists than the author of *Creative Evolution*. Yet, the similarity is undeniable: it is one of those unintentional agreements which sometimes occur, ironically enough, between two thinkers who, in other respects, may be radically different. In both Reichenbach and Bergson we have basically the same biological orientation in epistemology; the same opposition to the Kantian idea of the immutability of reason; the same belief that the human intellect in its classical form is a result of adaptation to a certain environment; finally, the same conviction that the nature of matter is profoundly different from its superficial macroscopic appearance. It is true that there were also important differences between both thinkers. Reichenbach did not draw all of the consequences from his radical statement quoted above. Furthermore, he was a positivist – and the historical association of positivism with physicalism prevented him from siding with Bergson on other important issues.

The most outstanding representative of the biological and genetic approach to the problems of epistemology is Professor Jean Piaget. Starting from the sound principle that "every response, whether it be an act directed towards the outside world or an act internalized as thought, takes the form of an adaptation or, better, of a readaptation",[8] he regards intelligence as "a generic term to indicate the superior forms of organization or equilibrium of cognitive structures". In other words:

It (i.e. intelligence) is the most highly developed form of mental adaptation, that is to say, the indispensable instrument for interaction between the subject and the universe when the scope of this interaction goes beyond immediate and momentary contacts to achieve far reaching and stable relations. But...it is an ultimate goal, and its origins are indistinguishable from those of sensory-motor adaptation in general, or even from those of biological adaptation itself.[9]

This passage is characteristic enough; it sounds as if taken from Spencer's *Principles of Psychology* or Mach's writings. It is hardly surprising that Piaget is resolutely opposed to the antigenetic views of Russell, the Vienna circle and logical positivism in general.[10] But besides his general biological and genetic attitude, Piaget's numerous and fascinating monographs have some implicit bearings on the problems of epistemology, in particular those that were raised by the discoveries of modern physics. In his book *La construction du réel chez l'enfant*[11] he showed that the belief in permanent objects gradually develops in a child's mind under the pressure of experience and that the elaboration of the belief in the reality of space takes place concomitantly. There is no place here to go into all the details of Piaget's investigations; let us only recall that in the first phase the child's universe consists of the images which are capable of recognition, but still lack substantial permanence as well as spatial organization; that only in the fourth phase the child begins to search for the objects which disappeared from the sensory field without taking their displacements into account; that in the fifth phase, the displacements are taken into account, but only if they are perceived; and that only in the sixth phase (12–18 months) is the belief in the substantial permanence of the objects existing outside of the sensory field established, even if their displacement was not perceived.

In his book *Le développement des quantités physiques chez l'enfant: Conservation et atomisme*[12], Piaget[12a] showed how the child gradually arrives at the idea of substantially permanent object on the microphysical scale. More specifically, he showed how the child arrives at the idea of quantitative conservation of substance, of its volume and of its weight at the same time as at the notion of atomic structure of matter. In the first phase, the dissolution of a piece of sugar in water is regarded as a sheer annihilation: neither its weight nor its volume and not even its substance is regarded as being conserved. (The child predicts that even the sweet taste of the solution will eventually vanish in the same way as odors eventually evaporate...) In the next phase, the child begins to believe in the preservation of substance (sugar) in the solution, but without weight and without occupying any volume; this type of permanence seems to be of a "potential" kind, similar to the way in which, according to Aristotle, the components are present in the mixture. This confirms another of Piaget's discovery that the original notion of identity or permanence is purely qualitative[13], i.e. free of any quantitative connotation: sugar persists, though neither its weight nor its volume... Only in the last phase, when the child realizes that the solution weighs more and permanently occupies a slightly larger volume after a piece of sugar has been dissolved, it arrives at the correct view: that the substance dissolved only *apparently* vanished, i.e. that it was merely divided into smaller and smaller and, eventually, invisible grains, scattered through the water and indestructibly persisting in it. Only then is the joint conservation of substance, weight and volume firmly grasped. In the same book he showed how the logico-arithmetical operations develop concomitantly with the operations performed with physical bodies and therefore do not constitute any inborn a priori functions of mind. This was already clear from his book *La genèse du nombre chez l'enfant*[14] when he showed how the belief in the elementary commutation law $a+b=b+a$ is impossible as long as the permanence of the objects in the aggregate symbolizing the number is not explicitly recognized. This was noticed earlier by Helmholtz who pointed out that objects, in order to be countable, must fulfil certain conditions: during the process of counting they must neither disappear, nor merge together; nor can any of them split in two parts, nor can any new one suddenly come into existence.[15] Now the permanence (i.e. indestructibility and uncreatibility) as well as distinctness and indivisibility are properties of the macroscopic solid bodies; in Piaget's terms, the logical and arithmetic operations are the results of the "internalization" of our manipulations of physical solids.[16] All this is in agreement with Bergson's view that our logic is "the logic of solid bodies", that *homo sapiens* should rather be called *homo faber*, and that "our intellect is characterized by the unlimited power of decomposing according to any law and of recomposing into any system".[17]

Piaget's view clearly implies that the traditional form of intelligence is a result of the interaction of human mind with the macroscopic bodies of our daily experience and as such it is not necessarily applicable *beyond* the limits of this experience. This is what contemporary physics clearly showed and this is what Bergson's epistemology anticipated. Piaget was fully aware of the agreement of his view with the implications of modern physics. He, who showed how the notion of permanent object is formed in early childhood by the pressure of our daily surrounding, grasped fully the significance of the fact that modern physics began to challenge the applicability of such notion on both the microphysical and megacosmic scale. In the second volume of his *L'introduction à la épistemologie génetique* Piaget pointed out that we still continue to speak of the universe as a whole, either finite or infinite, without realizing that we are illegitimately transferring our category of 'object' – which is a diaphanous replica of the macroscopic solid body – beyond the realm of its applicability.[18] He could have added one powerful argument based on the relativity theory: because of the non-existence of the objective cosmic 'Now', it is meaningless to treat the universe at large as an enlarged object of our macroscopic experience, consisting of simultaneously co-existing parts. In the same book and in his more recent article 'The Child and Modern Physics', Piaget pointed out how physicists gave up the concept of permanently existing object on the microphysical level. Thus "by an extremely curious coincidence it was found that a very young baby acts with regard to objects rather like a physicist."[19] This may sound paradoxical, but what appears strange becomes completely intelligible within the framework of the new genetic epistemology. There is no reason to fear that modern physics invites the physicist to re-descend to the mental level, pre-linguistic and pre-operational, of the infant. And when Piaget says that there is a kind of "return to the primitive" (*une sorte de retour au primitif*)[20] on the part of the physicist disbelieving in the permanence of micro-objects, he explicitly stresses that this return is conscious and deliberate, based on the effort "to re-create a mentality free of all preconceived notion." It can be succinctly summed up in the following way: the infant does not believe in permanent objects because it has not yet been exposed sufficiently long to macroscopic experience; the microphysicist does not believe in permanent objects because his experience *transcends* macroscopic experience.

But while Piaget was aware of the agreement of his epistemology with some results of present physics, he realized much less to what extent his views coincided with the biological epistemology of Bergson. It is true that he conceded that Bergson "saw very well the rôle of *homo faber* in the formation of reason", but he claimed incorrectly that Bergson failed to trace this formation to the sensory-motor level. Equally erroneously, he imputed to Bergson the view sharply separating "intellect" from "intuition".[21] Piaget's interest in Bergson's philosophy was rather limited, certainly and surprisingly smaller than in the idealistic epistemology of Brunschwicg. However, he conceded in 1959 that his reading of *Creative Evolution* made him realize the possibility of the biological theory of knowledge.[22]

An additional significance of Piaget's genetic epistemology is that it is free of the controversial assumption of the heredity of acquired characteristics. This view, proposed by Lamarck, was accepted by Spencer and was implied in Poincaré's view of space, quoted above. Piaget's monographs showed that the Lamarckian assumption is not necessary for a genetic approach to the cognitive functions of man. They showed that "the logic of solid bodies" or, in his terminology, atomism and the belief in conservation of substance, develop in man's mind under the continuous pressure of the macroscopic environment; hence, there is no need to make an appeal to inherited tendencies. It is true that despite the generally hostile attitude of the neo-Darwinian biologists, the problem of the heredity of acquired characteristic is far from being settled definitively; in truth, it is not taboo even to Piaget.[23] But this is an additional reason for not using such a controversial hypothesis as the basis for any epistemological theory. In any case, Bergson was not inconsistent when he combined his genetic approach to the structure of human intellect within his resolute denial of the Lamarckian assumption of the inheritance of acquired characteristics.

## NOTES

[1] Cf. my article 'The Development of Reichenbach's Epistemology' *The Review of Metaphysics* XI (1957), p. 42–67.

[2] H. Reichenbach, *Ziele und Wege der heutigen Naturphilosophie*, Leipzig 1931, p. 54.

[3] *Relativitätstheorie und Erkenntnis a priori*, Braunschweig 1920, p. 70. Cf. also *Atom and Cosmos* (transl. by E. S. Allen), G. Braziller, New York 1957, p. 293: "natural science taught us that reason is not a rigid chest of logical drawers, that thought is not the eternal repetition of inherited norms...."

[4] *Atom and Cosmos*, pp. 288–293. Cf. also A. March, *Die physikalische Theorie und seine Grenzen*, Braunschweig 1960, pp. 11–20.

[5] W. Finkelburg, *Atomic Physics*, New York, McGraw Hill, 1950, p. 245.

[6] *C.E.*, p. XIX.

[7] *The Rise of Scientific Philosophy*, Univ. of California Press, Berkeley, 1951, pp. 189–190.

[8] J. Piaget, *Psychology of Intelligence*, Littlefield, Adams & Co., Patterson, 1960, p. 4.

[9] Piaget, *op. cit.*, p. 7.

[10] *Ibid.*, pp. 18–21.

[11] Neuchâtel, 1937, pp. 11–96.

[12] Neuchâtel, 1962. Cf. also his *L'introduction à l'épistémologie génétique*, Presse Universitaire de France, Paris, 1951, II, pp. 146–152; and his article 'A propos de la psychologie de l'atomisme', *Thalès* V (1949) 3–7.

[12a] In collaboration with B. Inhelder.

[13] J. Piaget, *Genetic Epistemology*, Columbia Univ. Press, 1970, p. 53.

[14] Neuchâtel, 1950.

[15] Cf. Ch. 2, Note 2.

[16] *Genetic Epistemology*, p. 20.

[17] *C.E.*, p. 173.

[18] Vol. II, pp. 212–13.

[19] *Scientific American* **197** (March, 1957), pp. 46–51.

[20] *L'Introduction*, II, pp. 234–235.

[21] *Op. cit.*, II, 20; III, pp. 124–126.

[22] Piaget's letter in *Revue de Théologie et de Philosophie*, Lausanne IX (1959) 44.

[23] J. Piaget, *Biologie et connaissance*, Gallimard, Paris, 1967, pp. 313–314. Interesting facts are mentioned by Karl von Frisch, *Man and his Living World* (transl. by Elsa B. Lowenstein), Harcourt & Brace, New York, 1962, pp. 277–278.

# CHAPTER 9

# THE LOGIC OF SOLID BODIES
# FROM PLATO TO QUINE

In the last chapter of *Creative Evolution* Bergson tried to trace the persistent influence of "the logic of solid bodies" through the whole history of Western thought. His analysis is entitled "The Cinematographical Mechanism of Thought and the Mechanistic Illusion". It begins with the Eleatic school which identified Being with a solid immutable sphere, and ends with Herbert Spencer who, without trying to eliminate becoming altogether, still insisted on its subordination to Being when he tried to deduce "the law of evolution" from the conservation laws of energy and matter. It would be otiose to restate here Bergson's excellent and documented analysis. It is true that this analysis could be extended greatly in the sense that the persistent tendency to get rid of time and change could be illustrated by a far greater number of examples than those given by Bergson. For instance, Bergson omitted an analysis of ancient atomism which, because of its Eleatic roots and its tendency to admit change only in its most 'innocuous' form as a mere displacement of the eternally immutable bits of matter, would have provided him with the most graphic illustration of the "logic of solid bodies". Bergson also completely disregarded the medieval thought, which in its fascination with the static and timeless "Ens realissimum" clearly exhibited the persistent Eleatism of the human intellect.[1] In the modern era, both Schopenhauer and Nietzsche would have provided him with two other significant instances of "eternism" – to use A. O. Lovejoy's term; Schopenhauer, who explicitly assimilated his timeless "Will" with the Eleatic *One and All* (ἕν καὶ πᾶν) and Nietzsche, who despite his emphasis on evolution and his pronounced sympathy for Heraclitus ended with the almost mystical hymn to Eternity and its everlasting "ring of recurrence".

But, as Albert Thibaudet pointed out, Bergson's ambition was not the same as that of the historians of philosophy, such as Edward Zeller, for instance; he himself characterized the chapter mentioned above as 'a glance at the history of systems' or as a mere 'sketch of a criticism of philosophical systems'. Moreover, a more extensive application of Bergson's analysis was made by Émile Meyerson, who, though not a Bergsonian, traced various attempts to eliminate change and time in both philosophy and the sciences and who regarded the search for 'identity in time' as one of the most conspicuous and most constitutive features of classical explanation. Meyerson's book *Identity and Reality* appeared only a year after Bergson's *Creative Evolution* – ironically in the same year that J. M. E. McTaggart published his famous article 'Unreality of Time'[2], whose content has been discussed since then. We shall supply other illustrations of the fact that Eleatism, that is, 'the logic of solid bodies" in various degrees is as much alive in this century as it was in the past.

The case of McTaggart shows that Bergson's analysis of 'mechanistic illusion' does not apply only to mechanism in its original, physicalistic sense. It equally applies to all kinds of *static idealism* which tries to reconstruct concrete reality out of the elements which share their rigid immutability with the solid bodies of our macroscopic experience. In this respect, there is no basic difference between the material atoms and McTaggart's monads. McTaggart himself can appropriately be characterized as a *static pluralist*; he tried to combine the affirmation of genuine diversity with his static interpretation of Hegelianism and, consequently, with his denial of time.[3] By applying the idealistic terminology to static elements, their affinity with the macroscopic solids is merely hidden but not affected. This is what Bergson observed with respect to Plato:

Concepts, in fact, are outside each other, like objects in space; and they have the same stability as such objects, on which they have been modeled. Taken together, they constitute an "intelligible world," that resembles the world of solids in its essential characters, but whose elements are lighter, more diaphanous, easier for the intellect to deal with than the image of concrete things: they are not, indeed, the perception itself of things: but the representation of the act by which the intellect is fixed on them. They are, therefore, not images, but symbols. Our logic is the complete set of rules that must be followed in using symbols. As these symbols are derived from the consideration of solids, as the rules for combining these symbols hardly do more than express the most general relations among solids, our logic triumphs in that science which takes the solidity of bodies for its object, that is, in geometry. Logic and geometry engender each other, ... It is from the extension of a certain natural geometry, suggested by the most general and immediately perceived properties of solids, that natural logic has arisen; then from this natural logic, in its turn, has sprung scientific geometry, which extends further and further the knowledge of the external properties of solids. Geometry and logic are applicable to matter;* in it they are at home, and in it they can proceed quite alone.[4]

This passage clearly anticipated Piaget's discovery that the logical and arithmetical operations developed concomitantly with the manipulation of solids and Reichenbach's observation, quoted above, on the correlation of the corporeal substance with the two-valued logic. Even one of the severest critics of Bergson, René Berthelot, concedes that Bergson was correct when he pointed out the affinity of logical relations between concepts with the spatial relations of exteriority and inclusion.[5] This is true of the Aristotelian syllogistics as much as of the modern calculus of classes; the Aristotelian formal logic was not inappropriately called "the traditional calculus of classes." The fact that the relations between class-concepts can be symbolized by Euler's circles or Venn's diagrams shows clearly how closely *juxtaposition* and *logical distinction* are related. Thus the three possible relations between two concepts – identity, similarity (i.e., partial identity) and difference correspond to the three geometrical relations between two Euler's circles – a complete coincidence, a partial overlapping and a complete exteriority. No other relations are possible; the only difference between the situation just described and Venn's three-circles-diagrams is in the number of the classes considered. The fact that on the basis of Boolean algebra mechanical devices for performing the logical operations can be constructed is another striking confirmation of Bergson's and Piaget's claim about the close correlation of logical operations and mechanical manipulations.

But Berthelot was correct that the classical subject-predicate logic is not the only kind of logic; the existence of the logic of relations, which is irreducible to the classical syllogistic, should be a sufficient warning against the sweeping claim that human intellect is fated to think only in spatializing categories. But has Bergson ever claimed this? We dealt with this point in Chapter 7 of this Part and stressed that the misunderstanding was due to Bergson's too sweeping use of the terms 'intellect', 'geometry' and 'logic' while he meant 'the traditional intellect', 'traditional geometry', and 'traditional logic'. We must not forget that in 1907 the three-valued logic was still unknown and also that equally unknown – except among a few specialists – were geometries not requiring the rigid bodies. In truth,

* To wit: "*macroscopic* matter".

it was the mathematician-philosopher Bertrand Russell who shortly before that time insisted on the impossibility of spaces with variable curvature, on the ground that in such space "the axiom of free mobility", would not hold. But precisely this axiom affirms that neither the size nor the shape of the bodies are affected by their displacement; in other words, their rigidity is guaranteed.[6] And the attitude of Russell was still liberal compared with that of neo-Kantians who – like Louis Couturat – intransigently singled out the Euclidian space as the only possible one.[7] Yet, the names of both Russell and Couturat were made famous by their active participation in the modern development of logic!

Furthermore, René Berthelot himself conceded that 'at first glance' the modern arithmetization of mathematical analysis made by Krönecker, Dedekind, Tannéry and Cantor, seemingly confirmed Bergson's diagnosis.[8] This arithmetization can be characterized as a very refined and sophisticated renaissance of Pythagoreanism. The points of modern analysis are extensionless and they are infinite in number; but they are as *external* to each other as the Pythagorean 'units with a bulk'. Thus the logic of solid bodies is merely refined, but not basically changed; the extensionless points are as rigid and as discontinuous as the atoms of Leucippus who, according to the beautifully concise expression of John Burnet, "gave the Pythagorean monads the character of the Parmenidean one".[9] Russell approvingly quotes Poincaré who wrote:

The continuum thus conceived is nothing but a collection of individuals arranged in a certain order, infinite in number it is true, but external to each other. This is not the ordinary conception, in which there is supposed to be, between the elements of the continuum, a sort of intimate bond which makes a whole of them, in which the point is not prior to the line, but the line to the point. Of the famous formula, the continuum is unity in multiplicity, the multiplicity alone subsists, the unity has disappeared.[10]

We shall see in Part II that it is this intimate, dynamic bond of the successive continuum which escapes the logic of solids of which modern set theory is the most refined form. Cantor's definition of the set (*Menge*) as "a collection of definite, well discernible objects" (*die Zusammenfassung der bestimmten, wohl unterschiedenen Objecte*) shows clearly the *external* character of the relations in which its constitutive elements stand to each other – relations which are modeled after the relation of juxtaposition.[11]

A more recent, but equally striking, illustration of "the logic of solids"

is Wittgenstein's *Tractatus Logico-Philosophicus*. The following propositions from the first part of this book are certainly characteristic:

2.01.      A state of affairs (*der Sachverhalt*) is a combination of objects
           (*Gegenständen, Sachen, Dingen*).
2.013.     Each thing (*Ding*) is, as it were, in a space of possible states of
           affairs (*Sachverhalte*). This space I can imagine empty, but I
           cannot imagine the thing without space.
2.02.      Objects are simple.
2.021.     Objects make up the substance of the world. That is why they
           cannot be composite (*zusammengesetzt*).
2.026.     There must be objects, if the world is to have an unalterable
           form.
2.027.     'Unalterable' (*das Feste*), 'persistent' (*das Bestehende*) and
           'object' (*der Gegenstand*) are synonymous terms.
2.0281.    Objects are what is unalterable (*das Feste*) and persistent (*das
           Bestehende*); their configuration is what is changing and un-
           stable.
2.0272.    The configuration of objects produces states of affairs.
2.032.     The determinate way in which objects are connected in a state
           of affairs is the structure of the state of affairs.
2.04.      The totality of existing states of affairs is the world.

The shades of Democritus and Herbart are clearly discernible in these
passages. Obviously, the 'objects' of Wittgenstein are not material particles; but in their main characteristics they *are* the atoms of Democritus,
but like the "reals" of Herbart or the "elements" of Mach, transposed to
a higher level of abstraction. In other words, there is a perfect isomorphism
between physical atomism and the logical atomism of Wittgenstein: the
objects of *Tractatus* are as immutable, discontinuous, indivisible and
simple as the indivisible and homogeneous particles of classical physics. In
both kinds of atomism, change is reduced to the changing 'configurations'
(Wittgenstein even uses the same term) of these ultimate units. Similarly,
the affinity between the void of the atomists and 'the intelligible space' of
Herbart and 'the logical space' of Wittgenstein is unmistakeable. It is only
consistent for Wittgenstein to espouse the relational theory of time;
according to him, there is no 'passage of time', separable from concrete
processes (6.3611) and since the processes are reducible to the changing

configurations of the objects, this view bears a remarkable similarity to
Lucretius's claim that 'time is nothing by itself', that is, nothing separable
from the changing configurations of the unchangeable atoms.

Similar and even more radical trends are present in still more recent
English speaking philosophers. This has been pointed out in a documented
article of Nicholas Rescher, 'The Revolt against Process'.[12] While Wittgenstein still retained the relational theory of time in which time is sub-
ordinated to things without being eliminated, Donald Williams went so
far as to speak of 'the myth of passage', proposing to replace becoming
by 'the four-dimensional manifold of events'. His view and similar views
of this kind were admittedly influenced by a widespread misconception of
the relativistic space-time – a misconception which I have analyzed
critically several times before.[13] Nothing confirms Bergson's diagnosis of
the native Eleatism of human intellect more than this persistent tendency
towards 'treating time as spacelike', to use Professor Quine's words.[14] We
shall return to this problem in Part II. And although Professor Strawson
does not eliminate time, but merely subordinates it to things, his "cate-
gorial preference" for the three-dimensional material bodies bears such a
striking resemblance to the logic of solid bodies that it would be otiose
to dwell on it. Professor Quine on the very first page of his book which
has a characteristic title, *Word and Object*, concedes it quite candidly:

Entification begins at arm's length; the points of condensation in the primordial conceptual scheme are things glimpsed, not glimpses. In this there is little cause for wonder.
Each of us learns his language from other people. ... Linguistically, and hence conceptually, the things in sharpest focus are the things that are public enough to be talked
of often, and near enough to sense to be quickly identified and learned by name; it is to
these that words apply first and foremost.

This is undeniably true as a *statement of fact*, although as a *genetic
account* it should be slightly amended. (In the light of Piaget's researches,
it is the glimpses, not the objects, which come first, even though the elaboration of the idea of object comes very early in childhood.) In other words,
nobody denies that there is an "immemorial doctrine of ordinary enduring middle-sized physical objects" (Quine) or a "massive central core
of human thinking" (Strawson). But the epistemological significance of
this fact is, it seems to me, different from that assumed by Quine and
Strawson. No matter how important – both socially and biologically – the
macroscopic bodies of our daily experience are, they do not represent the

only furniture of the universe. There are very important ingredients of reality that are not necessarily the things "common and conspicuous enough to be talked of publicly, common and conspicuous enough to be talked of often, and near enough to be quickly identified by sense and learned by name". In truth, the most revolutionary discoveries in science were the discoveries of the entities which were not only inconspicuous, but not even perceptible by senses; and of which nobody talked about before. This is the reason why no familiar name could be applied to them – not even the word 'object' or 'thing' as the most recent development of physics shows. What if the very term 'object' is, according to Professor Gonseth's profound remark, nothing but a 'macroscopic prejudice'?[15] What if our object-oriented language and our object-oriented logic is merely a result of the long process of adjustment to the realm of middle dimensions? This possibility, clearly envisioned by Bergson, has been made highly probable by the evolution of physics in this century.

This is now being recognized more and more. Professor David Finkelstein in his recent paper[16] points out that in each physical theory there are three different levels of the increasing generality: (a) those statements dealing with the distribution of matter and energy, with what can roughly be called the theory of mechanics; (b) on the deeper level there are the statements dealing with the spatio-temporal structure of the world which, until the coming of the relativity theory, was regarded as rigid and independent of the concrete physical content with which the statements on the first level deal; (c) on the deepest level, which corresponds to the highest degree of abstraction, there are the statements rarely formulated explicitly in the physical theories since they belong to arithmetic, set theory, and propositional calculus. Now while the statements on the two previous levels lost their alleged a priori character – first physics, then geometry became empirical sciences contrary to the claims of Kant and neo-Kantians – the deepest logical level is still generally regarded as above any possible modification. Logic is not an empirical science – at least *not yet*. But Finkelstein raises the question whether the present upheaval in physics will not require a similar far reaching *widening* or *generalization* of logic, as revolutionary as the generalization of Newtonian mechanics and Euclidian geometry. Within such generalized logic the traditional two-valued logic would still retain its legitimate place, being applicable on the most superficial level of daily experience (which so unfortunately fascinates the ordinary language philosophers). Similarly, Euclidian geometry remains a useful approximation on the planetary scale, where space curvature is negligible, and Newtonian mechanics still holds very approximately on the macroscopic level of small velocities and a large number of quanta. Yet, traditional logic would lose its privileged, absolutist status since, like Euclidian geometry and Newtonian determinism, it would become incorporated into a larger and more flexible framework.

There is hardly need to stress how well this envisioned widening of logic fits into Bergson's biological theory of knowledge; for only in the genetic perspective of this theory does the *broadening* of physics, geometry, and logic become intelligible. Nietzsche's heretical statement that "logic (like geometry and arithmetics) only holds good of fictitious entities which we have created" and that "truth is that kind of error without which certain species of beings cannot exist", clearly anticipated the relativization of traditional logic which Finkelstein suggests now.[17] But his own epistemology, in spite of its biological orientation, remained too much colored by his commitment to certain dogmas of classical physics, to allow him to draw all the consequences from his own truly remarkable and revolutionary anticipations.[18] Similarly, Professor Quine recently conceded the importance of the evolutionary explanation of the origin of logic; logic itself, he conceded, may be "a result of natural selection".[19] But does he realize that a *biological* success does not necessarily mean a *perfect ontological fit* and that the successful survival of the traditional forms of thought, including the two-value logic, is due to their applicability to certain limited conditions, to our confinement to the world of middle dimensions in which we live? And now, when, technologically and intellectually, we are breaking out of this confinement, a broadening of classical logic, including of its refined forms of set theory and propositional calculus, becomes imperative?

The term 'broadening' or 'widening' or 'generalization' of logic is far less misleading than 'relativization'. The latter term – unlike the former – suggests a complete relativization of truth which – as we know from the time of Socrates' criticism of the sophists – is intellectually self-defeating. In other words, *becoming of logic requires the logic of that becoming*; but we may be sure that such logic of becoming will be quite different – more flexible and far more comprehensive than the provincial logic of solid bodies which itself is a local by-product of the evolutionary becoming of life.[20] It is the dynamic structure of becoming in general which will be the main object of Part II.

## NOTES

[1] Cf. my article 'The Doctrine of Necessity Re-Examined' *The Review of Metaphysics* V (1951), especially pp. 18–19.

[2] *Mind* 17 (1908).

[3] John M. E. McTaggart, *Studies in Hegelian Dialectic*, Russell & Russell, New York, 1964, Ch. V.

[4] *C.E.*, p. 177.

[5] René Berthelot, *Un Romantisme utilitaire*, II, Alcan, Paris 1913, pp. 190–193.

[6] B. Russell, *An Essay on the Foundations of Geometry*, Cambridge University Press, 1897, pp. 150–154.

[7] Louis Couturat's review of Russell's book mentioned above in *Revue de métaphysique et de morale* VI (1898), 354–380, esp. pp. 373–374.

[8] R. Berthelot, *op. cit.* pp. 179–180.

[9] John Burnet, *Early Greek Philosophy*, The Meridian Library, New York, 1957, p. 336.

[10] B. Russell, *The Principles of Mathematics*, Norton, New York, 1964, p. 347. The passage quoted is from *Science and Hypothesis* (*F.S.*, p. 43).

[11] A. Fraenkel, *Einleitung in die Mengenlehre*, Dover, New York, 1946, p. 4.

[12] 'The Revolt against Process', *The Journal of Philosophy* LIX (July 1962), 410–417.

[13] Donald Williams, 'The Myth of Passage', *Journal of Philosophy* 48 (1951) 457–472; my answer 'The Myth of Frozen Passage', *Boston Studies in the Philosophy of Science*, II, Dordrecht and New York, 1956, pp. 441–463 with the subsequent rejoinder by D. Williams; cf. also my article referred to above in Note 1 (pp. 40–45) and my book *The Philosophical Impact of Contemporary Physics*, Van Nostrand, Princeton, 1964, Ch. XI, XII, XIII.

[14] Willard Van Orman Quine, *Word and Object*, M.I.T. Press, Cambridge, p. 172.

[15] Ferdinand Gonseth, *Les mathématiques et la réalité*, Paris 1936, p. 158.

[16] Professor David Finkelstein in his paper 'Matter, Space and Logic' in *Boston Studies for Philosophy of Science*, Vol. V, Dordrecht and New York, 1969, pp. 199–215.

[17] *The Will to Power* (transl. by Anthony M. Ludovici), Russell & Russell, New York, 1964, II, § 493, 516.

[18] Nietzsche's belief in the law of conservation of energy clashed with his view that "substance" and "object" are mere fictions; his sympathy for the idea of process was more than neutralized by his cyclical theory of time and of the identical recurrence of events. The difference between Nietzsche's biological theory of knowledge and that of Bergson was well pointed out by R. Berthelot, *op. cit.*, pp. 72–82.

[19] In the discussion at the Boston Colloquium for the Philosophy of Science, November 21, 1967. Cf. also Quine's book *Ontological Relativity and Other Essays*, Columbia Univ. Press, New York, 1969, pp. 126–127.

[20] Cf. *C.E.*, p. XX, "... our thought in its purely logical form, is incapable of presenting the true nature of life... Created by life, in definite circumstances, to act on definite things, how can it embrace life, of which it is only an emanation or an aspect? Deposited by the evolutionary movement in the course of its way, how can it be applied to the evolutionary movement itself?"

# PART II

## BERGSON'S THEORY OF DURATION

CHAPTER 1

# THE MEANING OF IMMEDIACY

So far we have dealt only with the negative character of Bergson's epistemology. Its positive features are inseparable from Bergson's concrete views concerning the nature of mind, matter and organic life. His first problem was the nature of psychological existence; the solution which he proposed influenced in a decisive way the subsequent elaboration of his philosophy. Out of Bergson's psychology grew naturally what may be called the 'Bergsonian physics' and the specifically Bergsonian solution of the classical mind-body problem; out of his psychological and physical views sprung in an equally organic way his neo-vitalistic re-interpretation of evolution which later colored even his particular views concerning the nature of ethics and religion.

The problem of the nature of psychological existence was attacked by Bergson in his first book *Essai sur les données immédiates de la conscience*. Here his epistemological method, which we explained in the first part, found its initial application. Although a complete formulation of Bergson's biological theory of knowledge can be found only in *Creative Evolution* nearly two decades later, the same epistemological attitude is unmistakably, though less explicitly, present in his first book. Its original French title is in this respect more indicative than the title of the English translation. It indicates that the main theme of the book is the nature of the immediate introspective data. What is the nature of psychological existence when it is intuited in its immediacy and epistemological purity, i.e. freed from irrelevant and misleading associations? In this book Bergson himself characterized his epistemological attitude as a 'reversed Kantianism'.[1] A more adequate term would be a "completed Kantianism." Kant and the classical epistemology in general stressed the importance of the subjective elements in our sensory perception. According to nearly all classical philosophers the preliminary condition for acquiring a more adequate insight into the nature of external reality is to make a clean sweep of the subjectivity which enters our sensory perception. Thus Democritus regarded the secondary qualities as mere "psychic additions" which should be eliminated from our perception if we want to reach the true ontological core of matter. This theory of 'psychic additions', after its temporary eclipse in the Middle Ages, triumphantly returned with the birth of modern science and modern philosophy. A big problem was where to draw a dividing line between the 'psychic additions' and the primary qualities of matter. Descartes went farther than Democritus when he included into psychic additions even the impenetrability of matter; Berkeley prophetically challenged the whole distinction between the primary and the secondary qualities, claiming that they are both subjective; Kant went still farther when he claimed that even the spatial and temporal relations are supplied by the subject. While the true reality for Democritus was matter in space, for Kant it was a hypothetical 'thing-in-itself' outside of space and time, having nothing in common with the matter perceived by our senses; we are not even sure whether we should speak of it in a grammatical plural or singular since the categories of unity and plurality are, according to Kant, not applicable outside the phenomena.

It is not important for the time being to ask whether this progressive elimination of the allegedly subjective elements did not go too far; this question will emerge naturally later. What is important now is to be aware of the trend which classical epistemology exhibited; and this trend consisted in an increasing emphasis on the presence of subjective elements in our knowledge of the external reality. Bergson on the other hand called attention to the fact that the converse question has hardly been raised: to what extent is our knowledge of the 'internal world', i.e. our awareness of introspective data, colored and possibly even distorted by elements borrowed from sensory perception? This is the fundamental problem of Bergon's first book: are we aware of our 'inner states' immediately, or is our introspection stained by the admixtures unconsciously borrowed from our sensory experience? The probability that sensory associations do interfere with our self-perception is certainly not negligible if we consider the fact that the knowledge of the external world comes up considerably earlier than awareness of our self. Sensory perception by far *precedes* introspection. It is then only natural to expect that the latter is influenced by the former, as a new experience at first is almost invariably expressed in the terms of the older and more familiar one.

The most obtrusive sensory admixtures in our introspective awareness, which, according to Bergson, most seriously distort our introspection, are due to the habits of *spatialization and visualization*. They both are really two aspects of one and the same obsession: visualization is only a more concrete form of spatialization. Both habits are very obstinate and excessively difficult to eliminate; their force is strengthened by their never-ending successful application to the objects of our daily experience. Even outside of the boundaries of our spontaneous visual perception, spatial and visual models were for a long time the only possible ways of explanation. Their failure, due to a further widening of our experience, is only very recent and does not threaten their continued usefulness in the world of the middle dimensions. Thus it is only natural that our habits of spatialization and visualization continue to dominate our imagination, and are virtually present even under the disguise of apparently purely abstract concepts, and along with explicit verbal denials. The basic fallacy of all forms of materialism and even sensualism was the assumption that the whole of our experience, including the elusive introspective data, is interpretable in crudely sensory and even visual terms. What else is the claim of the materialists of all ages that what we call 'thought' is nothing but a movement of subtle material substance, whether this substance is constituted by the 'fire-atoms' of Democritus or by the sub-microscopic changes in the DNA molecules? What was even more serious is that even those thinkers who saw clearly the fallacies of physicalism have not entirely been able to escape from the sway of basically the same mental habits. The fact that this influence was of a less conspicuous and more subtle kind makes the corresponding 'epistemological psychoanalysis' only more difficult. This visual and spatializing sensualism in its conscious or semiconscious form was the main target of Bergson's criticism in his first book.

It is needless to emphasize how the biological orientation of Bergson's epistemology is discernible even in this early stage. Continuous and repeated warnings against spatialization and visualization, which is a *Leitmotiv* of the *Essai*, is only a negative aspect of the belief that reality is constituted by heterogeneous strata and that our mind is adapted only to *some* of them, but certainly not to the whole of reality; consequently, that it is illegitimate to apply the intellectual forms fitting one of these strata to all others. (It would be premature now to raise the question whether the heterogeneous strata of reality are absolutely foreign to each other or whether an attentive analysis may not discover a denominator common to all the strata. But we may say, anticipating our future discussion, that Bergson believed he had found such a denominator in his concept of temporal process or duration.) The question which is important now is: what were the positive results of Bergson's epistemological struggle against the fallacies of spatialization and visualization in psychology? In other words: what is the meaning and content of 'immediacy' and 'intuition'?

Both terms became the main source of misunderstanding of Bergson's philosophy, and this is one additional reason for giving their full explanation. First the notion of *immediacy*. Without exaggeration, there is hardly anything *less immediate*, less *given*, than Bergson's 'immediate data'. The word 'immediate' was ill chosen by Bergson because what he meant by it was certainly not 'immediate *de facto*' but rather 'immediate *de jure*'. In other words, the introspective datum is immediate when it is freed from irrelevant and extraneous elements which, so to speak, 'mediatize' it. A simple example will illustrate it: a musically educated person will almost certainly associate with his auditory perception of melody its graphical symbol, a score written on a sheet of paper, a visual image of the orchestra and, possibly, if he is a musician, vague tactile reminiscences of strings, bow, piano keys, etc. The cluster of these heterogeneous images is in a certain sense immediately present to his mind; yet, it would be wrong and misleading to confuse this kind of immediacy with the immediacy of musical experience which appears only when all accessory non-musical images and recollections are radically eliminated. By not doing it, we confuse the auditory data with the visual and tactile ones, and even run the risk of losing sight of the central nucleus of musical experience, although we may still continue to *talk* about it. It is true that the "audition colorée" may be very effective in poetry or even in subjective interpretations of musical experience, as the case of Rimbaud and Baudelaire clearly showed; but epistemologically it always means a translation of the auditory data into visual terms. In other words, it means a transition of the experience whose salient feature is temporality into terms which, if not entirely devoid of temporal character, are at least preponderantly spatial in nature. The search for immediacy here means a search for *epistemological purity*[2], i.e. an effort to avoid the confusion of heterogeneous strata of experience.

If the Bergsonian notion of *immediacy* is thus understood, there will be little danger of misunderstanding the Bergsonian *intuition*; both terms are almost synonymous. Bergson did not use this term in his first book and he would probably not have used it at all, had he anticipated all the misinterpretations and confusions which it aroused. The first reference to this term appeared in his *Introduction to Metaphysics* in 1903. There are sporadic references to this term in *Creative Evolution*, and the fact that the intuition is opposed to the analytical intellect easily aroused suspicion of Bergson's irrationalism. Only one whole essay is dedicated to this problem: the paper *L'Intuition philosophique*, read at the philosophical congress at Bologna in 1911. In this paper the following characteristic of intuition is given:

It seems to me that intuition often behaves in speculative matters like the demon of Socrates in practical life; it is at least in this form that it begins, in this form also that it continues to give the most clear-cut manifestations: it forbids. Faced with currently accepted ideas, theses which seemed evident, affirmations which had up to that time passed as scientific, it whispers into the philosopher's ear the word: "Impossible!" Impossible, even though the facts and the reasons appeared to invite you to think it possible and real and certain. Impossible, because a certain experience, confused perhaps but decisive, speaks to you through my voice, because it is incompatible with the facts cited and the reasons given and because hence these facts must have been badly observed, these reasonings false.[3]

There is perhaps an unfortunate ring of Rousseau in this passage which is apt to hide its true meaning, especially if the passage is taken out of the context of Bergson's other writings. Nevertheless, the passage does contain the essence of the Bergsonian intuition. This intuition begins with the attitude of distrust for the accepted modes of thought; there is no need, of course, to call this distrust by an unduly poetic and highly misleading term 'inner voice'. (Any 'whispering of inner voice' would only arouse the suspicions of contemporary naturally suspicious psychologists and philosophers who are used to psychoanalyzing all philosophical views but their own.) This attitude of distrust originates in a vague awareness of certain experiences incompatible with the accepted modes of thought; the essence of intuition is precisely to bring these vague and rather implicitly felt data into a clear focus and to show that the new forms of understanding thus created are superior to the old ones. Their superiority is precisely in *their greater flexibility*; not only do they take into account the aspects of experience neglected by previous theories, but they also explain with equal success the facts already known and claimed as the main basis of evidence for the previous theories. In other words, it is important to show that the old facts not only do not contradict a new theory, but that, when correctly interpreted, they follow as natural consequences from it.

It is thus clear then that what was designated by the word 'intuition' is a very complex process which had nothing in common with emotion and instinct, and which certainly does not go on effortlessly and passively. It is true that the incipient phase of this process is rather vague and devoid of explicit clarity; we must not forget, however, that the terms 'vague' and 'clear' are relative and that the feeling of vagueness is often an *effect of contrast* resulting from the deceptive clarity of the old ways of thinking. But as the process of reinterpretation goes on and as the inadequacy of the old ways of thought becomes more and more evident, the incipient vagueness gradually yields to the growing insight until eventually the situation is exactly reversed: a hazy anticipatory feeling is transformed into the clarity of understanding, although usually of a more complex type; while what previously was clear will finally appear as an oversimplification which, by its artificial simplicity, distorts and mutilates the data of experience. The ambiguity of the term "intuition" was due to the fact that Bergson applied it both to the initial anticipatory insight as well as to its terminal phase in which its content has already been purged of all parasitic elements. The analysis of all the outstanding revisions of established modes of thought would convincingly illustrate this gradual transition from a hazy anticipatory glimpse into the lucidity of final insight. This transition is, according to Bergson, a characteristic feature of every intellectual effort.[4]

## NOTES

[1] *Time and Free Will, An Essay on the Immediate Data of Consciousness* (transl. by F. L. Pogson), Harper Torchbooks, 1960, pp. 222–224. Hereafter referred to as *T.F.W.*
[2] This is what is repeatedly stressed by Vladimir Jankélevitch, *Henri Bergson*, Paris 1959, *passim*.
[3] *C.M.*, p. 129.
[4] Cf. Note 8 of Chapter 7, Part I.

# CHAPTER 2

# CONTENT OF THE BERGSONIAN INTUITION

Such was apparently the case with Bergson's own intellectual development. According to his own admission, an overwhelming feeling of inadequacy came to his mind when he was reading Herbert Spencer's *First Principles*, whose author he at first sincerely admired. This original feeling of inadequacy soon took a more precise form: the *awareness of the reality of time* contrasting with the apparent superfluity of time in the mechanistic and deterministic scheme. The awareness of this contrast became a source of the Bergsonian philosophy, its "dynamical scheme" (*"le schéma dynamique"*) out of which all main trends of his thought unfolded, including his philosophy of physics. The assertion of the reality of succession is an element common both to Bergson's psychology and to his philosophy of nature.

Such an assertion may appear a truism. Who else with the exception of some extreme idealistic philosophers denied the reality of succession? But the revolutionary character of Bergson's affirmation of time will become more evident only when we realize that the time thus asserted is considerably different from a vague and common sense notion which was more or less adopted by the majority of scientists. We have to bear in mind that Bergson asserted the reality of the temporal first on the psychological level. But even in his first book the problem of psychological time, or, as he preferred to call it, of psychological duration, began to merge with the problem of *duration in general*. This is why the understanding of the notion of *true duration ("durée réelle")* in his *Essai sur les données immédiates de la conscience* is a key to the understanding of the essence of his philosophy.

In view of the negative and inhibitory character of intuition which led Bergson to compare it to Socrates's *daimonion*, it is hardly surprising that the salient features of the intuition of duration are expressed in the following negative propositions.

(1) The dynamic continuity of psychological duration must not be conceived as a succession of clearly defined and mutually external units. The terms 'mental states', 'ideas', 'sensations', 'impressions', or 'elements' by which associationistic psychology from Hume and Condillac up to Mach and Bertrand Russell tried to reconstruct the dynamic continuity of introspective experience are borrowed from classical physics; by combination and recombination of such 'mental atoms' all changes and varieties of our psychological life were to be explained. But such isolable and permanent elements are merely fictitious entities, artificially carved out of the dynamic continuity of what James called the 'stream of thought', and artificially solidified by the same act of naming and conceptualization which separates them from their temporal context. This leads us to the second feature, the most important of all:

(2) The psychological duration is by its own nature forever incomplete, being always a *fait accomplissant* and never a *fait accompli*; in other words, it is a *continuous emergence of novelty* and can never be conceived as a mere rearrangement of permanent and pre-existing units. It never barely *is*, it always *becomes*. Consequently, it cannot be adequately represented by spatial diagrams for the simple reason that the diagram consists of *simultaneous* parts and thus erroneously suggests that the successive phases of duration coexist together; more specifically, that the future, though hidden to our subjective knowledge, coexists in all its specific details alongside the present and that no real succession and no genuine novelty ever occur.

(3) The reality of novelty implies the qualitative diversity of the successive phases: every duration is by its very nature *heterogeneous*. The so-called quantitative and homogeneous time which allegedly contains all changes and events is in truth nothing but a disguised space and thus it is inadequate in a twofold sense: by its static character it tends to obscure the incompleteness of any duration which it symbolizes; by its homogeneity it tends to suppress the qualitative diversity of successive phases.

(4) The negation of the homogeneity of duration implies also the negation of its mathematical continuity, that is, of its *infinite divisibility*. Again such divisibility belongs not to duration itself, but only to its spatial symbol. The negation of infinite divisibility logically eliminates the reality of durationless instants. The mathematical instant is another fictitious entity suggested by a wrong analogy between the temporal moment and

geometrical point. The psychological present, the experienced 'now', is always a *concrete quality* and never an infinitely thin mathematical point devoid of temporal thickness.

(5) The negation of the homogeneity of time implied for Bergson the elimination of the Newtonian time as an empty and inert receptacle, additionally filled up by concrete changes and events. In other words, there is no distinction between the duration itself and its content. Psychological events are not *in* time, since they in their ceaseless emergence *constitute true time itself*. This was one of the most far-reaching and boldest of Bergson's claims. It was clear that in analyzing the structure of psychological duration he was drawing the consequences for the nature of time in general. It must be admitted that there is no question that this aspect of Bergson's thought remained very obscure in its first phase. It led him in his first book to claim that true time is confined only to the psychological realm while the physical world is devoid of duration. In *Matter and Memory*, however, this untenable dualism of the temporal mind and timeless matter was given up, since becoming was reinstated into the physical realm. Thus psychological time ceased to be the only real time, although it was used as a model for comprehending the structure of objective duration. (But this will be dealt with in the third part of the book.)

These five negative statements are too concise to be convincing and therefore a more detailed exposition is needed. Although it is extremely difficult, if not impossible, to consider separately the five negative features of psychological duration, clarity of exposition makes it necessary. It will be precisely this step by step exposition that will elucidate more convincingly the organic connection between the individual features of duration, the connection which makes some repetition in the exposition unavoidable. Then it will become clear that it is almost childish to number each individual feature separately, since all of them are complementary and inseparable aspects of one single, though very complex, dynamic reality.

CHAPTER 3

# THE DYNAMIC CONTINUITY OF DURATION

By insisting on the continuity of psychological duration Bergson protested against the natural tendency to conceive of our mental states as
distinct and clearly separate entities which, precisely because of their
distinctness and separability, may be treated in arithmetical fashion.
This idea of distinct mental states was most clearly and explicitly formulated by the sensualistic associationism of the eighteenth century and its
nineteenth and twentieth century followers; but implicitly and half-
consciously it is present even in the minds of those who are vigorously
opposed to this trend and who would resent any suspicion of their
affinity with Hume or Russell.

Sensualism and associationism arose as a reaction against the inadequacies of Cartesian rationalism. Against the artificial and barren unity of
the Cartesian *res cogitans*, which is an immutable and impersonal
substratum of successive mental states, sensualism emphasized the rich
variety and continuous change of our inner experience. The bare Ego of
the rationalists and idealists is never introspectively perceived, since as
Hume showed, our psychological experience always comes in the form
of a concrete sensation, concrete image or concrete feeling. Berkeley's
attack against the reality of general ideas was inspired by the same conviction, to wit that only the distinct, particular mental states exist. This
emphasis on concrete experience and reluctance to accept any purely
verbal or postulated entities which are hypothetically invented, but never
intuited, is fully shared by Bergson. If he does not agree with the conclusions of associationism, it is because he does not believe that they are
*empirical enough*. Sensualistic associationism in the name of experience
rejected the idea of an unchanging empty Ego underlying the concrete
psychological events; but did it succeed in getting rid of the notion of
fictitious substance entirely? Are not Hume's "impressions", Condillac's
"sensations", Herbart's "*Vorstellungen*", Taine's "mental events", Mach's

"elements", etc. *minute substances* which on a reduced scale share all the essential features of the Cartesian substance, especially its independence and immutability? Hume was fully aware of it and explicitly admitted it.[1] The affinity of Herbart's psychological elements with substantial physical atoms is obvious. The same thing is true of the associationism of the later period. It is almost amusing to see Mach, who is so critical of the concept of atom in physics, accept unreservedly the atomistic conception of mental data. In recent British sensualism the term "atom" is consciously accepted and the atomistic view of mind is only a consequence of the atomistic view of reality in general (Russell, Wittgenstein).[2]

This atomistic approach to introspective data was evidently inspired by the spectacular success of atomism in physics. The temptation was great to interpret all psychological changes in terms of the combination and recombination of simple, self-identical elements. The names of these elements varied, but not their basic meaning. Physical atomism had different varieties; so did psychological atomism. Are the physical atoms all of the same kind? The answer of Democritus followed by Epicurus, Lucretius, Gassendi and classical atomism in general, was affirmative; Anaxagoras, Empedocles, followed in the modern era by Sennert, Berigard[3], John Dalton and nineteenth century chemistry in general, insisted on the irreducible qualitative differences between the ultimate units of matter. But around 1900 the electron theory resolved the allegedly irreducible qualitative differences between Dalton's atoms into differences in configuration of the homogeneous subatomic units. It is interesting that an analogous situation appeared in the history of psychological atomism. Although experience seemed to substantiate beyond doubt the irreducible qualitative differences between 'mental elements', the Democritian ideal of a basic homogeneity was never abandoned. Thus even in psychology there were persistent efforts to discover under the qualitative differences separating various sensory qualities the basic homogeneous 'sub-sensations'. The device was the same as in the physical sciences: to split the apparently indivisible psychological units into simpler homogeneous components in order to reduce the apparent qualitative differences to the difference of 'arrangement'. This view, the roots of which may be traced to Leibniz's idea of *"petites perceptions"*, was proposed in the last century by Hyppolite Taine, followed on this point by Herbert Spencer.[4] It was revived in this century by E. B. Holt and, most recently, by S. C. Pepper.[5]

These differences between qualitative and quantitative forms of atomism are less important in view of their basic agreements. These agreements were: (1) In the same way as the apparently inexhaustible variety of physical nature is reducible to a few kinds – or possibly one kind – of unchangeable physical units, the baffling variety of our inner psychological life is due to the differences in arrangement and complexity of relatively few, or possibly one, kinds of elements. (2) In the same way that physics placed the movement of physical corpuscles into the homogeneous and indifferent medium of space, the 'movement' of psychological atoms was supposed to take place in the equally homogeneous and inert medium of time. Although the Newtonian separation of time from events taking place in it was not explicitly listed among the presuppositions of the associationistic psychology, it was nevertheless always present implicitly.

Bergson challenged both these assumptions. He tried to show that both the postulate of permanent, self-identical entities as well as that of the passive, indifferent and container-like medium of time are nothing but clumsy and artificial constructions which have no counterparts in concrete, psychological experience. It is true that the solid bodies of our sensory experience have, superficially at least, well-defined outlines and permanence through time; but do we have the right to endow the alleged 'mental states' with the same properties? To what extent are we justified in borrowing corpuscular models from our daily physical experience and applying them to our introspective data? Or more concisely: is 'the logic of solid bodies' applicable beyond its natural limits, that is, to our introspective experience? Bergson claims that in the continuity of successive phases of our psychological life there are no sharp boundaries, no precise contours. What we call a 'mental state' is an entity artificially carved out of its dynamic context:

Sensations, feelings, volitions, ideas – such are the changes into which my existence is divided and which color it in turns. I change then without ceasing. But this is not saying enough. Change is far more radical than we are at first inclined to suppose. For I speak of each of my states as if it formed a block and were a separate whole. I say indeed that I change, but the change seems to me to reside in the passage from one state to the next: of each state, taken separately, I am apt to think that it remains the same during all the time that it prevails. Nevertheless, a slight effort of attention would reveal to me that there is no feeling, no idea, no volition which is not underlying change every moment: if a mental state ceased to vary, its duration would cease to flow...

This amounts to saying that there is no essential difference between passing from one state to another and persisting in the same state. If the state which "remains the same" is more varied than we think, on the other hand the passing from one state to another resembles, more than we imagine, a single state being prolonged; the transition is continuous. But just because we close our eyes to the unceasing variation of every psychical state, we are obliged, when the change has become so considerable as to force itself on our attention, to speak as if a new state were placed alongside the previous one. Of this new state we assume that it remains unvarying in its turn, and so on endlessly. The apparent discontinuity of the psychical life is then due to our attention being fixed on it by a series of separate acts; actually there is only a gentle slope; but in following the broken line of our acts of attention, we think we perceive separate steps... Now states thus defined cannot be regarded as distinct elements. They continue each other in an endless flow.[6]

This means that there is no radical distinction between *succession* and *duration*, although these two notions are conventionally considered to be irreducible and in a certain sense opposed to each other. But before we consider the identification of these two notions, the identification which is the very essence of Bergson's theory of duration, let us consider the first negative statement, that the plurality of psychological phases is not that of distinct and separable elements. It is from precisely this negation that the fusion of succession and duration inevitably follows. Simultaneously and independently of Bergson, the very same negation of psychological atomism was energetically put forth by William James. Although James's *Principles of Psychology* appeared only one year after Bergson's *Essai*, there is no evidence of James's dependence on Bergson; on the contrary, the fact that the philosophically most significant passages of James's book appeared considerably earlier in the British review *Mind* as well as in a French translation in *Critique Philosophique* seems to indicate the possibility of James's influence on Bergson rather than *vice versa*.[7] In any case, the term *stream of thought* was coined by James in 1884, five years before Bergson, in his article *On Some Omissions of Introspective Psychology*. In it the uninterrupted continuity of psychological life as well as the artificial and fictitious character of the concept of distinct "state" were vigorously emphasized. As James then wrote:

A permanently existing 'idea' or '*Vorstellung*' which makes its appearance before the footlights of consciousness at periodical intervals, is as mythological an entity as the Jack of Spades.[8]

The originality of Bergson lies elsewhere. Though he might have been and probably was influenced by James's criticism of early associationism, he went much farther than James, and from basically the same premises drew more radical conclusions. Although the denial of any precise contours separating successive 'states' of consciousness virtually implied a radical inapplicability of the *concept of number* to the introspectively perceived psychological diversity, James failed to see it or at least to state it explicitly. This is probably due to the preponderantly visual character of his imagination. We must not forget that he almost become a painter before turning philosopher. Consequently the natural human tendency toward spatialization and visualization was never attacked by him systematically nor with a vigour comparable to that of Bergson. It is extremely curious to see James, who was one of the first protagonists of imageless thought, use *visual* metaphors in describing the *non-sensory* data of consciousness: 'fringe', 'halo', 'places of flight', etc.! Even the term 'stream of thought' is visual in its nature and in this respect significantly different from the musical and kinesthetic metaphors of Bergson. The profound originality of Bergson was that he became clearly aware of all the risks involved in our unconscious tendency to visualize the data which by their own nature are basically recalcitrant to any such attempt, and that he tried to be on guard against these tendencies vigilantly and consistently. This implied for him a radical resignation to the use of the *concept of number* in psychology.

Arithmetical multiplicity presupposes the existence of distinct, mutually external units, which when juxtaposed along the ideal geometrical line called 'axis of numbers', form complex aggregates in a similar way as the atoms of Democritus by their juxtaposition formed complex material aggregates. This is more than an analogy; it indicates the profound affinity between the atomistic attitude of mind and the process of arithmetization. This affinity accounts for the simultaneous emergence of Pythagoreanism and atomism in Western intellectual history. We shall return to this point later. But if the continuity of the stream of thought defies any analytical subdivision into distinct and separable units, or, in other words, if psychological continuity is incompatible with any attempt of arithmetization, in what sense can we speak about the *diversity* and *heterogeneity* of our psychological life? In what sense are the mental events *many*, if they are not many in the arithmetical sense? How may they be *different* if they cannot be *counted*? How can we speak of their plurality without considering them to be made up of distinct and, consequently, *countable* units?

Bergson's answer that the multiplicity of psychological duration is of a different kind than numerical multiplicity, being a multiplicity of a *qualitative* kind, appears to be utterly unconvincing at first glance. What else may the concept of qualitative multiplicity be, we ask, than an *ad hoc* notion created out of embarrassment and a reluctance to apply the usual concept of plurality? Is this not contrary to the principle of economy of thought and Occam's razor? Before considering these doubts in detail, we have to remember that the continuity of duration is of a *dynamic* type. So far we have stressed the noun 'continuity'. Its full meaning cannot become clear without considering in detail its *dynamic character*. Only when we realize that it is far more correct to speak with Whitehead of the 'becoming of continuity'[9] rather than the 'continuity of becoming' can the notion of qualitative multiplicity be approached more fruitfully and without the usual prejudices.

## NOTES

[1] David Hume, *A Treatise on Human Nature*, Book I, Part IV, Section 5: "They (i.e. perceptions) *are therefore substances*, as far as this definition explains a substance." (Italics mine.) The definition Hume uses is altogether Cartesian; perceptions can be regarded as substances since "they are also distinct and separable... and *have no need of anything else to support their existence*". (Italics added.) Cf. my article 'Process and Reality in Bergson's Thought', *The Philosophical Forum* XVII (1950–1960) 30–31.
[2] Bertrand Russell, *The Principles of Mathematics*, Norton, New York, 1964, p. 466: "We are sometimes told that things are organic unities, composed of many parts expressing the whole and expressed in the whole.... The only kind of unity to which I can attach any precise sense – apart from the unity of the absolutely simple – is that of a whole composed of parts. But this form of unity cannot be called organic; for if the parts express the whole or the other parts, they must be complex, and therefore contain parts; if the parts have been analyzed as far as possible, they must be simple terms, incapable of expressing anything except themselves." The logic of atomism could not have been expressed with a greater clarity; and the affinity of Russell's thought on this point with Herbart's metaphysics of absolute simple units is unmistakable, though rarely mentioned. An exception in this respect was Jean Wahl's book *Les philosophes pluralistes d'Angleterre et d'Amérique*, Paris 1920 in which Russell's and G. E. Moore's unconscious Herbartism was clearly pointed out (p. 217).
[3] Cf. Kurt Lasswitz, *Geschichte der Atomistik vom Mittelalter zum Newton*, Hamburg–Leipzig 1890, I, pp. 434–450; 482–498.
[4] H. Taine, *De l'intelligence*, 16th Ed., Paris 1927, I, p. 201; Book III, Ch. I, II; H. Spencer, *The Principles of Psychology*, 3rd Ed., I, §60, in particular pp. 150–1.
[5] E. B. Holt, 'The Place of Illusory Experience in a Realistic World', in *The New Realism*, Macmillan, New York, 1912, p. 350; Stephen C. Pepper, 'A Neural Identity Theory of Mind', in *Dimensions of Mind* (ed. by Sidney Hook), Colliers Books, 1961, especially pp. 58–60.

[6] *C.E.*, pp. 3–5.

[7] Cf. my articles 'Stream of Consciousness and "Durée réelle"', *Philosophy and Phenomenological Research* X (1950) 331–353; and 'Process and Personality in Bergson's Thought', *The Philosophical Forum* XVII (1959–1960) especially pp. 32–33.

[8] The article quoted, *Mind* IX (1884) 6–12; *The Principles of Psychology*, I, p. 236.

[9] A. N. Whitehead, 'Time', in *Proceedings of the Sixth International Congress of Philosophy*, Longmans, Green, London, 1927, p. 64; reprinted in *Interpretations of Science* (ed. by A. H. Johnson), Bobb-Merrill, New York, 1961, pp. 241–247.

# CHAPTER 4

# THE INCOMPLETENESS OF DURATION:
## NOVELTY AND ITS DENIALS

Bergson's assertion that psychological duration is a *fait accomplissant* and not a *fait accompli* may sound like a truism; who would deny that psychological time as well as time in general is by its own nature *incomplete*, something which is not given at once, something which is gradually growing? Even those thinkers who most intransigently deny the objective reality of succession eventually concede its reality on the *psychological* or, as they say, on the 'phenomenal' level; otherwise it would be meaningless to speak of its illusory character. I may well deny the reality of change *outside* of my mind, but not *within* my own consciousness; otherwise how could I *have* the illusion of change?[1]

But by asserting the reality of psychological duration, Bergson went far beyond common sense belief in the reality of change and succession. His "true duration" is a continuous emergence of *genuine novelty*. In other words, Bergson's temporalism means *indeterminism*. This indeterminism is far from being absolute, as will be pointed out later; nevertheless it is resolutely opposed to the strict determinism of the classical period. Psychological atomism, which Bergson criticized in his first book, was almost always associated with determinism. In the nature of every kind of atomism inheres the tendency to reduce every change to a displacement of the immutable, permanent elements. Thus genuine novelty is virtually eliminated, since it is reduced to a 'new' arrangement of the pre-existing elements. But even this 'newness' of arrangement is entirely spurious because: (1) it is completely deducible from previous configurations; (2) there is no guarantee that it has not already occurred in the past.

Now, we say that a composite object changes by the displacement of its parts. But when a part has left its position, there is nothing to prevent its return to it. A group of elements which has gone through a state can therefore always find its way back to that state, if not by itself, at least by means of an external cause able to restore everything to its place. This amounts to saying that any state of the group may be repeated as often as desired, and consequently that the group does not grow old. It has no history. Thus nothing is created therein, neither form nor matter. What the group will be is already present in what it is, provided "what it is" includes all the points of the universe with which it is related. A superhuman intellect could calculate, at any moment of time, the position of any point of the system in space. And as there is nothing more in the form of the whole than the arrangement of its parts, the future forms of the system are theoretically visible in its present configuration.[2]

There is no need to dwell on the historical and logical connection of the atomistic view of reality with the idea of complete predetermination. I dealt with this point extensively in the first part of my previous book.[3] It is only natural that Bergson, in attacking atomism in psychology, at the same time challenged the atomistic belief that every change is explainable as a mere reshuffling of the pre-existing elements. According to him, there is an element of irreducible contingency or indetermination in each moment of our psychological history. Each moment of our psychological duration constitutes a genuine – and not merely spurious – novelty, indivisible and consequently unanalyzable into its components for the simple reason that there are no 'components' present, at least not components in the atomistic and arithmetical sense. In a similar sense White-head, three decades after Bergson, will speak of "the self-creativity of actual occasions".[4] This does not mean that the influence of past phases is absent from the novelty of the present phase. This would be altogether incompatible with Bergson's insistence on the dynamic continuity of the stream of thought, which is analytically unbreakable into the mutually external phases. On the contrary, the past is pervading the present, and subsequent discussion will show how such immanence of the anterior phases in the present moment should be understood. But for a while our attention will be focussed on Bergson's claim that the denial of genuine novelty is tantamount to the denial of succession in general; that without an element of indetermination there is no true duration, no succession, no change except in a purely verbal fashion.

This means that even in his first book Bergson's target was far wider than associationism and psychological determinism; it was *determinism in general*, whether of a physicalist or an idealist type. His original aim was to uphold relative human freedom or, more specifically, an element of irreducible novelty in each moment of psychological duration. But in trying to explore the nature of psychological duration, he was reaching a conclusion which went beyond the limits of psychology: without genuine novelty there is no real succession and *vice versa*. There is hardly a more

controversial part of Bergson's philosophy. Is it really true that rigorous
determinism is incompatible with temporality?

To this question the answer of common sense as well as that of an
average scientist is clearly negative. No concept is more familiar to us
than the idea of the process whose phases, though necessarily connected,
nevertheless distinctly *succeed each other*. All classical thinking in sciences,
not only in astronomy and physics, but also in the biological and social
sciences, is based on the idea of *necessary connection in time*. In such
connection Kant saw the very nature of causality. As Ralph Barton Perry
wrote in his polemic against Bergson, "it is entirely possible to maintain
the existential priority of time and to be a vigorous determinist as well".[5]
According to Perry, even the strictly determined mechanical system *ages*,
although it ages according to law. Thus the simple, mechanical movement
of a particle, all of whose future positions are completely and accurately
predictable, seemingly contradicts Bergson's assertion that real succession
and rigid necessity are incompatible.

Such is the widespread popular view, common both to scientific and
philosophical circles. Only a very careful and attentive analysis will
disclose the inherent difficulties of this belief. What is the precise meaning
of the necessary causal connection between two successive events, $a$ and
$b$? It means that all the features of the event $b$ – even those which are the
most individual and apparently most contingent – are logically deducible
from the event $a$. Astronomy and mechanics provide us with an unlimited
number of illustrations, and it was from rational mechanics that Perry
borrowed his example allegedly illustrating the compatibility of succession
and strict causal determination. The situation is not essentially different,
if instead of a single material particle, we consider a whole system of
them, no matter how large it may be: given their positions and velocities
at a given instant of time, all their future states, that is their future positions and velocities, are in principle derivable. Only our purely human
limitations prevent us from extending accurate, mathematical anticipation to the whole universe and its future history. Laplace's famous
passage calling the astronomer's mind a mere faint replica of the universal, omniscient mind embracing in the same huge formula all successive
states of the universe, is sufficiently known and quoted too many times to
be repeated again.[6] But it is less known that a quarter of a century before
Laplace, Immanuel Kant, the alleged defender of human freedom, in

one passage of his *Critique of Practical Reason* applied the Laplacean formula not only to the human body, but to the human mind as well:

It may therefore be admitted that if it were possible to have so profound an insight (*so tiefe Einsicht*) into a man's mental character as shown by internal as well as external actions, as to know all its motives, even the smallest, and likewise all the external occasions that can influence them, we could calculate a man's conduct for the future with as great certainty as a lunar or solar eclipse.[7]

This passage will appear less paradoxical if we remember, first, that the quoted view of Kant followed inevitably from his insistence that the category of causality should apply without any restriction to the phenomenal world; freedom was banished by Kant into the "intelligible realm", beyond the limits of the phenomenal world. (We shall return to this grotesque Kantian solution of the problem of freedom later.) Second, we have to remember that it would be a mistake to associate rigorous determinism only with the mechanistic or naturalistic view of nature. Long before Laplace, Leibniz was convinced that a sufficiently powerful and deep insight into "the inner parts of Being" would provide us with the ability to see the future in the present "as in a mirror".[8] Even Fichte, the philosopher of "Absolute Ego" and of "Freedom", did not hesitate to assert a complete predictability of the inner states as well as their complete determination:

Thus my connection with the whole of nature is what determines everything that I was, what I am and what I shall be, and the same mind could infer from any possible moment of my being what I was before and what I shall be after. All that I am and that I shall be, I am and I shall be in a necessary fashion, and it is impossible for me to be something else.[9]

It is then merely consistent for Fichte to say that "for pure reason everything exists at once; time exists only for imagination".[10] Similarly, Hegel is no less deterministic when he insists in his *Wissenschaft der Logik* that "the effect does not contain anything which was not contained in the cause and *vice versa*".[11] The same thing is true of William Hamilton, who claimed that there is "absolute tautology" between cause and effect.[12] Friedrich Paulsen, a representative of neo-Kantianism, apparently disrespectfully, but consistently, applied the Laplacean pattern of explanation to the mind of his master himself:

The omniscient physiologist ... would explain the author of the *Critique of Pure Reason* just as he would explain a clockwork. In consequence of this particular arrangement of the brain-cells and of their interconnection with each other and the motor-nerves, certain stimuli exciting the retina and the tactile nerves of the fingers had to occasion certain movements which are in no wise different from those of a writing automaton or a music-box.[13]

Not long before this, John Tyndall in his famous Liverpool address did not say anything different when he said that "not alone the more ignoble forms of animalcular or animal life, not alone the nobler forms of the horse and lion, not alone the exquisite mechanism of the human body, but that the human mind itself – emotion, intellect, will, and all their phenomena – were once latent in a fiery cloud". Consistently he drew the conclusion that "all our philosophy, all our science, and all our art – Plato, Shakespeare, Newton, and Raphael – are potential in the fires of the sun".[14] This clearly shows that as far as complete determinism was concerned, there was no disagreement whatever between the idealists and the naturalists of the last century; both accepted the idea of a mechanistic predetermined nature. The only difference between them was the claim of the idealists that a mechanistic nature is a mere "phenomenal manifestation" of the ideal reality. Pragmatically speaking, the difference was non-existent and this explains two apparent paradoxes: an easy conversion of the determinist Hegelianism into the most intransigent form of materialism, and the steady alliance of Neo-Kantianism and positivism.

It would be easy to multiply such illustrations and to show, more specifically, how the doctrine of absolute necessity, which implies the integral predetermination of the future, penetrated even into literature where the theme of the inevitability of personal as well as of collective destiny became very popular in the naturalistic and psychological novel. Let us mention just two examples: Tolstoi and Anatole France. The philosophical epilogue of *War and Peace* is pervaded by the same idea of universal necessity as the philosophical discourses of Dr. Socrates in the *Histoire comique*, when he insists that the whole cosmic past has, as it were, conspired to make Chevalier's suicide inevitable. "Even when the solar system was only a pale nebula with a radius a thousand times greater than that of Neptune", says Anatole France, speaking through Dr. Socrates, "the actions of all men, including this particular and tragic one of Chevalier, were already inexorably predetermined – for the human mechanism is only a special case of the universal mechanism".[15]

How do we explain the fact that determinists of all ages had this tendency to deny – most often only implicitly, but sometimes even explicitly – the reality of succession? Was it not because the causal order of classical science and classical philosophy was essentially timeless and thus by its very nature implied a radical denial of succession? A short glance at the history of Western thought will convince us that this correlation of the deterministic and the timeless view has deep historical roots. Laplace's and Kant's 'omniscient mind' is only a metaphorical expression for the causal order immanent in nature; we may also say that it is simply the omniscient God of St. Thomas and St. Augustine depersonalized and secularized.[16] The connecting link between theological and naturalistic determinism was the pantheism of Bruno and Spinoza, who identified divine predestination with the immanent determinism of nature. Theological determinism, pantheized in this way, has found itself in natural agreement with the determinism of modern science; let us not forget that Bruno was an older contemporary of Galileo and that Spinoza was a contemporary of Newton. Divine Providence was superseded by the impersonal order of nature; but they were both equally deterministic and, ultimately, equally timeless. If we go still further back in the past, we see that the roots of theological determinism lay in the fact that the primary reality was conceived in the ancient Greek philosophy as transcending succession and change. This was true of the Eleatic Being, Plato's Ideas, Aristotle's God, the Plotinian One; while the Stoics by equating Necessity and Providence, ἀνάγκη and πρόνοια, very probably influenced St. Paul's doctrine of predestination. The inherent logic was always the same: the timelessness of the Supreme Being implies the timelessness of its insight; its complete foreknowledge, in making the future present, abolishes succession, change, time. In the last chapter of *Creative Evolution* Bergson convincingly pointed out the persistence and continuity of the Eleatic tradition through the history of Western thought.

## NOTES

[1] Cf. the words of Hermann Lotze: "We must either admit becoming or else explain the becoming of an unreal appearance of Becoming." Quoted by G. J. Whitrow, *The Natural Philosophy of Time*, Thomas Nelson & Sons, London, 1961, p. 311.
[2] *C.E.*, p. 11.
[3] *The Philosophical Impact of Contemporary Physics*, in particular Ch. IX.
[4] *Process and Reality*, Macmillan, New York, 1930, p. 228: "An actual entity is at once the product of the efficient past, and is also, in Spinoza's phrase, *causa sui*." P. 339: "All actual entities share with God this characteristic of self-causation."

5 Ralph Barton Perry, *Present Philosophical Tendencies*, Longmans, Green, New York, 1916, pp. 251–252. Cf. a similar view in C. F. von Weizsäcker, *The History of Nature*, Univ. of Chicago Press, 1950, p. 137.

6 Pierre Simon de Laplace, *A Philosophical Essay on Probabilities* (transl. by F. W. Truscott and F. L. Emory), J. Wiley & Sons, New York, 1902, p. 4.

7 Immanuel Kant, *Critique of Practical Reason and Other Works on the Theory of Ethics* (transl. by T. K. Abbot), Longmans, Green, London, 1909, p. 193.

8 G. W. Leibniz, 'Vom Verhängnisse', quoted by E. Cassirer, *Determinism and Indeterminism in Modern Physics*, Yale University Press, New Haven, 1956, p. 12.

9 J. G. Fichte, *Die Bestimmung des Menschen, Sämmtliche Werke*, Berlin, 1845, II, pp. 182–183.

10 J. G. Fichte, *Grundlagen der gesammten Wissenschaften, Sämmtliche Werke*, I, p. 217.

11 *Wissenschaft der Logik, Sämmtliche Werke* (ed. by G. Lasson), Leipzig 1923, IV, pp. 190–191.

12 William Hamilton, *Lectures on Metaphysics*, Boston 1859, p. 533.

13 F. Paulsen, *Introduction to Philosophy* (transl. by F. Tilly), H. Holt, New York, 1912, p. 88.

14 John Tyndall, 'Scientific Use of Imagination', in *Fragments of Science*, Appleton, New York, 1872, p. 159.

15 *A Mummer's Tale. The Works of Anatole France in English Translation* (ed. by J. Lewis and B. Niall), Vol. XXIX, p. 118.

16 Cf. M. Čapek, 'The Doctrine of Necessity Re-Examined', *The Review of Metaphysics* V (1951) esp. pp. 18–19, 24–27.

CHAPTER 5

# SUPERFLUITY OF SUCCESSION IN THE
# DETERMINISTIC SCHEMES

Such historical connection of rigorous determinism and the timeless view was hardly accidental. For we have seen that for both naturalistic and idealistic (or theological) determinism the future is *logically implied* by the present or by any previous moment. Every logical implication is admittedly timeless. It is a commonplace in logic to distinguish between the implication itself which is beyond time – 'tenseless' as it is fashionable to say today – and the psychological process of inference with its distinctive successive phases. The words 'antecedent', 'consequent', 'it follows', etc. are essentially metaphorical and misleading because of their obvious temporal connotations. The simultaneity of the conclusion with the premises may be illustrated and even visualized by analyzing the traditional categorical syllogism: All $M$ are $P$, all $S$ are $M$; *therefore* all $S$ are $P$. By drawing conventional Euler's diagrams it becomes immediately obvious that the inclusion of the class $S$ into $P$ *coexists* with two previous inclusions, $M$ in $P$ and $S$ in $M$. There is no succession here, not a trace of any movement, except the shifting movement of our attention which, after first noticing the first two inclusions, *perceives* finally the simultaneous inclusion of $S$ into $P$. In other words, the conclusion does not follow from the premises in a temporal sense; on the contrary, it is tenselessly *contained* in them.

The same pre-existence of the valid conclusion in the premises is found also in mathematical reasoning, indeed in every reasoning in general. When we write the left side of the equation while having its right side blank, we know that the solution exists *prior* to our finding of it; the right side is blank only to our temporary ignorance. This is why we say that we *discover* the truth instead of *creating* it. The belief that reality itself, including all its future successive phases, is nothing but a single huge, though tremendously complex, pattern of pre-existing truth which is gradually and laboriously unveiled to our imperfect intellect is at the very bottom of classical determinism. In such a view the future is comparable to the right side of the algebraic equation which appears blank only to our limited knowledge, but which is unambiguously and completely contained in the left side of the equation. Under such conditions succession, duration and change are mere illusions, and coming into being is, according to one recent expression, merely "coming into our awareness".[1]

But then the most embarrassing question comes: where does succession come from? Why is the future development of the universe, though already given and necessarily implied by the present, not present *now*? If the future logically pre-exists in the present state of the universe, why does it require a certain time to become present? In the completely determined universe successive states are comparable to the images placed side by side along the cinematographic film prior to its unrolling. Then, Bergson asked: "Why, then, the unrolling? Why does reality unfurl? Why is it not spread out? What good is time?"[2] Or in the words of *Creative Evolution*:

Therefore the flow of time might assume an infinite rapidity, the entire past, present and future of material objects or of isolated systems might be spread out all at once in space, without there being anything to change either in the formulae of the scientist or in the language of common sense. The number *t* would always stand for the same thing; it would still count the same number of correspondences between the states of the objects or systems and the points of line, ready drawn, which would be then the "course of time".

What is it that obliges us to wait, and to wait for a certain length of psychical duration which is forced upon me, over which I have no power? If succession, in so far as distinct from mere juxtaposition, has no real efficacy, if time is not a kind of force, why does the universe unfold its successive states with a velocity which, in regard to my consciousness, is a veritable absolute? Why with this particular velocity rather than any other? Why not with an infinite velocity? Why, in other words, is not everything given at once, as on the film of the cinematograph? The more I consider this point, the more it seems to me that, if the future is bound to *succeed* the present instead of being given alongside of it, it is because the future is not altogether determined at the present moment, and that if the time taken up by this succession is something other than a number, if it has for the consciousness that is installed in it absolute value and reality, it is because there is unceasingly being created in it ... something unforeseeable and new.[3]

Referring to the Laplacean determinism, Bergson says:

In such a doctrine, time is still spoken of; one pronounces the words, but one does not think the thing. For time is here deprived of efficacy, and if it *does* nothing, it *is* nothing. Radical determinism implies a metaphysics in which the totality of real is postulated complete in eternity, and in which the apparent duration of things expresses merely the *infirmity of a mind that cannot know everything at once.* (Last italics added.)[4]

Bergson probably was not entirely correct in claiming that a determinist pronounces the word 'time' without thinking of real succession. The state of mind of an average determinist is more complex, and it was correctly analyzed by Bergson in his first book. There he pointed out that the belief in necessary causal connection is an association of two plainly incompatible ideas, that of mathematical necessity, implying the preformation and even pre-existence of a 'future' whose 'futurity' becomes merely apparent, and that of a temporal process with distinctly successive phases called 'cause' and 'effect' respectively.[5] This association is so intimate and automatized by habit and, consequently, so familiar to the average mind, that its absurdity is not noticed. It requires a considerable effort to separate the incompatible elements fused into the deceptive unity of our instinctive belief. But only thus can progress in redefining or rather widening the concept of causality be achieved; only by insisting on the genuine succession of the causal links can we free the concept of causality of its traditional, static, Laplacean character, by which the whole world is transformed, as Poincaré said, into "an immense tautology", or as William James said, into a "block universe".[6] If duration is real, future as future is unreal, i.e. not yet present, and its coming into being is a *genuine* and *not merely apparent* novelty.

In the above quotation Bergson expresses this view in another way, by saying that time is causally efficacious. This is what causal determinism denies. Its concern is "to transform relations of succession into relations of inherence, to eliminate the action of duration and to replace apparent causality by a 'basic identity'," to wit, by the identity of the successive phases.[7] In his classical works Émile Meyerson showed convincingly that the leading idea of mechanism and determinism in general is "identification in time", which in its ultimate consequences leads to a complete elimination of time. However such an operation is never completely successful, since the irreducibly successive character of reality stubbornly resists its own elimination. This resistance of reality to the identifying schemes was called by Meyerson "*l'irrationnel*". (Such a term is hardly appropriate since the irreducible character of succession appears irrational *only* within the Spinozistic-Laplacean necessitarian scheme, and the assumption that the timeless determinism is the only type of rationality is nothing but a *petitio principii*.) It is hardly surprising that Bergson was one of the first to recognize the significance of Meyerson's work.[8]

Let us conclude: if the rigid form of determinism claims to account for *all* the features of any future event, and if, at the same time fails to account for its most fundamental one, its 'posteriority', its 'futurity', its 'not yet' or, briefly, for *the delay in its actualization,* then it certainly does not merit its name. Furthermore, if the same doctrine tries to incorporate into its own body the element which is *ex definitione* excluded from it, – and we have seen that succession is excluded as long as causation is equated with logical implication – then it ceases to be logically coherent. Genuine succession *does* imply genuine novelty.

In his first book Bergson tried to establish the reality of novelty on the psychological level, since his main topic was *psychological* duration. But, as already mentioned, his conclusions, even then, virtually went beyond psychology; for, even if novelty were confined to the introspective realm, it would become *a part of the total reality* and the problem of how it is related to the alleged absence of novelty in the physical world inevitably arises. As we shall see, this point was far from clear to Bergson himself. This is hardly surprising. To speak of genuine novelty, or, which is the same, of the causal efficacy of time, in the physical world of 1889 would have required enormous intellectual courage. It required a nearly equal courage even seven years later when Bergson in *Matter and Memory* accepted the implications of his analysis of duration for the physical realm as well. Only then could Perry's objection, based on the idea of "the physical process aging according to the strict law," have been convincingly answered. In truth, a convincing answer for physicists was not possible prior to 1927.

In a few instances determinist philosophers became aware of the incompatibility between time and necessity. Thus Hans Driesch who, while being a vitalist, was nevertheless a determinist of the most uncompromising type, after stating his Laplacean belief in the predetermination of the universe asked the following question: Why do the timeless entelechies unfold their activities *in time*? In other words: why the development of the individual from the ovum to its adult form is not realized in *one single and complex act*? He frankly admits: "To this, we have no answer."[9] It is obvious that the question about the legitimacy of becoming can arise only within the timeless necessitarian scheme. On the other hand, Hyppolite Taine, who became famous by applying strict, Spinozistic determinism to the history of literature and the theory of art, nevertheless refused to follow his master in eliminating temporality when he replaced the static geometry of the *Ethics* by his "living geometry" (*la géométrie vivante*)[10] without realizing that he was juxtaposing two incompatible words. It is unnecessary to repeat what we said about the tenseless character of any logical, and therefore also geometrical, implication.

Most frequently, the incompatibility between the affirmation of real novelty and deductive necessity was only vaguely felt; but this vague feeling was sufficient to find expression in certain peculiarities of language which were invented to reconcile the necessitarian claim of the logical equivalence of cause and effect with the vague but strong feeling of their genuine differences resulting from their real succession. William James pointed out the function of the word '*quatenus*' by which Spinoza tried to reconcile eternity and the unity of substance with the realm of manifold and changing modes: "*Deus quatenus infinitus est*" is different from "*Deus quatenus humanam mentem constituit*". James also pointed out that determinism, while insisting on the identity of cause and effect, nevertheless makes a verbal concession to the vaguely felt difference between them in admitting that their identity is only 'basic', i.e. not altogether complete, and that the cause contains its effect only "virtually".[11] But if we concede a genuine difference between the successive phases of every process, no matter how minute, do we not admit an element of novelty by which the effect is differentiated from its antecedent?

Becoming, instead of being completely denied, was banished from the metaphysical realm of true reality and modestly lodged in the region of 'phenomena'. In other words, ultimate reality was placed outside of time and only its phenomenal aspect – a 'surface aspect' – was regarded as unrolling in time. It matters little if this true Being was the Sphere of Parmenides, the Matter of Democritus, the *Ens realissimum* of the Scholastics, the Substance of Spinoza, the *Ding-an-sich* of Kant, the Unknowable of Spencer, the Absolute of Bradley or the impersonal order of nature symbolized by the 'Omniscient Mind' of Laplace. The conclusion always remained the same: time, succession, and becoming do not belong to 'reality itself', but to the semi-real region of phenomena. However, when one admits the dichotomy of a 'reality in itself' which is outside of time, and the 'region of becoming' in which the phenomena succeed each other, he has merely stated the problem without even attempting to solve it. Since Plato's time the same question has not been answered: why is reality cut into two regions – that of the immutable and 'perfect' and that of change and the 'imperfect'? Or, more pointedly, how what is supposedly one and the same reality can appear in two such thoroughly incompatible aspects? William James asked this question before Bergson in reference to Hegelianism, but his question also concerned all static monisms:

Why, if one act of knowledge could from one point take in the total perspective with all mere possibilities abolished, should there ever have been anything more than that act? Why duplicate it by the tedious unrolling, inch by inch, of the foredone reality? No answer seems possible.[12]

The superfluity of time in the necessitarian scheme could hardly have been stated more clearly and concisely. In this quotation, James pointed out that in the rigorous, deterministic scheme genuine possibilities do not have place; they are literally 'abolished', since there is no middle ground between necessity and impossiblity. Whatever happens is necessary; whatever does not occur is impossible. Possibility, as determinists from Democritus to the modern opponents of microphysical indeterminism have never tired of repeating, is merely a name for human ignorance. Not only, whatever is real is necessary, but also, whatever is necessary is real. Since the future in the deterministic scheme possesses the character of absolute necessity, it acquires *ipso facto* the status of reality, of something actually – 'tenselessly' – existing, which only our limited 'timebound' insight fails to perceive. "Future" is merely a label given to the unknown part of reality which *coexists* with our present moment in a sense similar to that in which hidden, distant scenery coexists with our limited visual field. Long before Laplace, Spinoza in the 29th proposition of the first book of the *Ethics* saw clearly that determinism in its rigorous form implies the denial of the objective character of possibility or contingency as well as the abolition of succession. Only the objectively ambiguous future, that is, the future charged with various possibilities is a genuine future. Aristotle had a correct glimpse of it long before Bergson when he rejected the idea of a future completely stripped of its contingency. For such a "future" loses its character of futurity and becomes a mere name for an undisclosed part of the huge, cosmic tautology. It is clear that one must choose one of two assertions: *either* real succession with an element of real contingency, *or* complete determinism with the total absence not only of possibilities, but of succession as well.

## NOTES

[1] A. Grünbaum, *Philosophical Problems of Space and Time*, A. A. Knopf, New York, 1963, p. 329.

[2] *C.M.*, p. 109.

[3] *E.C.*, p. 12; pp. 368–369.

[4] *Ibid.*, pp. 44–45.

[5] *T.F.W.*, pp. 215–218.

[6] Cf. Chapter 2, Part I, Note 40; W. James, 'The Dilemma of Determinism', in *The Will to Believe and Other Essays in Popular Philosophy*, Longmans, Green, New York, 1915, p. 150.

[7] *T.F.W.*, pp. 209.

[8] Cf. 'Rapport sur un ouvrage de Émile Meyerson, *Identité et réalité*, 23 janvier 1909', in H. Bergson, *Écrits et paroles* (ed. by R. M. Mossé-Bastide), Presses Universitaires de France, 1959, II, pp. 311–312; also André Metz, 'Bergson et Meyerson', in *Bergson et nous. Actes du X⁰ Congrès des Sociétés de Philosophie de Langue française. Bulletin de la Société Française de Philosophie* **53** (1959), 235–237.

[9] Hans Driesch, *Philosophie des Organischen*, Leipzig 1928, Chapter 'Entelechie und Zeit', pp. 325–326.

[10] H. Taine, *Histoire de la littérature anglaise*, 12th Ed., Paris, 1905, IV, p. 389.

[11] W. James, *A Pluralistic Universe*, Longmans, Green, London, 1947, p. 47; *Some Problems of Philosophy*, Longmans, Green, London, 1940, pp. 192–194.

[12] W. James, 'On Some Hegelisms', in *The Will to Believe*, p. 271.

CHAPTER 6

# THE LEIBNIZ-FOUILLÉ ARGUMENT
## FOR THE COMPATIBILITY
## OF SUCCESSION AND DETERMINISM

In the mind of an average, classical determinist, succession and logical implication were simply juxtaposed and their incompatibility veiled by the ambiguity of the words "follows", "antecedent", and "consequent", which were understood in both a temporal and logical sense. As long as he realized that temporal and logical sequence were two different things, he was satisfied to say that they were both *empirically* correlated; in other words, that the coexistence of causal necessity and succession was a datum of experience which did not need to be explained. But in a few instances an attempt was made to make this correlation intelligible, in other words, to *explain* it. Such was the case of the once famous French philosopher and psychologist, Alfred Fouillé. He wrote explicitly:

We live in time and we reason in time. Well, in time it would be contradictory to say that the future exists and exerts influence because in such a case I would be at the same time alive and dead, truly alive and truly dead, my future death being already *real* as much as my present life. Such a theory would mean an elimination of all possible thought and of all possible experience, since thought cannot admit the simultaneous presence of contradictories and because experience cannot grasp the present and the future simultaneously.[1]

The argument is probably borrowed from Émile Boutroux, for whom it was equivalent to a rejection of necessitarianism[2]; but not so for Fouillé, who on the same page did not fail to add that contingency is absent from the past, the present and the future. According to Fouillé, we have only to replace static determinism by what he called "*le déterminisme dynamique*", which accepts the genuine reality of succession. What is interesting in Fouillé's position is that temporality is not only accepted as authentically real, but is to be *explained logically* as an unavoidable consequence of the law of contradiction. *For it is the logical incompatibility of successive phases of the universe which prevents them from coexisting simultaneously.*

The argument is extremely interesting, and if valid, it would provide the answer to Bergson's question, the one that was raised before him by Lequier, Delboeuf, Boutroux and James[3]: Why is the already developed film of reality only gradually unrolled instead of being spread out at once like a map? Fouillé's answer was virtually present in Leibniz's definitions of space and time: "Extension is the order of possible coexistences, time is the order of *inconsistent possibilities* which nevertheless are connected. Thus the first type of order concerns things which are simultaneous or existing together, the second type of order characterizes those which are incompatible *and which are nevertheless considered as existing: and this makes them successive.*"[4] (Italics mine.) Again, succession is regarded as a logical consequence of the incompatibility of certain predicates. Thus one cannot be alive and dead simultaneously, but only in succession. Yet the passage quoted shows clearly how divided Leibniz' mind was on this point. While he claimed that it is the logical incompatibility of successive events which prevents them from occurring simultaneously, he at the same time stressed their causal connection; and since, like Spinoza, like Laplace, like any other classical determinist, he identified causation and logical implication, he became entangled in a hopeless contradiction. For how can we assert that two events, *a* and *b*, are incompatible (and for this reason can occur *only one after another*), when at the same time we claim that *a* implies *b*? How can the effect be *incompatible* with its cause, if the former is *logically deducible* from the latter? Leibniz was undoubtedly correct in insisting on the incompatibility of successive events; but he contradicted himself when he assimilated the causal order to the logically implicative order.

The very same difficulty appears in Leibniz' view of substance. Substance, according to him, contains *all* its future states as its own predicates. Thus the substantial monad of Alexander the Great contains in itself all the incidents of his life, his childhood, his triumphs and finally his death by poisoning.[5] It is clear that these various events are both successive and incompatible; but how then can they be assimilated to various predicates *inhering* in one subject? And, conversely, how can the incompatibility and succession of the predicates be deduced from their logical inherence in the subject? For logical inherence is essentially timeless, as Leibniz himself concedes. Thus despite his honest effort to do justice to the reality of succession, Leibniz was definitely leaning toward the static view. The same was true of Fouillé as well. While he insisted on the reality of change, he implicitly denied it when in his polemic against Renouvier he compared causal necessity to the necessity of mathematical reasoning: "When two and three are present, five can be neither absent *nor delayed* (*"ni en retard"*); ...it is also immediately present." (Italics mine.)[6] This is true of mathematical equations, but not of real changes in which effects are indeed irremediably *after* their causes. It is this *delay* of the events with respect to their antecedents which the logico-implicative theory of causality fails to account for.

The Leibniz-Fouillé argument, in a less precise form, occurs very frequently, especially in preponderantly deterministic historical and sociological reasonings, and leads to the same difficulties. For example, was the French revolution possible after the death of Louis XV in 1715? The answer which we would hear would be: 'Certainly not, because not all necessary conditions were present.' But although it was impossible in 1715, was it not necessary *already then* as a future event in 1789? Certainly it was. No other answer is possible for any consistent determinist. Thus we have two opposite claims, and determinists seem to be committed to both of them: (a) the French revolution was necessary from the beginning, in truth from beginningless eternity; and (b) the French revolution was impossible until 1789. – Let us go further: Was the revolution possible in 1788? For an unprejudiced historian *it was*. At that time the revolutionary propaganda was already extensively active and effective; the authority of the aristocracy, hierarchy and dynasty was seriously undermined, the sight of the nobility riding in carriages already drew threatening gestures from the people, while Marat was publicly reciting Rousseau's *Contrat Social* on the boulevards of Paris. Yet, for a rigid determinist the outbreak of the revolution was still impossible in 1788 for the simple reason that *it did not occur*. For him to speak about the *possibility* of the revolution in 1788 is nothing but a vague and metaphorical language *because no real possibilities exist*. Whatever does not happen is impossible and we speak of "possibility" only because of our ignorance. We can apply the same reasoning to the history of any individual person. For instance, the murder committed by Raskolnikov was impossible up to the last moment. As a matter of fact, it was impossible even in the last second when he was already raising the hatchet with his hands half-paralyzed by a strong, unconscious resistance to the coming crime. More generally, it is the same impossibility which prevents the successive stages of one process from being simultaneously present.

But if this is so, is it still meaningful to speak of *predetermination*? Would it not be logical to follow this reasoning up to the end and say with the author of *Time and Free Will* that the event is necessary only when it is actually present and that the whole thesis of "dynamic determinism" can be reduced to a simple truism: "What is being done is being done"?[7] For no matter how narrow the time interval separating two successive events becomes, their incompatibility is *the very opposite* of the iron-like, necessitating link which Leibniz, Fouillé and other determinists postulated between them. Only when the future event becomes present does what may be metaphorically termed the 'hesitation of reality between several possibilities' come to an end; and as Bergson observed, "*time* (to wit: the time interval separating the present from the future) *is this very hesitation or it is nothing at all*".[8] By becoming actual the formerly future event excludes all rival possibilities for this particular moment. Thus the present moment is indeed the only possible event, but it becomes so *only* by the very act of its own actualization, after it became present. But it was *not* the only possible event prior to its actualization, when it was a mere possibility competing with other possibilities for its own actualization. In ignoring these other possibilities, or, what is the same, in declaring them to be illusory and non-existent, rigorous determinism assimilates the future to the present by retrospectively projecting the definiteness of the latter into the intrinsic indefiniteness of the former. In Whitehead's words, written about the same time as those of Bergson quoted above: "Whatever is realized in any one occasion of experience necessarily excludes the unbounded welter of contrary possibilities. There are always 'others' *which might have been and are not*."[9] (Italics added.)

The true significance of Bergson's indeterminism and, in particular, of its *relative* character cannot be fully grasped, if we overlook his emphasis on the *mnemic immanence of the antecedent phases of every process in its present moment*. This immanence of the past in the present follows from his criticism of the atomistic, associationistic psychology, or rather from Bergson's generalization of this criticism. The continuity of psychological duration is used by him as a model by means of which the structure of *any* duration can be understood. The particular synthesis of the novelty of the present with the persistence of the past is the next topic of our discussion.

## NOTES

[1] A. Fouillé, *La pensée et les nouvelles écoles anti-intellectualistes*, Paris 1911, p. 140.

[2] Emile Boutroux, *The Contingency of the Laws of Nature* (transl. by F. Rothwell), Open Court, Chicago, 1920, pp. 21–22.

[3] See the references to these authors in Fouillé's polemical articles in *Revue philosophique* **14, 15, 16** (1882–1883): 'Les nouveaux expédients en faveur du libre arbitre', 'Les arguments métaphysiques en faveur du libre arbitre', 'Le libre arbitre et la contingence des futurs', and 'Causalité et liberté'.

[4] G. W. Leibniz, *Philosophische Schriften* (ed. by C. J. Gerhardt), Berlin 1875–1890, IV, p. 568. (Hereafter referred to as *G.*)

[5] *Le discours métaphysique*, §6.

[6] A. Fouillé, 'Les nouveaux expédients en faveur du libre arbitre', *Revue philosophique* **XIV** (1822) 607.

[7] *T.F.W.*, p. 182.

[8] *C.M.*, p. 109. Cf. the whole subsequent passage in which Bergson describes how he realized the superfluity of time in Spencer's and, more generally in any mechanistic scheme: "Nevertheless, I said to myself, time is something. Therefore it acts. What can it be doing? Plain common sense answered: *time is what hinders everything from being given at once.* It retards, or rather it is retardation... Would it not then be a vehicle of creation and of choice? Would not then the existence of time prove that there is indetermination in things? *Would not time be that indetermination itself?*" (Italics added.)

[9] A. N. Whitehead, *Adventure of Ideas*, Macmillan, New York, 1933, p. 356.

# CHAPTER 7

# THE HETEROGENEITY OF DURATION:
# LOVEJOY-USHENKO'S OBJECTIONS

If true succession implies the appearance of genuine novelty and contingency, this means that there is an irreducible, qualitative difference between successive phases of any temporal process. It is precisely this difference which accounts for the element of indetermination and creativity which constitutes the essence of the present moment. In other words, true duration is by its own nature heterogeneous; homogeneous time is merely a practically convenient but theoretically inadequate fiction. As we shall see, it is not quite adequate even as far as physical reality is concerned, though it has a maximum convenience and practical applicability on the macroscopic level. The affirmation of the qualitative heterogeneity of successive phases of duration will throw additional light on Bergson's notion of "qualitative multiplicity", which he opposed to the arithmetical multiplicity of distinct juxtaposed units. Bergson's emphasis on the heterogeneity of successive moments may at first glance appear to contradict his insistence on the introspective continuity of duration. Thus the late Andrew Ushenko wondered about the following passage of *Durée et Simultanéité*:

A melody to which we listen with our eyes closed, heeding it alone, comes close to coinciding with this time which is the very fluidity of our inner life; but it still has too many qualities, too much definition (*determination*), and we must first efface the difference among the sounds, then do away with the distinctive features of sound itself, retaining of it only the continuation of what precedes into what follows and the uninterrupted transition, *multiplicity without divisibility and succession without separation*, in order finally to rediscover basic time. Such is immediately perceived duration, without which we would have no idea of time. (Italics mine.)[1]

Incidentally, hardly any other passage shows more graphically how little immediate the immediacy of Bergson's duration is and how much of an abstracting effort is inherent in it. It shows that Bergson's qualitative multiplicity is essentially of an *imageless kind*, transcending and, in a sense, independent of the kaleidoscopic variety of the sensory qualities which so exclusively attracted the attention of associationistic psychologists from Hume to Russell. We shall later return to this point; for the present let us turn to Ushenko's comment, which at first glance has a great deal of plausibility:

This passage shows very clearly the tendency to strip *durée* of all qualitative differences which allow for the possibility of transition. Instead of giving an account of such differences, emphasis is laid on the identical element underlying the various phases of a process ... Thus the denial of discrete occurrences precludes historical development. For with the elimination of discreteness novelty is impossible. All the future is swallowed by the indifferent continuity. The sameness of continuous duration erases from its own face all characteristic features: the multiplicity and variety of phenomena vanish. And the very creativity or transition to a new phase is no longer possible.[2]

Ushenko's comment is especially interesting as it shows how easily misunderstandings about the nature of duration arise. Ushenko's criticism is basically the same as that of Lovejoy, who claimed that Bergson's insistence on the indivisible continuity of duration is equivalent to the abolition of any distinction between the successive phases of duration. Thus, according to Lovejoy, Bergson unwittingly tends toward Eleaticism and monism![3] Ushenko apparently overlooked two crucial phrases: "*multiplicité sans divisibilité*" and "*succession sans séparation*"; both are synonymous with "qualitative multiplicity", which now has to be elucidated. Then and only then will it become clear that indivisible continuity in Bergson's sense is *not* the timeless identity in which every novelty would be condemned to disappear; nor is the heterogeneity of successive phases comparable to discreteness in the arithmetical sense. The misunderstanding of the last point was the source of Lovejoy's misunderstanding of Bergson's thought. It is true that Bergson – like James, like any opponent of psychological atomism – denied the presence of any definite boundaries separating successive phases of psychological duration; but this does not mean that he denied the *qualitative diversity* of these successive phases. Lovejoy's and Ushenko's error was to confuse *the absence of separation* with the absence of *qualitative difference*, two very different things. Related to this confusion was the gratuitous assumption of both Lovejoy and Ushenko that the only type of unity is the undifferentiated *static* unity and that succession is an *external* relation. Since Bergson denied that succession is an external relation, he must be guilty of unconscious Eleaticism. But was this accusation justified? Is there no middle term between the homogeneous static unity and the arithmetical multiplicity? In other words, is Bergson's term "continuous emergence of novelty", which fuses two apparently opposed notions into a single one, empirically justified and logically coherent? This is the fundamental question which we face now.

The question of empirical justification can be solved only by an appeal to introspective data freed from the irrelevancies and admixtures of misleading and adventitious associations. By analyzing musical experience Bergson shows that the continuity and heterogeneity of successive phases are *not* mutually exclusive, both being two complementary aspects of one and the same temporal process, which is neither a bare unity nor a sheer multiplicity. Is the intuitive fact of melody a single whole or a sum of distinct parts? Is it a unity or plurality? It is both of them at the same time, *or rather neither of them*, at least as long as both terms have their usual arithmetical connotations. It is the continuity of the temporal whole which nevertheless is differentiated into the diversity of successive tones; but this diversity is *not* that of a multiplicity of distinct terms, otherwise the perception of melody would never arise; instead we would have a series of separate sensations, each corresponding to one, single tone and each being immediately forgotten after being replaced by the subsequent one. This means that the individual tones dynamically *cohere* in spite of their qualitative differences, and this coherence, far from being static and timeless, as Mr. Ushenko believes, is inseparable from novelty. In Whitehead's happy words, it is the "becoming of continuity rather than the continuity of becoming". In this case, it is nothing but another term for the dynamic growth of melody itself, for the continuous emergence of new tones. This is the reason why Bergson distinguishes the numerical multiplicity of terms which are externally added to one another from the 'qualitative multiplicity' where terms interpenetrate one another. Perhaps the term 'qualitative diversity' or 'heterogeneity' would be less misleading, as the word 'multiplicity' almost inevitably conveys the association of numerical units.

Thus the problem of the unity and multiplicity of successive phases is inseparable from the problem of the relation between the past and the present; as a matter of fact, it is only another name for the same problem. As early as 1889 Bergson insisted that the relation between the past and the present cannot be that of mutual externality; the successive moments of duration penetrate each other without ceasing to be different.

This is what the *experience of transition* consists of. But Lovejoy denies there is any such thing:

What reason is there for maintaining that we have any direct experience of transition as such? Suppose that when Bergson invites us to concentrate our thought *"tout entier sur la transition et, entre deux instants, chercher ce qui se passe"*, he is inviting us to look for something which is not there – something which very naturally baffles intellect, for the simple reason that it is at once an unreality and absurdity! To this question, at any rate, concerning the actual verifiability of the occurrence of an experience of *pure* transition – as distinct from the experience of a sequence of discrete momentary states, each of which contains as a part of its content memory and anticipation and the past-present-and-future schematism, – the issue respecting the value of the fourth of Bergson's arguments reduces....[4]

Lovejoy thus claims that (a) there is no evidence of the experience of transition as being essentially different from the experience of discrete successive moments and (b) that furthermore, such experience is a contradiction in terms. His argument is thus allegedly both empirical and logical. We intend to show not only that there is such a thing as the experience of transition, but also that without such experience no perception of succession could ever arise.

Re (a). There *is* experience of transition, but it is of a very elusive kind. For this reason it easily escapes crude and unrefined introspection. This was true of the classical atomistic or associationistic psychology, whose attention was captured exclusively by what William James called 'the places of rest' or 'substantive parts' of our stream of thought, which because of their sensory vivacity are more accessible to observation. "The resting places", says James, "are usually occupied by sensorial imaginations of some sort, whose peculiarity is that they can be held before the mind for an independent period of time, and contemplated without changing". But besides these sensory nuclei there are also "places of flight ... filled with thoughts of relations, static and dynamic, that for the most part obtain between the matters contemplated in the periods of comparative rest". (Note an important adjective used by James: *'comparative* rest'.) The function of these 'places of flight', 'transitive tracts' or 'feelings of relation' is to lead from one substantive tract to another. "Like a bird's life, it (i.e. conscious life) seems to be made of an alternation of flights and perchings. The rhythm of language expresses this, where every thought is expressed in a sentence, and every sentence closed by a period".[5] This peculiar character of the feelings of tendency accounts for the extreme difficulty of observing them introspectively. By focussing our attention on the feeling of tendency (which is another name for the experience of transition) we arrest it in flight and transform it into something fixed and crystallized, which always has the character of a sensory image. In other words, we transform the transitive tract into the substantive sensory nucleus – the very opposite of what we wanted to achieve. Thus by the very act of introspective focussing the dynamic nature of the feeling of transition is destroyed in the same way, says James, as the mobility of Zeno's arrow is frozen into a momentary static position.[6]

It is obvious that James was one of the first discoverers of so-called "imageless thought', which so powerfully stimulated the introspective investigation of the logical processes of thought as it was carried on later, especially by Alfred Binet in France and by the Würzburg school in Germany (Külpe, Messer, Ach). To the old sensualism of the Hume-Condillac and Mill-Taine variety the logical thought was a simple succession of ideas, which were conceived as mere faint replicas of sensory images. The difficulty of observing the vague and imageless 'feelings of tendency' favored and apparently justified the atomistic and sensualistic prejudice in psychology. Thus the dynamic unity of thought was artificially reduced to a kaleidoscopic mosaic of 'sensory elements' which differed only by a lesser sensory vivacity from sensations. Needless to say, this position was very difficult to uphold consistently. Thus David Hume, after resolving personal identity into mutually external atomic 'impressions', conceded that the very same feeling proceeds "entirely from the smooth and uninterrupted progress of the thought along a train of connected ideas".[7] Apparently he did not realize that he was contradicting himself, since 'the smooth and uninterrupted progress of the thought', which is clearly another name for James's 'feeling of tendency', does not have any place in his atomistic psychology. For Hume's 'impressions' are mutually external, as he himself stressed when he wrote that "they are also distinct and separable, and may be considered as separately existent, and may exist separately". He even applied the Cartesian definition of substance to them, since they "have no need of anything else to support their existence".[8] Thus instead of eliminating the Cartesian substance, Hume merely cut it into tiny pieces called 'impressions' or 'perceptions', which nevertheless retained a substantial, thing-like character, similar to the sense in which Democritus's atoms retained the static character of the Parmenidean Being. Perhaps such a crude atomization of the dynamic continuity of our experience was excusable at the time of Hume; but basically the same kind of atomization persisted one century after Hume in Spencer, Taine and Mach, and two centuries after Hume in certain writings of Russell and Wittgenstein. Lovejoy, perhaps unwittingly, trod the same path.

Re (b). What is even more serious is that if we deny any experience of transition, we are in no position to account for our experience of succession at all. In the passage quoted above, Lovejoy claims that the experience of succession is not different from that 'of a sequence of discrete and momentary states'. But by this he evidently does not mean that in each particular moment we are only aware of one, single state. For in that case no experience of sequence would arise; succession of perception is not perception of succession. He claims that each momentary state 'contains an aspect of its content memory and anticipation and the past-present-and-future schematism' and it is this complexity of a single, momentary state which accounts for our experience of succession. Thus the present momentary state is colored by two coexisting images: one which is an effect of the immediately preceding state, and one by which the forthcoming future state is anticipated. But both the effect or the trace of the past state and the anticipated image of the immediately future state are *equally present*; they literally *coexist* with the present, momentary state, being only the "signs" of events not simultaneous with the present.

This is the famous theory of 'temporal signs', by which the experience of succession is supposedly explained or rather *explained away*. According to this view what we really experience when we speak of the perception of succession or transition is the superposition of the *present* sensation and of the *present* traces of the past sensations. This is usually interpreted physiologically: while the fading traces of the past cerebral processes are still going on, the present one is reaching its maximum intensity. These differences in intensity of the cerebral processes are translated by the mind into the differences of temporal order. Although Lovejoy probably had reservations about this physiological model of the perception of succession, he certainly agreed with the view that the awareness of succession is an introspective translation of the *present* differences in intensity into the differences of temporal order. In other words, there is, according to him, *no direct perception of the past*, no matter how immediate it may be. That this theory of temporal signs is hardly satisfactory was felt even by Theodor Lipps, one of those who upheld this view. He frankly conceded that mind's translation of the order of intensity into the order of succession remains entirely inexplicable.[9] There would be no possibility of differentiating between our experience of time sequence and our experience of coexistence. A melody in which the tones succeed each other would be indistinguishable from the chord in which the same tones sound simultaneously; the visual perception of motion, like for instance that of a shooting star, would be indistinguishable from the perception of the luminous tail which the same meteorite leaves behind it on the sky and which consists of juxtaposed, luminous points. Thinkers so widely different as Bergson, Royce and Russell agree entirely on this point.[10]

The theory of temporal signs, which reduces the awareness of succession to the coexistence of unequally intense sensations or images, has still another difficulty, of which Lovejoy apparently was not clearly aware. If, in keeping the difference between the past and present as sharp as possible, we confine the awareness of succession to a single, present moment, we still assume that this moment *endures* and as such contains a certain portion of the past. The only alternative which would avoid this would be to reduce the present moment to a single, mathematical, durationless point. Only a few philosophers go so far when the psychological present is concerned; but we shall return to this question soon. As long as we accept the temporal extension of the present moment, we also accept immediate memory, *coextensive, so to speak, with its duration.* In other words, we do accept a direct perception of the past, no matter how immediately anterior this past was. In conceding this, the direct experience of transition is accepted, however reluctantly.

In conclusion, we can say that, according to Bergson, there is such a thing as the perception of succession; but it is neither an atomic succession of perceptions, nor does it reside in the last term of the sequence. It is *coextensive* or rather *contemporary* with the succession itself, being both indivisible and differentiated. This is the reason why Bergson insists so strongly on both the indivisibility and heterogeneity of duration. But how can two such apparently antagonistic features be united?

## NOTES

[1] *Duration and Simultaneity* (transl. by Leon Jacobson), Bobb-Merrill, New York, 1965, pp. 44–45. Hereafter referred to as *D.S.*

[2] Andrew Ushenko, *The Logic of Events. An Introduction to a Philosophy of Time*, Univ. of California Press, Berkeley, 1929, pp. 120–121.

[3] A. O. Lovejoy, 'The Problems of Time in Recent French Philosophy', *The Philosophical Review* XXI (1913) 328–329. Lovejoy quotes A. E. Taylor who claimed that Bergson is "at heart as much of an Eleatic as Mr. Bradley." (!)

[4] Lovejoy, *loc. cit.*, p. 338.

[5] W. James, *The Principles of Psychology*, I, 243.

[6] *Ibid.*, p. 244. Cf. my article 'Stream of Consciousness and "durée réelle"', *Philosophy and Phenomenological Research* X (1950), especially pp. 331–338 about the affinities of James's and Bergson's views on this matter, – the affinities of which Bergson was unaware.

[7] D. Hume, *A Treatise of Human Nature*, Book I, Part IV, Section 5.

[8] D. Hume, *ibid.* On the substantial character of Hume's impressions cf. Arthur E. Murphy, 'Substance and Substantive', in *University of California Publications in Philosophy*, Vol. 9, p. 64; M. Čapek, 'The Reappearance of the Self in the Last Philosophy of William James', *The Philosophical Review* 62 (1953) 534.

[9] Quoted by W. James, *op. cit.*, p. 632 n.

[10] *D.S.*, pp. 50–51; Josiah Royce, *The World and the Individual*, Dover, New York, 1959 II, pp. 115–116; B. Russell, 'On the Experience of Time', *The Monist* XXV (1915) pp. 225–227. Cf. also M. Čapek, 'Time and Eternity in Royce and Bergson', *Revue Internationale de Philosophie*, No. 79–90 (1967) 22–33.

# CHAPTER 8

# THE DEEPER MEANING OF THE 'INDIVISIBLE HETEROGENEITY' OF DURATION

It has already been pointed out that the alleged contradiction between the indivisibility and the heterogeneity of duration is due to a semantic misunderstanding. As long as we understand, on the one hand, the terms "indivisibility" or "unity" in a rigid and static sense – we might say in an Eleatic sense – while, on the other hand, the term "heterogeneity" is made synonymous with "arithmetical multiplicity," the contradiction is obvious and unavoidable. As soon as we drop these two antagonistic terms, the contradiction disappears. But we have the right to drop them only if we convincingly show that there are certain types of experience which resist translation into arithmetical terms; in other words, to which neither the concepts of abstract, undifferentiated unity nor of abstract, atomic multiplicity apply. It was already pointed out that feelings of transition in general and the perception of melody in particular are such experiences. But these examples require further analysis, since they are so easily misinterpreted. Feelings of transition are easily misunderstood as being located "between" the substantive sections of our stream of thought, and as such are easily misconceived as other quasi-atomic items, existing *between* or *above* the atomic sensations or images whose connections they were supposed to restore. Thus arises the idealistic notion of the *unifying act*, which, when it was placed by Kant completely outside of concrete and moving experience, acquired the barren and homogeneous unity of the transcendental Ego, a unity as pure and absolute as that of a geometrical point. On the other hand, in the experience of melody our attention is almost inevitably focussed on the sensory diversity of the tones, and this diversity is naturally misconceived in the sense of an *arithmetical* multiplicity. The example of Herbart, that forgotten ancestor of Russell's and Wittgenstein's logical atomism, shows it quite convincingly. The passage of Bergson quoted above and incriminated by Ushenko shows that the qualitative diversity of duration means *something else* than a mere diversity of sensory qualities; "we must first efface the differences among the sounds, then do away with the distinctive features of sound itself, retaining of it only the continuation of what precedes into what follows and the uninterrupted transition, multiplicity without divisibility and succession without separation. ..." For this reason the perception of melody may be, in one sense, misleading because the diversity of the successive tones may hide from us *the imageless character* of the qualitative diversity of psychological duration. This is why Bergson invites us to consider the duration of one *apparently unchanging* mental state:

Let us take the most stable of internal states, the visual perception of a motionless external object. The object may remain the same, I may look at it from the same side, at the same angle, in the same light; nevertheless the vision I now have of it differs from that which I have just had, even if only because the one is an instant older than the other. My memory is there, which conveys something of the past into the present. My mental state, as it advances on the road of time, is continually swelling with the duration which it accumulates: it goes on increasing – rolling upon itself, as a snow ball on the snow.[1]

Or more concisely and without metaphors borrowed from our macro-scopic experience:

There is no consciousness without memory, no continuation of a state without the addition, to the present feeling, of the memory of past moments. This is what duration consists of.[2]

It is thus evident that what constitutes the novelty of the present moment is precisely the recollection of the immediately preceding moment; but this recollection, or as it was named by Richet and James, 'immediate memory', or by Stout, 'immediate retentiveness', is at the same time a dynamic link between the present and the past. Thus although the quality of novelty is a feature which differentiates *two* successive phases of the process, it is incorrect to say that it *separates* them; on the contrary, it binds them as well. *It has rarely been sufficiently stressed that the opposition between the novelty of the present and the persistence of the past is only apparent, since they both are merely two complementary aspects of one single, dynamic fact.* In other words, the traditional distinction between 'change' and 'duration' is not justified by attentive experience. This is why psychological time, and as we shall see, time in general, is the "indivisible and indestructible continuity of a melody where the past enters into the present and forms with it an undivided whole which remains undivided and even indivisible in spite of what is added at every instant *or rather thanks to what it is added*".[3] Without exaggeration, there are only a few other sentences in Bergson's writings which are equally significant in giving us a clue to understanding the paradoxical fusion of *the emergence of the present with the survival of the past*.

If we listen to a single, continued tone, we discover the same pattern. Although the sensory characteristics of the tone remain the same – its pitch, timbre and intensity – there is an indefinable and elusive change due to the simple fact that it *endures*. A tone which endures for two seconds is slightly different from what it was a second ago, even when its pitch, timbre and loudness remain the same; its present phase differs from its antecedent phase by the mere fact of being older. What makes its present phase *novel*, that is, *richer* with respect to its immediately anterior phase? Precisely the fact that the antecedent phase is *still remembered*; in other words, it is an immediate recollection which accounts for the qualitative difference between the present and the past and which we know by the name 'novelty'. In other words, the novelty of the present requires the persistence of the past as a necessary, contrasting background. But conversely, the pastness of the previous moment is impossible without the novelty of the present; it is a new moment, which, metaphorically speaking, 'pushes' it into the past. Our language, with its static and discontinuous nouns, is ill fitted to express the indivisible complexity of the intuitive fact; the nouns 'past', 'present' and 'moment' suggest static and discontinuous things moving in an imaginary space. Even verbs are far from adequate. Such verbs as "move" and "push" are imbued with kinematic and mechanical associations, wrongly suggesting the images of bodies pushing each other in space. It would be more accurate to say that the former present acquires the character of pastness by virtue of the emergence of a new present; it is only this emergence which really makes it 'previous', i.e. which deprives it of the character of presentness.

But conversely it is equally true that it was the passing of the former present (that is, the fact that the former present fades into the past, thus acquiring the character of pastness) which brought forth a new and immediately subsequent moment. There is no novelty without immediate memory and *vice versa*. They are not two different facts, but one single process considered from two different perspectives, prospectively and retrospectively, or using Husserl's language, "protentionally and retentionally".[4]

Let us be constantly on guard against the language we use! There is a definite danger in using the word "perspective," since it may wrongly suggest that the duality in question is merely subjective and thus spurious, and that the simplicity of the underlying fact is absolute, comparable to the bare simplicity of the atom or mathematical instant. Such a wrong interpretation is further strengthened by the words 'one single process', and the cases of Ushenko and Lovejoy show how easily the most serious and honest minds are misguided in this respect. The perspectives referred to above are nothing *external* to the dynamic fact of duration. They intrinsically belong to the very process itself. Their duality is *inherent* within the unity of temporal pulsation without ever becoming the duality of mutually separate units. A complete separation of successive phases would destroy their succession; but so would their complete, undifferentiated fusion. We shall treat this particular point in detail later; for our present purpose let us note as a sort of *resumé* another significant passage from Bergson:

Everyone will surely agree that time is not conceived without *before* and *after* – time is succession. Now we have just shown that where there is not some memory, some consciousness, real or virtual, observed or imagined, actually present or ideally introduced, there cannot be before *and* after; there is one *or* the other, not both; and both are needed to constitute time.[5]

Or in the words of Bergson's answer to Lovejoy:

The difficulties which you find in my description of duration are doubtless due to the fact that it is hard, if not impossible, to express in words a thing which is repugnant to the very essence of language. I can only attempt to *suggest* it. Duration is indivisible; but this nowise implies that the past and the present are simultaneous. On the contrary, duration is essentially succession; only it is a succession which does not imply a "before" and "after" external to one another.... It is only by an effort of reflection that, subsequently turning back upon this indivisible whole once constituted, you represent it to yourself as a simultaneity, because of its indivisibility; – which leads you to have a *spatial* image of it, capable of being cut up into distinct terms, decomposable into a "before" and "after", which then would be juxtaposed.[6]

Two terms are needed for the relation of succession, since at least two terms are needed for any relation. But we must not forget that the relation in question is quite different from other relations, since it has the following properties: (a) it is a dynamic relation, that is, such that by its own nature *becomes* and both of whose terms are created by this very becoming. This means that the qualitative difference which is the very essence of succession is not a relation between two *coexisting* terms, but two successive terms. It is almost preferable to avoid the term 'between' altogether, as it suggests the usual connotation of the spatial separation of two simultaneous terms. (b) It is not an *external* relation, since both terms, despite their succession and despite their difference – *or rather because of it* – are not separated. The novelty of the present is *within* this relation, since it is created by the very act of becoming; but the same is true of the antecedent phase whose very pastness would become utterly meaningless without the contrasting background of the emergent novelty of the present. Thus, paradoxically speaking, the qualitative difference "separating" two successive moments at the same time *joins* them. Consequently, two successive moments are never "two" in the usual arithmetical sense like two bodies in space or two ordinal numbers located on the axis of real numbers. Nor is it unqualifiedly *one* in the sense of the bare, undifferentiated unity of a single point. "The 'passing' moment is", as James observed, "the minimal fact, with the 'apparition of difference' inside of it as well as outside".[7]

It is only fair to stress that Bergson, in spite of his radical opposition to the classical, intellectualistic tradition, was not without predecessors, even though they were not as numerous as those who either completely or half-heartedly sided with the Eleatic or atomistic tradition. Heraclitus in his 'obscure' language, stressing the apparently contradictory character of perceptual change, 'the unity of opposites', the absence of distinction between change and duration as beautifully expressed in fragment 83 ("*It rests by changing.*") – in all this he anticipated Bergson's struggle against the inadequacy of language and traditional conceptualization when they are applied to the reality of time. The same is true of Hegel as far as the dynamic aspects of his thought are concerned. While Bergson, according to his own admission, like Whitehead, never studied Hegel[8], and therefore was not aware of the resemblance, William James definitely was when he wrote (not without humor): "Not only the Absolute is its own other, but the simplest bits of immediate experience are their own others, if that Hegelian phrase be once for all allowed." Needless to stress that it is the Hegel of Croce and Findlay, not the Hegel of McTaggart, who is akin to Bergson's view of reality *sub specie durationis*; in the concluding part of his book on Hegel Croce explicitly stressed this affinity.[9] The words 'by being their own others' suggest the same '*pénétra-*

*tion mutuelle'* of Bergson, the same *'durcheinander perceptual'* of James[10], which defies any artificial dissection into atomic bits externally separated and internally undifferentiated; in other words, it defies any arithmetization, which is so successful in the area of macroscopic, visual experience.

We can see why Bergson insisted on the radical difference between homogeneous time and concrete duration. Homogeneous, mathematical time, symbolized by a geometrical line, misrepresents the nature of concrete duration in several ways. By suppressing any qualitative difference between successive moments it does not do justice to their inherent heterogeneity; by symbolizing them by juxtaposed 'points-instants' located on 'the axis of independent variables' it wrongly suggests that succession is an *external* relation and thus destroys the dynamic indivisibility of duration. Finally and most seriously, the geometrical line that is *already drawn* to symbolize 'the course of time' tends to hide the most essential character of becoming – its incompleteness. Even if we try to do justice to the incompleteness of duration by imagining 'the line of time' as being incomplete and gradually extending itself in the direction of the future, the inadequacy of the spatial symbol is hardly removed. For the extreme point of the line, symbolizing 'Now', gradually occupies the originally unoccupied points in the future; and since in our geometrical diagram these points virtually exist prior to their occupation by the 'Now', the idea of the pre-existence of the future is again wrongly suggested. Briefly, spatialized time, whether in the form of one-dimensional space in psychology or four-dimensional space in physics, is radically incompatible with the essential features of becoming.

## NOTES

[1] *C.E.*, p. 4.
[2] *C.M.*, p. 211 (the essay 'Introduction to Metaphysics').
[3] *C.M.*, p. 83.
[4] E. Husserl, *Zur Phänomenologie des inneren Zeitbewusstseins (1893–1917)* (ed. by Rudolf Boehm), Martinus Nijhoff, The Hague, 1966 [Husserliana, 10], especially pp. 52–53.
[5] *D.S.*, p. 65.
[6] Reprinted first in A. O. Lovejoy, *Bergson and Romantic Evolution*, Univ. of California Press, Berkeley, 1914, p. 9; more recently in A. O. Lovejoy, *The Reason, the Understanding and Time*, The Johns Hopkins Press, Baltimore, 1961, pp. 185–187.
[7] W. James, *A Pluralistic Universe*, p. 283.
[8] Cf. Jules Chaix-Ruy, 'Vitalité et élan vital: Bergson et Croce', in *Études berg-*

*soniennes*, Vol. V, Paris 1960, p. 145. Croce reported his conversation with Bergson in *Quâderni della Critica* (November 1949).

[9] B. Croce, 'Cio che è vivo è cio che è morte della filosofia di Hegel', in *Saggi philoso-phici*, Vol. III, Bari 1948, p. 141.

[10] *Some Problems of Philosophy*, Longmans, Green, London, 1940, p. 199.

# THE UNREALITY OF DURATIONLESS INSTANTS: BECOMING NOT MATHEMATICALLY CONTINUOUS

The heterogeneity of time logically implies the denial of another cherished dogma of classical thought: *the concept of the infinite divisibility of time.* This concept is closely connected, though not entirely identical, with the concept of instant. Every interval of time, no matter how small, contains, according to the classical view, its component sub-intervals, which in their turn can still be subdivided, and so on *ad infinitum.* There are no indivisible intervals of time; the only truly indivisible elements of time are durationless instants with no temporal thickness. In this respect there was a remarkable agreement among classical thinkers, and the terms 'homogeneity', 'infinite divisibility' and 'mathematical continuity' were altogether synonymous. But psychological experience is decisively and unambiguously opposed to the concept of infinitely divisible time and the correlative concept of a durationless instant. William James was certainly not the first thinker to question the infinite divisibility of time and the concept of the mathematical present, but he gave to this denial probably the clearest expression in a passage which has since become a classic:

In short, the practically cognized present is no knife-edge, but a saddle-back, with a certain breadth of its own on which we sit perched, and from which we look in two directions into time. The unit of composition of our perception of time is a *duration*, with a bow and a stern, as it were – a rearward- and a forward-looking end. It is only as parts of this *duration-block* that the relation of *succession* of one end to the other is perceived. We do not first feel one end and then feel the other end after it, and from the perception of the succession infer an interval of time between, but we seem to feel the interval of time as a whole, with its two ends embedded in it. The experience is from the outset a synthetic datum, not a simple one; and to sensible perception looking back its elements are inseparable, although attention looking back may easily decompose the experience, and distinguish its beginning from its end.[1]

Some of James's metaphors may be misleading because of their crudely visual character; it is only too obvious that he remained a *painter* when he tried to describe the immediate data of experience. But what he tries to describe is essentially the same basic, dynamical fact of 'two-oneness', of the indivisibility of temporal relation and the inseparability of 'after' from 'before', which is so repeatedly emphasized by Bergson. If psychological duration is by its own nature an indivisible synthesis of the novelty of the present with the memory of immediate recency, no demarcation line can be drawn which would cut the present from the past, since both, in spite of their qualitative difference, remain *internally* related. In other words, the novelty of the present is a *concrete quality* and as such it cannot be a bare, mathematical point, especially since the psychological origin of the latter concept is, as we shall point out soon, too suspicious to be trusted. But this quality is a *quality of difference* with respect to the immediately recent past which, consequently, must not be forgotten. Otherwise no feeling of qualitative difference, the very essence of succession, would arise. Once again we return to the apparently baffling though undeniable fact that it is the qualitative difference between 'two' successive moments which joins and separates them at once.

At once? Do you not, it may be objected, surreptitiously re-introduce the rejected concept of instant in using the expression 'at once'? Not as long as I do not translate the immediately experienced quality of pre-sentness into the language of spatial symbolism from which the concept of mathematical instant arises. It is a prejudice, though a very natural one, to conceive each unity in the sense of the bare and undifferentiated unity of the mathematical point. The novelty of the present is an indivisible datum, *complex but not composed*[2], whose complexity is due to its indivisible correlation with the anterior moment. Let Bergson speak for himself:

I make no hypothesis, I do not call in aid a mysterious entity, I confine myself to observation. For there is nothing more immediately given, nothing more evident than consciousness, and mind *is* consciousness. Now, consciousness signifies, before everything, memory. At this moment that I am conversing with you, I pronounce the word "conversation". Clearly my consciousness presents the word all at once, otherwise it would not be a whole word, and would not convey a single meaning. Yet, when I pronounced the last syllable of the word, the three first have already been pronounced; they are past with regard to the last one, which must then be called the present. But I did not pronounce this last syllable "tion" instantaneously. The time, however short, during which I uttered it is decomposable into parts, and all of these parts are past in relation to the last among them. This last would be the definitive present, were it not, in its turn, decomposable. So that however you try, you cannot draw a line between the past and the present, nor consequently between memory and consciousness.[3]

Here, based on a direct appeal to introspective evidence and freed from encumbering visual metaphors, the non-instantaneous nature of the psychological present is asserted. There is a haziness of expression here, however, when reference is made to the "decomposability" of the experienced present. As the whole passage suggests, the time of the psychological present is *not* divisible; what is divisible is the *contemporary interval of physical* time which classical physics believed to be divisible *ad infinitum* and in this respect was, from the practical point of view at least, not wrong. We make this point explicit now, though it will be treated extensively in the third part.

Yet serious objections arise and in anticipating them it is exceedingly important to try to make some additional points clear. We shall consider them step by step, though they are merely various and complementary aspects of the same, though tremendously complex, problem – the structure of time. The questions we are going to consider are: (1) the apparently convincing character of the notion of the instantaneous present; (2) the relation of psychological continuity to mathematical continuity; and (3) the apparent contradiction between the dynamic continuity and the pulsation-like character of time. Some repetition will be unavoidable in view of the fact that the aforementioned topics are closely related and they even overlap to a certain extent.

Re 1. The infinite divisibility of time seems to be an unavoidable logical consequence of the *continuity of change*. Succession and change must be present even within the smallest intervals of time; "if a mental state ceased to vary, its duration would cease to flow",[4] as Bergson himself insists. But if this is so, no matter how small an interval of time may be, it always contains a 'before' and 'after'. In other words, it is always possible to discern successive subintervals within it. This necessarily leads to the idea of mathematical continuity, i.e. the infinite divisibility of time. Upon this conclusion the agreement among classical thinkers was almost universal. Kant expressed this view in precise language in one passage of his *Transcendental Analytic*, according to which time can consist of times only, and, consequently, there are no *parts* of time which are not themselves temporal. Kant then adds a very significant sentence which is an almost exact anticipation of Bergson:

Points and instants are only limits, that is, mere positions which limit space and time. But positions always presuppose the intuitions which they limit or are intended to limit;

and out of mere positions, viewed as constituents capable of being given prior to space or time, neither space nor time can be constructed.[5]

The existence of instants and, by implication, the existence of the instantaneous present seemed to be resolutely denied. Yet, if each fraction of time is divisible into smaller and smaller intervals, *ad infinitum*, are we not driven to an inescapable conclusion that the only true *parts* of time which are *not* divisible are durationless instants, in direct contradiction to the passage just quoted? A. N. Whitehead pointed out that Kant virtually contradicted what he'd said before in another passage:

I call an extensive quantity that in which the representation of the whole is rendered possible by the representation of its parts, *and therefore necessarily preceded by it*. I cannot represent to myself any line, however small it may be, without drawing it in thought, that is, without producing all its parts one after the other, starting from a given point, and thus, first of all, drawing its intuition. The same applies to every, even the smallest portion of time. I can only think in it the successive progress from one moment to another, thus producing in the end, by all the portions of time, and their addition, a definite quantity of time.[6]

From this point of view durationless instants seem to be the basic elements of time, as extensionless points are the fundamental elements of space. It is hardly surprising, then, that Kant explicitly accepted the instantaneous mathematical present. In this respect Kant is representative of the whole classical tradition in which the concepts of infinite divisibility and durationless instant were correlated.

What remains paradoxical is that the reasoning, whose inspiring motive was the conviction that time *cannot* be built out of parts which are not themselves temporal, ends with an explicit admission that time *does* consist of timeless parts! Nothing could illustrate more convincingly the lasting conflict between a *direct intuition* and its *spatializing symbol*. For it is obvious that in this case a serious effort to express adequately *the radical fluidity of time* only resulted in its clumsy imitation by a multiplicity of instants, static and discontinuous in nature. It is less obvious that such a mistranslation of the temporal intuition into a statically distorted picture is due to our unconscious habit to represent 'the flight of time' by a geometrical line whose individual points allegedly represent the successive moments. The infinite divisibility of this line is wrongly attributed to time itself. In Bergson's words:

But we naturally form the idea of instant, as well as of simultaneous instants, as soon as we acquire the habit of converting time into space. For, if duration has no instants, a line terminates in points. And, as soon as we make a line correspond to a duration, to portions of this line there must correspond "portions of duration" and to an extremity of the line, an "extremity of duration"; such is the instant – something that does not exist actually, but virtually. The instant is what would terminate a duration if the latter came to a halt. *But it does not halt. Real time therefore cannot supply the instant; the latter is born of the mathematical point, that is to say, of space.*[7]

On this point – the unreality of durationless instants in psychological duration, there is no disagreement between Bergson and one of his most hostile and ill-tempered critics, Bertrand Russell. Russell explicitly conceded that it is absurd to claim that one experience can exist only in one mathematical instant.[8] It is of secondary importance whether Russell realized the extent of his agreement with Bergson or not.

What is far more important is that Bergson's and Russell's disbelief in the durationless psychological moment is shared by psychologists in general. Though estimations of the minimum temporal interval perceptible by our consciousness vary, there is a fairly general agreement that it is not of a smaller order than one thousandth second. Bergson himself accepted James's value of 0.002 sec.[9] To speak about *psychological* intervals *smaller* than this interval is an epistemological misunderstanding. It would imply that the indivisibility and simplicity of psychological pulsation is only deceptive and that what we call the minimum temporal interval is 'in reality' an aggregate *composed* of a considerable, perhaps even infinite, number of successive subpulsations which are much shorter or even without any temporal length altogether. The belief that conscious sensations are made up of subconscious subsensations of which we are not aware was stated for the first time by Leibniz, but became very popular in the second half of the nineteenth century with the arrival of the positivistic physiological and associationistic psychology. The expression 'illusion of consciousness' occurred very frequently in books of psychology at that time. It was emphasized again and again that the introspective psychologist should not be deceived by the "apparent" simplicity of psychological events and that he always has to search for the *truly* simple and *truly* fundamental elements underlying the spurious indivisibility of mental wholes. This view was meaningful within the framework of psycho-physiological parallelism or epiphenomenalism, according to which psychological events are merely fragmentary aspects of physiological processes in the neural tissues. Consequently, psychological duration is merely *an epiphenomenon of objective physical time* within which the only 'truly real' events of our nervous system take place; and since this physical time was believed to be infinitely divisible, the temporal indivisibility of mental wholes must be only a spurious 'surface-phenomenon'. As they were unable to find any introspective evidence for "elementary sub-sensations" or "petites perceptions," like Leibniz, they simply *postulated* them. Such was the case with Herbert Spencer and Hyppolite Taine, and more recently, with E. B. Holt and Stephen Pepper.[10]

Since the time of Descartes it should be known that I may be wrong in *interpreting* the data of experience, but I cannot be wrong about *having* them. For instance, I may be wrong when I believe that the rainbow is localized on the sky or that the echo is coming from the direction from which it seems to be coming, but I *cannot* be wrong in my belief that I really *experience* the visual sensations of spectral colors or the auditory sensations indicating a certain direction is space. It is true that the interpretation of the data is often so automatic and unconscious that it appears to be a part of the immediate experience, but here lies precisely the main task of epistemological analysis – to disentangle indisputable data from all inferential and interpretative elements. The question is: What are the *indisputable data*? Is the temporal indivisibility of mental events one of them? Is it possible to deny that the temporal indivisibility of, for instance, the sensation of redness is as genuine a datum as the redness itself? Only by inference from complicated, physical experiments do we *infer* that the objective counterpart, i.e. the stimulus, of the sensation of redness is a very complex fact consisting of an enormous number of successive electromagnetic vibrations. But this does not mean that the sensation of redness itself is compound, since there is absolutely nothing in its quality which would suggest its complexity. There is nothing illusory in the temporal and qualitative indivisibility of psychological events; the illusion arises only when this indivisibility is illegitimately projected into the physical world. Only in this carefully redefined sense can the expression 'illusion of consciousness' remain meaningful.

The contrast between the indivisible, psychological quality and the enormous complexity of its physiological counterpart is one of the most serious difficulties of the identity or 'double-aspect' theory of the mind-body relation. As William James observed:

It is hard to imagine that "really" our own subjective experiences are only molecular arrangements, even though the molecules be conceived as being of a psychic kind.

A material fact may indeed be different from what we feel it to be, but what sense is there in saying that a feeling, which has no other nature than to be felt, is not as it *is* felt?[11]

Whatever the structure of *physical* duration is, there can be, then, no question that *psychological* duration is *not* infinitely divisible. Its nature is such as it is experienced, and if it is experienced as a pulsational flow in which the novelty of the present comes up as a concrete quality and not as an ideal, infinitely thin instant, then it *is* a pulsational flow and nothing else. From this point of view the term 'specious present' is highly misleading, as it wrongly suggests that the psychological present may be *spurious* and that the 'true present' should be looked for in the durationless instants of the physical world. The very opposite is true; the psychological present is genuine and real while the mathematical present is *specious*, not only in psychology, but as we shall see, in physics as well.

Re 2. It is evident that psychological continuity is altogether different from mathematical continuity, which is only another name for infinite divisibility. It is highly unfortunate that one and the same term is associated with such different and even antagonistic meanings. While dynamic continuity emphasizes the internal relations between successive phases of duration (Bergson's '*interpénétration mutuelle*', James's 'stream of thought'), "the intellectual representation of continuity is negative, being at bottom, only the refusal of mind, before any actually given system of decomposition to regard it as the only possible one".[12] Mathematical continuity is thus nothing but *discontinuity endlessly repeated* which ignores the natural articulation of duration into indivisible, temporal pulsations. It is very significant, though it may appear paradoxical at first, that in the idea of "mathematical continuity" all real continuity is gone and nothing is left except the discontinuity of isolated elements; in this observation Bergson is not alone, as his view is shared by the mathematicians Poincaré and Weyl. "The continuum so conceived", says Poincaré, "is only a collection of individuals ranged in a certain order, infinite in number, it is true, but *exterior* to one another. This is not the ordinary conception, wherein is supposed between the elements of the continuum a sort of intimate bond which makes of them a whole where the point does not exist before the line, but the line before the point. Of the celebrated formula, 'the continuum is unity in multiplicity', *only the multiplicity remains, the unity has disappeared*." (Last italics added.)[13]

Herman Weyl is equally, perhaps even more, emphatic:

We must nor forget that in the "continuum" of real numbers the individual elements are as completely isolated as the whole numbers...

Certainly: the intuitive and the mathematical continuum do not coincide; there is a deep chasm which separates them.

It was a merit of Bergson's philosophy to point out emphatically this deep discrepancy between the conceptual world of mathematics and the immediately experienced continuity of phenomenal time (la durée).

The conception of the flux (Ablauf) consisting of points and thus *being disintegrated into points*, proves to be mistaken. Precisely that by which continuity is made escapes us, the overflow (Überfluss) of one point into another, that which continually lets the enduring present glide away into the ever deepening past.[14]

This substantiates the epistemological conclusion of *Creative Evolution*, the book to which Weyl explicitly refers: "Of the discontinuous alone does the intellect form a clear idea."[15] Needless to repeat again that by the word 'intellect' is meant here only one of its petrified, though historically most important and still widely prevailing forms, i.e. "the logic of solid bodies." In applying this spurious "mathematical continuity" to real temporal processes, we become entangled in Zeno's paradoxes, which all result from the confusion of real duration with its underlying spatial symbol. Real time is not a succession of infinitely thin instants, but grows, as James says, "dropwise, by discrete pulses of perception".[16] This is true of psychological time, but we shall see that Bergson generalizes the pulsational structure to *any type* of change, including the physical processes.

## NOTES

[1] *The Principles of Psychology*, I, pp. 609–610.

[2] Cf. Vladimir Jankélevitch, 'De la simplicité', in *Henri Bergson. Essais et témoignages*, Neuchâtel, 1943: "Car, 'complexe' et 'compliqué' font déux." (Pp. 170–171.)

[3] *M.E.*, p. 69.

[4] *C.E.*, p. 4.

[5] Immanuel Kant, *Critique of Pure Reason* (transl. by Norman Kemps Smith), Macmillan, London, 1953, p. 204.

[6] A. N. Whitehead, *Science and the Modern World*, Macmillan, New York, 1953, p. 184. (Hereafter referred to as *S.M.W.*); Kant, *op. cit.*, p. 198. (Whitehead quotes Max Müller's translation.)

[7] *D.S.*, p. 53.

[8] B. Russell, 'On the Experience of Time', *The Monist* XXV (1915), p. 217.

[9] *The Principles of Psychology*, I, 613–614; *M.M.*, p. 202. It is the minimum interval determined by Exner as necessary for perceiving two auditory sensations as being successive.

[10] Cf. Chapter 3 of this part, Note 4 and 5.

[11] *Some Problems of Philosophy*, p. 151. Cf. a nearly identical passage in *The Principles of Psychology*, I, p. 163.

[12] *C.E.*, p. 170.

[13] 'Science and Hypothesis', in *F.S.*, p. 43.

[14] H. Weyl, *Das Kontinuum*, Zurich 1917, Ch. II, § 6, especially pp. 69–71.

[15] *C.E.*, p. 170.

[16] *A Pluralistic Universe*, p. 231; cf. also *Some Problems of Philosophy*, Ch. XI 'The Novelty and the Infinite. – The Perceptual Point of View'.

# CHAPTER 10

# THE INADEQUACY OF
# THE ATOMISTIC THEORY OF TIME

But does not this last sentence indicate the acceptance of the atomic structure of time with all its absurdities? This is the third possible objection to the denial of mathematical, point-like instants. We have to remember that the concept 'atom of time' is basically self-contradictory because (a) it implies a surreptitious return to the concept of mathematical instant, and (b) because it seems to imply that the temporal process is *made up* of parts which themselves are devoid of temporality. A thinly disguised return to the concept of mathematical instant is implied by the assumption that time is a succession of temporal segments, each having a finite length; but does not the idea of 'segment' imply the presence of both initial and final extremities by which each segment is delimited from the anterior and subsequent ones? And how else can this boundary be conceived except as instantaneous, without temporal thickness, at least as long as an infinite regress is to be avoided? The second difficulty seems to be equally serious: How can we speak about the *duration* of the time-atom, if there is no possibility of discerning *two* successive moments within it? Is this possibility not excluded by the alleged *indivisibility* of the corresponding 'atomic' interval? Not much is gained, if we claim that subsequent temporal segments have no 'sharp edges' and that they 'compenetrate each other' like liquids in osmotic solutions. Bertrand Russell would undoubtedly object that this is nothing but a 'thick fog'[1] and that even the thickest fog, when microscopically analyzed, would appear as made of discontinuous, tiny drops; then the very same problem would arise again.

This shows too clearly that no solution of these difficulties can be gained as long as we, consciously or unconsciously, are confined within the limits of visual and spatial imagery. In Bergson's own words, temporal processes cannot be reconstructed out of static and discontinuous elements, no matter whether these elements are punctual instants of finite, geometrical segments. It may be objected that the theory of finite, temporal segments was inspired precisely by the desire to *avoid* the clumsy construction of time by static durationless instants; the idea of finite segment *symbolizes* a finite, indecomposable stretch of duration. There is no question as to what the motive of the atomistic theory of time was; but this again shows the startling difference between the inspiring motive and the results obtained. It is very, very easy to forget the symbolic nature of spatial imagery and, whether consciously or unconsciously, take the idea of finite, geometrical segment literally. As soon as this happens, the idea of boundary-points inevitably creeps in and with them the idea of the mathematical instants which were meant to be avoided. Our imagination, in contemplating a static, geometrical segment and forgetting that this segment symbolizes *the indivisibility of a process* and not *the indivisibility of a thing*, will find it difficult to stop the operation of subdivision which can legitimately be applied to every static line. By using the word 'drops' as James does, a certain advantage is gained in that it does not suggest so obtrusively the idea of the rigid, static character of the component, temporal elements. But even liquid drops have surfaces! Therefore, as stated above, the basic difficulty would remain; visual pictures or diagrams are not made more adequate by dimming their contours and making them more 'impressionistic'. Even an impressionistic painting is made of *juxtaposed*, simultaneous parts and thus sets essentially the same trap for our understanding the nature of time. For this reason we have to reject the idea of the atomistic structure of time. If Whitehead uses the term 'atomic succession', he does it only in a metaphorical sense; otherwise his criticism of "the fallacy of simple location in time" would lose its meaning.

According to the atomistic theory of time the succession of the atomic blocks of duration is conceived as a purely *external* relation; each segment is "after" the immediately preceding one, but this "after" is understood in the sense of sheer externality and independence. This emphasis on independence is not entirely wrong, because it rightly focusses attention on the feature of *novelty* occurring in every temporal process. But it is *one-sided* in neglecting the complementary feature of dynamic continuity by which successive moments cohere, and we have already seen that without the mnemic survival of the past no novelty would be possible. By isolating novelty from its mnemic, temporal content as the atomistic theory of time does, we break the relation of succession into two separate elements, which, in virtue of their isolation, lose their temporal character altogether. In plain language, we would have 'after' without 'before'. When Lovejoy wrote: 'Only by conceiving succession as made up of units in some sense static can we conceive of it as made of discrete units',[2] he was undoubtedly correct, although he failed to realize how baleful the implications of such a view are for his own alleged temporalism. If we bear in mind that the novelty of each present is possible only against the backdrop of the fading past, we see that no moment of succession can be conceived in separation from its antecedent, temporal context. Does this then mean that it is therefore *identical* with its predecessors? This was the view which both Lovejoy and Ushenko imputed to Bergson; wrongly, of course, because the very essence of the novelty of the present is its qualitative *difference* from its immediately anterior past. It is this qualitative difference – the difference which both separates and unites two successive moments – which Bergson stressed untiringly when he insisted on the *heterogeneity* of duration.

Unfortunately, words like 'novelty', 'present', 'past', 'immediate memory', etc., are abstract words which can acquire their full meaning only when they are continually referred back to concrete experience. As Whitehead wrote several decades after Bergson: "*The sole appeal is to intuition*",[3] i.e. to the epistemologically purified datum of temporal experience. Let us consider such a concrete experience: the experience of melody which is relatively lightly pervaded by obtrusive, visual associations and for this reason was used by Bergson more than once to illustrate the general pattern of duration. It also applied in his answer to Lovejoy's criticism. But Lovejoy apparently remained unconvinced. He claimed that he never experienced a melody in which the notes had no "distinct and numerical multiplicity". He apparently did not realize that were his claim *literally* true, he would not have perceived any melody at all, but only each individual tone, immediately forgetting its predecessors. But he did *not* mean it literally, as is obvious from his own words: "I seem to myself, indeed, to hear each separate note, one after the other, though *while hearing each, I may be continuously aware of the total musical unit, or pattern, of which it is a part*".[4] (Italics mine.) In the italicized words Lovejoy explicitly surrenders his original claim that the succession of the tones is "numerical multiplicity." It is true that each individual tone retains its specific individuality, but this individuality is bathed, so to speak, in the *antecedent* melodical context. Numerical multiplicity is a mere summation of the individual terms – addends – which remain *externally* related in the sense that each of them remains unmodified by being added to the others. But melody is a temporally extended *Gestalt* in the sense of Ehrenfels, whose unity persists despite the individual specificity of its individual tones. And precisely because it is not a mere addition of the tones, it is misleading to call the individual tones 'parts' of it. As Kurt Koffka observed, consciously adopting Bergson's language, melody never *is* in any of its tones, but *passes* through them in a way similar to that in which a moving body never *is* at any one of its positions, but merely passes through them.[5]

The reason Lovejoy and other critics of Bergson fail to see this is due to the fact that their introspective analysis of duration is simply not subtle enough, not *empirical* enough. It is easy to jump from the fact that each tone retains its specific, qualitative individuality to the conclusion that each is *externally* related to the others like the addends in a sum. If it were so, an individual tone preceded by a silence would be undistinguishable from the same tone preceded by the succession of different tones. Although the sensory quality of this tone is the same, there is a subtle, non-sensory difference due to the difference in the antecedent temporal context. William James was clearly aware of this fact when he wrote:

Into the awareness of the thunder itself the awareness of the previous silence creeps and continues; for what we hear when the thunder crashes is not thunder *pure*, but thunder-breaking-upon-silence-and-contrasting-with-it. Our feeling of the same objective thunder, coming in this way, is quite different from what it would be were the thunder a continuation of previous thunder. The thunder itself we believe to abolish and exclude the silence; but the *feeling* of the thunder is also a feeling of the silence as just gone; and it would be difficult to find in the actual, concrete consciousness of man a feeling so limited to the present as to not have an inkling of anything that went before.[6]

James ends by pointing out the inadequacy of our language. We label each 'section' of our stream of thought by a different word, often neglecting the more elusive, imageless 'fringes' or 'halos' of which the 'feelings of relations' are constituted, and then wrongly inferring from *the discontinuity of the words* the discontinuity of the content which they designate. James himself was occasionally prone to this type of fallacy. When, in another passage of his *Principles*, he spoke of "elementary sensations of duration", he was, as D. S. Mackay pointed out[7], unconsciously slipping into the atomistic and associationistic language which he himself so effectively criticized. But that was merely an occasional slip and only a few lines after the misleading expression, 'the units of composition of our perception of time', was used, he stressed that "the experience is from outset a synthetic datum, not a simple one".[8] This is what he more felicitously described above: a concrete specimen of the heterogeneous continuity of duration, of the dynamic process which remains indivisible while being differentiated, which is neither barely *one* nor barely *many* in the strict arithmetical sense and which, as such, transcends the artificial dilemma of 'atomicity versus continuity', of 'unity versus multiplicity'.

## NOTES

[1] B. Russell, *Introduction to Mathematical Philosophy*, Allen & Unwin, London, 1924, p. 105.
[2] A. O. Lovejoy, 'The Problem of Time in Recent French Philosophy', *The Philosophical Review* XXI (1915) 533.
[3] *Process and Reality*, Macmillan, New York, 1930, p. 32.
[4] A. O. Lovejoy, *The Reason, the Understanding and Time*, p. 192.
[5] Kurt Koffka, *Principles of Gestalt Psychology*, Harcourt & Brace, New York, 1935, pp. 434–435.
[6] *The Principles of Psychology*, I, pp. 240–241.
[7] D. S. Mackay, 'Succession and Duration', in *The Problem of Time. University of California Publications in Philosophy*, Vol. 18 (1935) p. 194.
[8] W. James, *op. cit.*, I, p. 610.

CHAPTER 11

# THE UNITY AND MULTIPLICITY OF DURATION:
## BERGSON, RUSSELL AND BROUWER

We are now in a better position to understand why Bergson so persistently warns against confusing 'numerical multiplicity' which, according to him, is basically of a spatial nature, and the 'qualitative multiplicity' of duration. It is obvious that the notion of 'qualitative multiplicity' or 'heterogeneous continuity' is a difficult and complex one, being as far as possible from the mental "thick fog" to which Bertrand Russell ironically compares it.[1] In the basic intuition of duration we cannot neglect one of the complementary features without getting a distorted or one-sided notion of it. If we insist exclusively on the multiplicity and diversity of successive moments, we obtain what Bergson calls '*poussière des instants*', which are mutually external and separate, no matter how close they are imagined; if we insist on the unity of duration, we easily forget the heterogeneity of its successive phases and get the idea of empty, homogeneous, container-like time which in virtue of its homogeneity has the character of static eternity rather than of an incomplete growing process.

If I consider duration as a multiplicity of moments bound to one another by a unity which runs through them like a thread, these moments, no matter how short the chosen duration, are unlimited in number. I can imagine them as close together as I like; there will always be, between these mathematical points, other mathematical points, and so on, *ad infinitum*. Considered from the standpoint of multiplicity, duration will therefore disappear in a dust of moments not one of which has duration, each one being instantaneous. If on the other hand I consider the unity binding the moments together, it is evident that it cannot have duration either since, by hypothesis, everything that is changing and really durable in duration has been put to the account of the multiplicity of the moments. This unity, as I examine its essence, will then appear to me as an immobile substratum of the moving reality, like some intemporal essence of time: that is what I shall call eternity, – eternity of death, since it is nothing else than movement emptied of the mobility which made up its life. Examining closely the opinions of the schools antagonistic to the subject of duration, one would see that they differ simply in attributing to one or the other of these two concepts a capital importance. Certain of them are drawn to the point of view of multiplicity; they set up as concrete reality the distinct moments of a time which they have, so to speak, pulverized; they consider as being far more artificial the unity which makes a powder of these grains. The others, on the contrary, set up the unity of duration as concrete reality. They place themselves in the eternal. But as their eternity nevertheless remains abstract, being empty, as it is the eternity of the concept which by hypothesis excludes the opposite concept, one cannot see how this eternity could allow an indefinite multiplicity of moments to coexist with it. In the first hypothesis one has a world suspended in mid-air which would have to end and begin again by itself each instant. In the second, one has an infinitely abstract eternity of which one can say that it is especially difficult to understand why it does not remain enveloped in itself and how it allows things to coexist with it.[2]

This oscillation between the point of view of multiplicity and that of unity characterizes the whole of classical thinking about time, particularly in physics and the philosophy of science. *Les extrêmes se touchent*; the Zenonian tendency to break temporal intervals into an infinite number of durationless instants eliminates succession and duration as effectively as the opposite tendency to fuse successive moments into the unity of a homogeneous qualitatively undifferentiated medium. If under such conditions time is still imagined 'to flow', it is, as Bergson says, only because "reality obtains this sacrifice from the two doctrines by taking advantage of the inadvertence in their logic".[3]

Sometimes these two opposite – or rather apparently opposite – tendencies are united within one and the same thinker or even within one and the same book. Thus Bertrand Russell accepted the view of Zeno when he wrote in his *Principles of Mathematics*:

Weierstrass, by strictly banishing all infinitesimals, has at last shown that we live in an *unchanging world*, and that the arrow, at every moment of its flight is *truly at rest*. The only point where Zeno *probably* erred was in inferring (if he did infer) that, because *there is no change*, therefore the world must be in the same state at one time as at another. (All italics mine.)[4]

Yet, in spite of this assertion of the complete externality of subsequent events, in other sections of the same book he maintained the strict causal implication of one event by another and obviously adhered to the Laplacean view. This Laplacean view became even more obvious later. While in *Principles of Mathematics* he still insisted that the causal proposition "A implies B" means "A's being at *this* moment implies B's being at a *subsequent* moment" (Italics added.)[5] – not realizing that the temporal delay of the event B with respect to A becomes unaccountable if causal relation is assimilated to tenseless, logical implication – in *Our Knowledge of the External World* he wrote that "nothing of any scientific importance depends upon its (i.e. effect's) being after the cause". And even more strongly: "it is a mere accident that we have no memory of (the) future (!)."[6] It is then understandable that in the same book he concluded that "to realize the unimportance of time is the gate to wisdom".[7] But it is far less understandable why in the same book Russell ridiculed Kant's view about the subjective ideality of time.[8] On the other hand when in his *Mysticism and Logic* he declared the law of causality to be a "relic of by-gone age",[9] he only realized the lurking Humean implications of his belief in the complete externality of events without, however, realizing the impossibility of accepting Hume's and Laplace's positions at the same time. Thus Russell's hesitancies and discrepancies provide us with beautiful illustrations of two complementary fallacies, pointed out by Bergson, concerning the nature of time.

If duration is neither a bare unity nor a sheer multiplicity of external units, is it perhaps describable as a 'synthesis of both unity and multiplicity'? But Bergson does not believe that by combining and manipulating two inadequate concepts we could obtain anything but a clumsy imitation of the concrete reality of duration. In language strongly reminiscent of Hegel and Heraclitus, Bergson says:

There is scarcely any concrete reality upon which one cannot take two opposing views at the same time and which is consequently not subsumed under two antagonistic concepts. Hence a thesis and an antithesis that it would be vain for us to try logically to reconcile, for the simple reason that never, with concepts or points of view, will you make a thing. But from the object, seized by intuition, one passes without difficulty in a good many cases to the two contrary concepts, and because thesis and antithesis are seen to emerge from the reality, one grasps at the same time how this thesis and antithesis are opposed and how they are reconciled.[10]

All we may say then, is that both multiplicity and unity are *complementary aspects* of duration without being their *constituting parts*. The distinction between 'component part' (*partie composante*) and 'partial expression' (*expression partielle*) is the central idea of Bergson's epistemological essay *L'Introduction à la metaphysique* and the warnings against confusing these two notions are present in all his writings. Consistently with this view he says:

Every moment of our life presents two aspects, it is actual and virtual, perception on the one side and memory on the other. Each moment of life is split up as when it is posited. Or rather, it consists in this very splitting, for the present moment always going forward, fleeting limit between the immediate past which is now no more and the immediate future which is not yet, would be a mere abstraction were it not the moving mirror which continuously reflects perception as a memory.[11]

Here we find a concrete application of the distinction between *fait accomplissant* and *fait accompli*. Durational process in virtue of its dynamic nature is a continuous act of *splitting up* into two subsequent moments, but it never is *split up*; and this is why the temporal process is never *composed* of two – or more generally many – units in their usual arithmetical sense. Nor is the continuity of this splitting the bare undifferentiated unity of the Cartesian substance or of Kant's Transcendental Ego. If the passage quoted above may appear unduly metaphorical, let us compare it with the following passage of the mathematician L. E. Brouwer who characterized his own 'neo-intuitionistic' philosophy of mathematics in the following way:

This neo-intuitionism considers the falling apart of the moments of life into qualitatively different parts, to be reunited only while remaining separated by time, as the fundamental phenomenon of the human intellect, passing by abstracting from its emotional content into the fundamental phenomenon of mathematical thinking, *the intuition of the bare two-oneness*. This intuition of two-oneness, the basal intuition of mathematics, creates not only the numbers of one and two, but also all finite ordinal numbers, inasmuch as one of the elements of two-oneness may be thought of as a new two-oneness, which process may be repeated indefinitely...[12]

This is exactly what Bergson asserts; the past and the present are neither one nor two, but they are dynamically separated as well as united in the intuition of bare two-oneness. More adequately, they are *being united* and *being separated*! A great number of fallacies concerning the nature of time would disappear if we were constantly aware of the exceedingly important differences between *the participle present* and *participle past*.

Needless to say that Brouwer's intuitionism and the intuitionism of Bergson are two very different things; the former is far more limited in its scope, being a *philosophy of mathematics* only, in contrast to Bergson's more comprehensive and more ambitious thought. Yet, the comparison of both quotations above clearly shows that Bergson's and Brouwer's intuition have more in common than a mere name. Not only did both thinkers stress the essentially temporal character of mental activity, but they even agreed on the way in which this activity transcends the concepts of unity and multiplicity in their usual arithmetical sense: on this point Brouwer's language is as Heraclitean as that of Bergson. We shall see later some other similarities between Bergson's philosophy of mathematics and that of Brouwer.

## NOTES

[1] Cf. Note 1 of the previous chapter.
[2] *C.M.*, pp. 219–220.
[3] *Ibid.*, p. 220.
[4] B. Russell, *The Principles of Mathematics*, The Norton Library, New York, 1964, p. 347. Originally published in 1903.
[5] Russell, *op. cit.*, p. 476.
[6] *Our Knowledge of the External World*, Unwin & Allen, London, 1914, pp. 226, 234.
[7] *Ibid.*, p. 167.
[8] *Ibid.*, pp. 116–117.
[9] *Mysticism and Logic*, Pelican Books, 1953, Ch. IX 'On the Notion of Cause', p. 171.
[10] *C.M.*, p. 208.
[11] *M.E.*, p. 135.
[12] L. E. W. Brouwer, 'Intuitionism and Formalism', in *Bulletin of the American Mathematical Society* XX (1913) 81–96.

# CHAPTER 12

# IMMORTALITY OF THE PAST:
# BERGSON AND WHITEHEAD

We have previously pointed out the extreme difficulty of separately dis-
cussing various aspects of the problem of time. The unitary and, in a sense,
indivisible insight into the nature of duration is far from being undifferentiated; it is a kind of *Gestalt* which, in Jankélevitch's happy words, is
"complex, though not *composed*"; like any other *Gestalt* it is *not* built out
of atomic, independent, mutually separable components. But as soon as
one aspect of it is treated in isolation – and this happens almost inevitably
when we begin to *talk* about it – such treatment tends to be misleading;
we unconsciously confer the discontinuity of the discourse onto the
complex indivisibility of the referent. We pointed out how the element of
novelty is inseparable from the "mnemic" character of duration, that is,
from the survival of the immediate past in the present; how the inseparability of these two aspects accounts for the inapplicability to duration of
either the concept of arithmetical unity or multiplicity; how for the
same reason duration is neither infinitely divisible nor discontinuous in
the sense that it is *composed* of atomic units. We are now approaching
the last couple of correlated aspects – the irreversibility of duration and
the immortality of the past.

If we accept the impossibility of conceiving duration as a multiplicity
of arithmetical units, then 'the growth of duration' is *not* comparable to
the additive process by which a certain sum increases by having discrete
units added to it. Bergson's metaphors suggesting anything of this sort –
like, for instance, his metaphor of a growing snowball – are not particularly happy. If terms from botany can be borrowed, we may say that
'the growth of duration' is by 'intususception' rather than by 'apposition',
that is, than by the addition *ab extra* of discrete layers. Growth by intususception suggests an *internal* relation between the present and the
past; the present is *incorporated* into the past rather than *externally* added
to it and conversely, the past is not absent from the present. A happier

metaphor of Bergson's is that of biological ripening ('mûrissement'); it suggests this incorporation, 'intususception' of the present into its antecedent context better than the images borrowed from inorganic nature.[1] But then a difficult and very important question inevitably arises: *is real time divisible at all, and if it is, in what sense?*

The answer that real time, i.e. duration, while being qualitatively differentiated is, properly speaking, *not divisible at all*, at least not in the sense that its successive phases are in mutually external relations, sounds very paradoxical and even contrary to experience. It is certainly strange to claim that even the remotest past is *never entirely external* to the present. Yet, complete externality of the present with respect to the past or to some portion of it would imply a *complete independence* of successive phases of duration. Nothing would then happen to the character of the present if that particular 'portion' of the past were entirely eliminated; the quality of the present moment – in virtue of its *complete* independence from the hypothetically removed past – would remain entirely unaffected. But a complete destruction of the past is an intrinsic impossibility. What happened, happened irrevocably and cannot be undone, cannot be made as *not* having happened; although the effects of any past events can be checked or neutralized by the effects of subsequent events, *the event itself* cannot be erased from the past. *Praeterita autem non fuisse contradictionem implicat*, wrote St. Thomas, echoing Aristotle in this respect.[2] It is precisely this that constitutes *the immortality of the past.*

Curiously enough, this term was coined by Bertrand Russell[3] who failed to draw consequences from it; otherwise he would not have claimed that the hypothesis about the universe being created five minutes ago is intrinsically irrefutable.[4] Such a claim can be made obviously *only* if absolute externality of the past with respect to the present or – what is the same – its destructibility – is assumed. Alfred North Whitehead apparently took over the term 'immortality of the past' from Russell; but for him, unlike Russell, this concept become one of the corner-stones of his organic view of nature. A negative aspect of this idea is his criticism of 'the fallacy of simple location in time', the criticism which excluded the external relation of the past to the present. In a well written passage of *The Concept of Nature* Whitehead insisted:

The passage of nature which is only another name for the creative force of existence has no narrow ledge of definite, instantaneous present within which to operate. Its operative presence which is now urging nature forward must be sought for throughout the whole, *in the remotest past as well as in the narrowest breadth of any present duration.* (Italics added.)[5]

Bergson's own concept of the immortality of the past was not as general as that of Aristotle, St. Thomas or Whitehead. This was due to the fact that in his first two books he was concerned exclusively with the problem of the *psychological* past and its relation to the present. Yet even in his first book he began to use his concept of psychological time as a model for understanding time in general; his insistence on the immortality of the *mental* past even then implied the affirmation of the *total* survival of the past in general. But it is true that he was never as explicit on this point as Whitehead, C. D. Broad or Hilda Oakley.[6] A comparison with the views of other thinkers, in particular of James, will provide us with the best clues for understanding Bergson's own.

## NOTES

[1] *T.F.W.*, p. 176.

[2] Aquinas, *Summa Theologica*, Q 25, art. 4; Aristotle, *Ethica Nicomachea*, Book VI, Ch. 2.

[3] *Mysticism and Logic*, the essay 'A Free Man's Worship', p. 59.

[4] B. Russell, *Human Knowledge. Its Scope and Limits*, Simon and Schuster, New York, 1962, p. 212.

[5] A. N. Whitehead, *The Concept of Nature*, Cambridge University Press, 1930, p. 73.

[6] C. D. Broad, *Scientific Thought*, Routledge and Kegan, New York, 1923, p. 66; H. Oakley, 'The Status of the Past', *Proc. Aristot. Soc.* XXII (1932) 227–250. Cf. also my article 'The Elusive Nature of the Past', in *Experience, Existence and the Good. Essays in Honor of Paul Weiss* (ed. by Irwin Lieb), Southern Illinois University Press, 1961, pp. 126–142.

# CHAPTER 13

# JAMES'S AND BERGSON'S VIEWS
# OF THE PAST COMPARED

For James, as well as for Bergson, the present moment is not an infinitely thin instant, but a temporally thick pulsation tinged with 'immediate recency'. According to James, "the distinctly intuited present merges into a penumbra of mere dim *recency* before it turns into the past ... which is simply reproduced and conceived".[1] In other words, there are two kinds of past for James: one *directly intuited* inside of the 'sensible present'; the second outside of the volume of the sensible present, which can be only indirectly *reproduced*, not immediately grasped. From this point of view recollection is not a past state at all, but merely a *present* state symbolizing or representing an event forever gone.[2] In this sense Taine called memory "*l'illusion vraie*", because, though it deceives us in presenting a present mental state as an equivalent of the past one, it nevertheless gives us an indirect knowledge, often quite accurate, of the past.[3] This distinction between *immediate* memory (merging with the "specious" present) and *indirect* recall was generally accepted by psychologists.

It is true that James was aware that the volume of the psychological present *varies*, that it is different for different categories of sensation; the above quotation shows that he was aware that the 'backward edge' of the psychological present is dim and that no sharp cut separates immediate memory from the absolute past. He, however, believed that beyond this dim rim of 'immediate memory' which, according to him, can span no more than twelve seconds of 'public time'[4], the past hopelessly escapes the direct reach of consciousness. While an impression, or, to speak in a less atomistic way, a certain mental quality which occurred eleven seconds ago, still dimly persists on the hazy pastward edge of our mental present, all that is beyond that rim is literally *dead* and perished forever. This is what James called a 'genuine past'; although it can be symbolically represented by a present state which is *similar* to it, it can never be *re-presented*, that is, 'presentified' – *vergegenwärtigt* as Germans say – in a literal sense. No matter how close the similarity of the present recollection to the past state is, they are both always absolutely *external* to each other.

It may be pointed out, as I have more than once[5], that under such conditions there would be no possibility whatever of differentiating between false and true recollection and, more generally, between myth and history. For if the faithfulness of recollection is measured by the degree of similarity between it and the original past state, and if, *ex hypothesi* this past state is absolutely dead, i.e. *unreal*, how can we say that it resembles the present state which supposedly represents it? How can we compare the existing present state with something which is no longer in existence? How can the relation of similarity hold between two terms, one of which does not exist at all? *Either*, then, we do not take the unreality of the past in such an absolute sense as we pretend to *or* there is no difference between history and myth; then our memory would be, to use Taine's language, not a 'veridical illusion', but illusion pure and simple.

Bergson did not use this argument, even though it is implicitly present in his claim that *forgetting rather than remembering* needs an explanation. His main explicit argument is that the difference between immediate memory and indirect reproduction is merely *that of degree* rather than that of nature. Once we concede that there is such a thing as a direct perception of the immediate past, – and this concession is inherent in any consistent denial of the instantaneous, durationless present, – then the main objection against the possibility of the perception of the past, no matter how distant, is removed. The only other alternative which can be consistently upheld is to deny radically *any* perception of the past, no matter how close or "immediate" it may be; in other words, to claim that what we are aware of is nothing but an instantaneous, point-like present, devoid of any duration. While even Bertrand Russell regarded such a view as absurd, Alexius Meinong and C. A. Strong went that far. According to the latter, even what we call 'immediate memory' is nothing but an illusion:

To take this illusion seriously is to be guilty of a sort of naive realism in the field of time. The impossibility of such a direct consciousness of past time appears, further, from the consequences to which it should lead if true. If we can be directly conscious of a feeling that occurred half a second ago in spite of the fact that the feeling is now past and gone, why not also of a feeling that occurred a whole second ago, or an hour or a day or a week? The consciousness would be in no wise miraculous. *Why cannot*

*we be directly conscious of any past experience, no matter how remote*? If such consciousness is not to be thought of, then for the same reason the direct consciousness of a half
a second ago is not to be thought of... Our consciousness of even the nearest past must
be ideal, not actual; representative, not intuitive. (Italics added.)[6]

Bergson formulated the problem in the same way; in truth the italicized
question is the same that he raised. But his answer was diametrically
opposed to that of Strong. He agreed with Strong that the difference
between immediate memory and indirect recall is merely that of degree;
but instead of denying immediate memory as Strong did, he not only
accepted it as the datum of experience, but regarded it *as a clue to understanding the nature of memory in general*. For Bergson, consciousness is
by its own nature memory; we already quoted the passage of his in which
he analyzed what we experience in pronouncing the word 'conversation'
and in which he pointed out that no definite demarcation line can be
drawn between the present and the past. On this point, Bergson's severe
critic, Arthur O. Lovejoy, unwittingly agreed with him when he wrote
that Descartes "could by simply keeping his attention fixed upon his
consciousness, have, so to say, seen *cogito* transforming itself into *memini*
before his eyes",[7] To the question italicized in C. A. Strong's passage
Bergson's answer was unhesitatingly affirmative:

I should no more seek the explanation of the integral preservation of this entire past
then I seek the explanation of the preservation of the three first syllables of "conversation" when I pronounce the last syllable. Well, I believe that our whole psychological
existence is something like this simple sentence, continued since the first awakening of
consciousness, interspersed with commas, but never broken by full stops. And consequently, I believe that our whole past still exists.[8]

The difference, not only from the point of view of Strong and Meinong,
but from that of James as well, is quite obvious; it is due to Bergson's
more radical attempt to purify our notion of time of all visual and
geometrical imagery. James, who rejected the artificial concept of a
point- like instant unlike Kant, Meinong and Strong, did not succeed
in getting rid of all subconscious visual representations. His view of
the present moment retained definite geometrical characteristics; it is
something comparable to *a linear segment*. He did not go as far as Lovejoy,
who replaced the succession of instants by the succession of contiguous,
temporal, sharply separated intervals; nevertheless, in introducing
absolute externality between more distant moments of time he also yielded
to an obtrusive, geometrical analogy. His "perishing pulses of thought"

are in this respect analogous to Lovejoy's temporal segments; the only difference is that James admitted not only the "contact", but the overlapping of the immediately successive moments. But only this kind of "next-to-next" continuity is recognized by him; the moments separated by more than an interval of twelve seconds are absolutely external to each other. More accurately, we must say that between such moments there is no relation whatever, not even that of mutual externality, *since the previous moment completely vanished.* Even when James stresses the impossibility of estimating *when* the past becomes absolutely 'dead', he does not give up his basic assumption that such a death really occurs.[9]

From Descartes to Meinong the present was regarded as a mathematical point; Lovejoy extended this point up to a definite temporal length; James insisted that this length has no constant and sharp edges; but all these three views agreed in considering the present moment as *the only reality* hovering between two abysses of non-being or, in Lotze's words, "wavering between a darkness of the past which is done with and no longer at all, and a darkness of the future, which is also nothing".[10] Time thus conceived is a "perpetual perishing" in which a spark of the present moment is continually extinguished in order to be replaced by another spark, equally ephemeral. Such was the instantaneous world of the Arabian atomists, of Descartes and of Leibniz, the world perpetually perishing and perpetually recreated by God, its evanescent events lost forever in the abyss of non-being. The world of Heraclitus was no different.

The main difference between such a view of time and the Bergsonian view consists in a different view of *the status of the past.* The difference in structure between two types of time can be illustrated graphically. It is true that any attempt to illustrate the temporal relations graphically involves a danger of inconsistency; are not all visual symbols of becoming basically inadequate? Yes, they are; however, if we keep in mind that different parts of a diagram are *not simultaneous, but successive stages* of temporal process, the danger of misinterpretation is considerably, though not completely, reduced. The following two pictures show approximately the differences in structure between the 'stream of thought' and '*durée réelle*'.

$p_1$, $p_2$, $p_3$... represent successive specious presents; $m_1$, $m_2$, $m_3$... represent the links of immediate memory (not recognized by Lovejoy, but admitted by James under the name 'feelings of transition' or 'feelings of relation'). It is evident that in the 'stream of thought' or in the Heraclitean flux in general the successive moments $p_1$ and $p_3$ are *completely external* each to the other, being linked only by the intermediate terms. Bergson's duration shows a much more complex structure. Besides the links of 'immediate memory' jointing together two immediately successive terms, there are a number of relations, or using Whitehead's term, 'prehensions', binding together temporal terms which are not 'contiguous'. These links are more and more tenuous as the difference in date grows larger; we might say that the intensity and the span are 'indirectly proportional'. Essentially, however, there is only a difference of degree and not that of kind between immediate memory and more tenuous mnemic links joining temporally noncontiguous moments. In Bergson's duration the whole past persists, though its different parts persist in different degrees; for James only the 'immediate past', perceived on the 'rearward edge' of the specious present, is real.

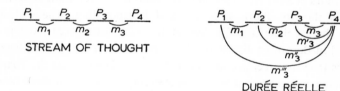

STREAM OF THOUGHT

DURÉE RÉELLE

This is the main difference between the 'next-to-next' continuity of the stream of thought and the total continuity of *durée réelle*. While in James's view the distant phases of the past are completely separated from the present moment, – Bergson would say 'separated by full stops', – the Bergsonian past is *indivisibly* immanent in the present 'occasion'. Within the dynamic totality of the past there are no gaps, no 'full stops'. But such indivisibility does not mean *undifferentiatedness*. On the contrary, the specific characteristics of the past moments find their expression in an enormous number of temporal links of unequal span and unequal intensity, superposing each other and ending, provisionally, in the actual moment. For this reason it is more appropriate to speak of "many-oneness" rather than the "two-oneness" of each moment of duration; interestingly enough, it is also the term which L. E. Brouwer eventually adopted.[11] It is highly misleading to speak of the relation of the present to the past as a dyadic, two-term relation; this is due to our habit of speaking of the past in a *grammatical singular* as well as to the related fact that we do not have a comparative of the adjective "past" – "paster" – to express the different degrees of pastness. When we try to do it, we resort to spatial metaphors and we speak of different events being *more or less remote* from the present. This is what our diagram even suggests and this is why it *may* be misleading. We must not forget that the diagram is still a clumsy representation of the extremely complex network of dynamical relations, a network which, unlike the diagram itself, cannot be decomposed into distinct and juxtaposed parts. Yet, misguided by the diagram and interpreting it literally and using the present tense ('past *persists*'), we tend to juxtapose the past *alongside* the present and in this way we eliminate its pastness.

Bergson never tired of warning against such confusion. It is therefore amusing to see Russell accusing him of committing it. According to Russell, Bergson made an 'elementary blunder' in confusing the past event with its present recollection.[12] It would have been difficult to distort Bergson's thought more seriously and irresponsibly. The distinction between the recollection of a past event and a past event itself is fully preserved in Bergson's distinction between '*souvenir-image*' and '*souvenir pur*'. It is true that recollection itself ('*souvenir-image*') has the character of an image and as such it is a *present* event; but this is a *mere terminal phase of the process of remembering* by which the past event ('*souvenir pur*') prolongs itself into the present moment and this 'prolongation of the past into the present' is the very essence of duration. Without its own root in the subsisting past the recollection would be indistinguishable from a faint present image; and its character of pastness, or using the language of phenomenology, its intentional reference to a definite past occurrence, would be completely unintelligible. This is why Bergson so persistently attacks the Humean and, more generally, sensualistic identification of recollection with a mere *faint present image*. The past, while ceasing to be present, did not cease to exist, or rather *subsist*; in Dewey's words, it is *present in its absence*.[13] This is rather an odd way of putting it and merely shows how exceedingly difficult it is to convey the meaning of the virtual subsistence of the past by means of our object-oriented, macroscopically conditioned language. It is beyond the scope of this chapter to discuss even cursorily the main ideas of *Matière et mémoire*, in particular the complex relations between the virtually subsisting *souvenirs purs* and the actually existing images or sensations; also how out of the cooperation – and antagonism – of these two factors *souvenir image*, that is, a concrete recollection itself, arises. Suffice it to recall what was said at the beginning of this chapter, that to equate the past with a sheer non-being is self-contradictory and that without affirming the existence of the past in some way, as Aristotle, St. Thomas, C. D. Broad, H. Oakley, Bergson, Whitehead and Hartshorne did, no memory, no history is possible; nothing else would be left than Santayana's "solipsism of the present moment" in which fact and fancy would exist, for a moment, before joining the yawning chasm of the non-being in which the past not only "sinks", but also completely vanishes. As Russell wrote relatively recently:

Leibniz did not carry his monadism far enough, since he applied it only spatially. Not only is a man private from other people, but he is also private from his past and future selves. It is not "here" alone that is private, but also "now."[14]

The difference between James's and Bergson's view of the past came up very distinctly in their early correspondence at the beginning of this century. It concerned the question of the subconscious, which is only a different aspect of the problem of the past. James, denying any ontological status to the distant past except the physiological one, admitted the psychological existence of only *one present state*, the present being understood as having a certain temporal breadth. Bergson expressed his disagreement in his letter of March 25, 1903:

If we reduce the memories to the categories of things, it is clear that there is no mean for them between presence and absence; either they are unqualifiedly present to our mind, and in this sense conscious; or, if they are unconscious, they are absent from our mind and should no longer be considered as present psychological realities. But in the world of psychological realities I do not believe that there is occasion for presenting the alternative "to be or not to be" exclusively. The more I try to grasp myself by consciousness, the more I perceive myself as the totalization or *Inbegriff* of my own past, this past being contracted with a view to action. "The unity of Self" of which philosophers speak, appears to me as a unity of an apex of a summit to which I narrow myself by an effort of attention – an effort which is prolonged during the whole of life, and which, as it seems to me, is the very essence of life.[15]

In other words, like James, Bergson rejected the static, substantial Ego of traditional idealism; but, unlike James, he did *not* reduce the self to a mere 'vanishing pulse of thought' which, according to James, is the only 'Thinker'.[16] This difference is again due to the Heraclitean character of the Jamesian 'stream of thought' where the perpetual perishing is over-stressed and the mnemic persistence overlooked. John Dewey's observation of a 'vanishing subject' in James's psychology could have hardly been more accurate.[17]

But Dewey did not take into account the last stage of James's development in which his view of the past, under the admitted influence of Bergson, was substantially modified. This change showed up distinctly in his letter of June 13, 1909 commenting on *Creative Evolution*:

The position we are rescuing is "Tychism" and a really growing world. But whereas I have hitherto found no better way of defending Tychism than by affirming the spontaneous addition of *discrete* elements of being (or their subtraction), thereby playing the game with intellectualistic weapons, you set things straight at a single stroke by your fundamental conception of the continuously creative nature of reality.[18]

And in a more detailed way in a letter of June 27, 1909:

I think that the center of my whole *Anschauung* since years ago I read Renouvier, has been that something is doing in the universe and that *novelty* is real. But so long as I was held by the intellectualist logic of identity, the only form I could give to novelty was tychistic, i.e. I thought that the world in which discrete elements are annihilated and others created in their place, was the best descriptive account we could give of things... This sticks in the human crop – none of my students became good tychists! Nor am I any longer, since Bergson's synechism has shown me another way of saving novelty and keeping all the concrete facts of law-in-change.[19]

Obviously, Renouvier's idea of discrete creation and discrete annihilation, adopted also by Lovejoy, accounts to a large extent for James' "perishing pulses of thought." There is no real perishing, no real annihilation in the philosophy of *durée réelle*, which James finally adopted under the name of "synechism." Nor is there any abrupt, *ex nihilo* creation. Or, in the language of *A Pluralist Universe*:

(Their) changes are not complete annihilations followed by complete creations of something absolutely novel. There is a partial decay and partial growth, and all the while a nucleus of relative constancy from which what decays drops off, and which takes into itself whatever is grafted on until at length something wholly different has taken its place.[20]

In other words, as pointed out above, the perishing of the past and the emergence of the present are two aspects of a single dynamic fact of which the concepts of complete annihilation and *ex nihilo* creation are two artificially disjoined and distorting snapshots. The quoted passage from James clearly suggests that if we ever decide to use the term 'growth'

metaphorically for describing the reality of time, it is 'growth by intususception' rather than by 'apposition', by an addition *ab extra* of discrete
units.

## NOTES

[1] W. James, *The Principles of Psychology*, I, p. 636 n.
[2] *Ibid.*, p. 627: "The feeling of past time is a present feeling."
[3] H. Taine, *De l'Intelligence*, 16th Ed., Paris, 1927, II, p. 49.
[4] W. James, *op. cit.*, I, pp. 612–613.
[5] Cf. M. Čapek, 'The Elusive Nature of the Past' (quoted in Chapter 12, Note 6), p. 136; 'Memini ergo fui', in *Memorias del XIII Congreso Internacional de Filosofia*, Mexico 1964, Vol. V, pp. 415–426, esp. pp. 419–420.
[6] C. A. Strong, 'Consciousness and Time', *The Psychological Review* III (1896) 156. Cf. also A. Meinong, 'Das zeitliche Extensionsprinzip und die sukcessive Analyse', *Z. f. Psychologie und Physiologie der Sinnesorgane* VI (1890). In the same journal (XIII (1897) 326–349) L. William Stern in the article 'Die psychische Präsenzzeit' effectively criticized Strong's defense of the point-like present.
[7] A. O. Lovejoy, *The Revolt against Dualism*, Open Court, Lasalle, Ill., 1967, p. 381.
[8] *M.E.*, p. 70.
[9] Cf. my article 'Stream of Consciousness and "durée réelle"', in *Philosophy and Phenomenological Research* X (1950) esp. 338–346. The main ideas of this article were incorporated into this chapter.
[10] H. Lotze, *Metaphysics* (transl. by B. Bosanquet), Oxford 1884, II, 268.
[11] E. W. Beth and J. Piaget, *Épistémologie mathématique et psychologie*, Presses Universitaires de France, Paris, 1961, p. 117.
[12] Bertrand Russell, 'The Philosophy of Bergson', *The Monist* XXII (1912) 342–343.
[13] John Dewey, 'Realism Without Monism or Dualism', *The Journal of Philosophy* XIX (1922) 354.
[14] B. Russell, *Human Knowledge, Its Scope and Limits*, p. 90.
[15] Ralph B. Perry, *Thought and Character of William James*, Harvard University Press, Cambridge University Press, 1948, II, 611.
[16] W. James, *op. cit.*, I, p. 342.
[17] John Dewey, 'The Vanishing Subject in the Psychology of James', *The Journal of Philosophy* XXXVII (1940).
[18] Perry, *op. cit.*, II, pp. 618–620. I dealt with the last phase of James's thought which was so distinctly influenced by Bergson in the article 'The Reappearance of the Self in the Last Philosophy of William James', *The Philosophical Review* LXII (1953) 526–544.
[19] Perry, *op. cit.*, II, p. 656.
[20] *A Pluralistic Universe*, p. 258.

CHAPTER 14

# THE IRREVERSIBILITY OF DURATION:
# THE COMMENTS OF ROYCE AND INGARDEN

The correlation between the immortality of the past and the irreversibility of duration was pointed out by Bergson in one of the initial paragraphs of *Creative Evolution*:

From this survival of the past it follows that consciousness cannot go through the same state twice. The circumstances may still be the same, but they will act no longer on the same person, since they find him at a new moment of his history. Our personality, which is being built up each instant with its accumulated experience, changes without ceasing. By changing, it prevents any state, although superficially identical with another, from ever repeating it in its very depth. That is why our duration is irreversible. We could not live over again a single moment, for we should have to begin by effacing the memory of all that had followed. Even could we erase this memory from our intellect, we could not from our will.[1]

In other words, it is the total preservation of the past which implies the irreversibility of duration. The fact of forgetting is not regarded by Bergson as incompatible with the immortality of the past, since, according to him, every forgetting is, so to speak, superficial and does not affect the totality of our mental past which persists subconsciously, even when it is not explicitly remembered:

But, even though we may have no distinct idea of it, we feel vaguely that our past remains present to us. What are we, in fact, what is our character, if not the condensation of the history that we have lived from our birth, – nay, even before our birth, since we bring with us prenatal dispositions? Doubtless we think only with a small part of our past, but it is with our entire past, including the original bent of our soul, that we desire, will and act.[2]

Moreover, Bergson pointed out, the phenomena of hypermnesia, which may be either spontaneous or artificially induced in hypnosis, show that the inability to recollect distinctly certain phases of our past may be only temporary. In the language of *Matter and Memory* this happens when 'the prolongation of the past into the present' does not arrive at its terminal phase in which the explicit memory-image appears; and this happens when the present state of the brain prevents the consummation of this final phase in the process of remembering. The exposition of the central ideas of *Matter and Memory* is beyond the scope of this book, but let us say at least this: Bergson, while being critical of the 'double-aspect' theory, according to which there is an exact parallelism between the physical and mental events, does *not*, however, deny that every sensory image has its physiological counterpart. He does *not* deny the reality and the importance of 'engrams' or physiological traces. It is '*souvenir pur*', but not '*souvenir-image*' which exists or rather *subsists* independently of the brain processes; but no distinct and explicit recollection is possible if '*souvenir pur*' cannot 'actualize' itself in '*souvenir-image*'. Penfield's recent experiments by which the reactivation of some apparently vanished engrams can be induced by physical means would not embarrass the author of *Matter and Memory* at all; he would rightly point out that this substantiates his claim that the process of recall is consummated only when the re-emerging '*souvenir pur*' can prolong itself into the corresponding '*souvenir-image*' correlated with certain present processes in the brain. The difference between hypnotically and electrically induced hypermnesia is merely that of degree, not of kind. Unfortunately, space is lacking here to extend this rather unsatisfactorily concise digression.

It is important to realize that the immortality of the past *cannot* be established empirically beyond doubt. No matter how impressive the phenomena of hypermnesia are, they are clearly insufficient to justify any absolute certainty in this respect. Like all limited empirical evidence, they cannot serve as the basis of unchallengeable generalization; while making the immortality of the mental past probable, they cannot make it certain. Even far less certain, then, is the claim for the total preservation of *any* kind of past – *unless we base this claim on grounds other than empirical.* This is what first Aristotle and then St. Thomas did; they claimed, and rightly so, that the notion of the destructibility of the past is *inherently contradictory*. If Bergson would have adopted such a logical approach, his claim about the immortality of the *mental* past would have been far more convincing, since it would be implied in the immortality of the past in *general*. But this he has not done, though it was implied by his own view of duration.

This is what was pointed out by Josiah Royce in his article 'The Reality of the Temporal' in 1910. He pointed out that the uniqueness and novelty which Bergson claims for each particular event *cannot be directly experienced*, but presupposed: "For how shall I immediately feel or see or otherwise sense the truth, if it be indeed a truth, that this fact of sense was never a fact for me, or for anybody else before?" Royce did not deny the novelty and uniqueness of events; he only claimed that such features cannot be verified empirically, but must be what he called 'supersensuous truth'.[3] Royce's argument applies to the immortality of the past as well, since such immortality cannot exist without novelty and vice versa. In the passage of *Creative Evolution* quoted above the correlation between the irreversibility of time and the integral preservation of the past was established; it is the latter which guarantees the former. In other words, it is the total survival of the past which prevents a recurrence of an identical state, that is, its complete repetition. Since an identical recurrence of a certain event means *the absence of novelty*, we may say that the total survival of the past *excludes the absence of novelty*. Can we also say that conversely the presence of novelty requires the immortality of the past? This is far less obvious, but is nevertheless true. For, as we pointed out above, the novelty of the present is possible only on the contrasting background of the persisting, or rather subsisting, past; if any portion of the past could be destroyed, a complete recurrence of a certain event would be possible. But an event completely identical with some previous event would lack the quality of novelty, since novelty in the radical Bergsonian sense means "to be different from the *whole* antecedent context of its past."

Could it then be maintained that two *completely* identical events are still *numerically* different, as when some Stoics claimed that another Socrates appearing in the next identical cycle of the cosmic recurring history would be *numerically* different from the Socrates of our cycle?[4] From the Bergsonian point of view it is impossible that two successive events, no matter how 'distant' from each other they are, were merely *numerically* different; every *genuine* temporal difference implies difference of quality. This is why from this point of view the successive instants of mathematical, homogeneous time are only *nominally* successive; from the same point of view Nietzsche was only consistent when he insisted that an *altogether complete* recurrence would abolish not only all qualitative differences between the original and the repeated event, *but even their numerical difference*. A radical, integral recurrence of the past would necessarily eliminate the whole interval separating the original event from its alleged identical replica. Identity in time – which should be carefully distinguished from the mnemic, heterogeneous continuity of duration – can belong *to one single moment only*, never to two genuinely successive moments. Immortality of the past, the irreversibility of duration and the novelty (uniqueness) of each of its phases are three aspects of one and the same complex truth.

Now this truth is, according to Royce, 'supersensuous', and as such cannot be perceived or intuited as Bergson claimed. Roman Ingarden, whose relatively little known study is one of the most detailed, most attentive and subtlest analysis of Bergson's thought [5], would agree with the first part of Royce's statement, not with the second. Royce's argument was as follows: the true essence of duration cannot be grasped within any particular perceptual datum, since such datum, be it a sensation or sensory image, cannot exhibit in virtue of its particularity the universal character of duration, including its irreversibility and novelty. "For how shall I immediately feel or see or otherwise sense the truth, if it be indeed a truth, that *this* fact of sense never was a *fact* for me, or for anybody else?" In other words, novelty and uniqueness of each moment must be "presupposed", never intuited. But Royce's reasoning clearly assumed that the Bergsonian intuition is a 'bare particular', an atomic datum externally related to its temporal context and which cannot reveal anything beyond its own isolated content. But is it fair to impute to Bergson any view *even remotely similar* to psychological atomism? Did anybody else, except William James and perhaps James Ward, more vigorously attack the associationistic psychology, even before the anti-atomistic Gestalt psychology came into being? Is not Bergson's untiring stressing of the continuity of duration thoroughly incompatible with what James appropriately called the 'mind-dust theory'? Royce clearly regarded experience – any kind of experience – in an atomistic fashion; under the influence of Kant and classical rationalism in general he regarded experience as a sheer *manifold*, that is, as the diversity of externally related items; according to him, as with Kant, any unifying aspect of experience must be supplied *ab extra*, by the synthesizing activity of reason. In assuming this Royce had an easy task to show that no experienced moment of time can disclose anything beyond its own confined content; consequently, its novelty cannot be perceived, it must be 'presupposed', and its truth must be of a 'supersensuous' kind. Apparently Royce also equated, at least in this context, experience in general with sensuous experience only; consequently, 'supersensuous' would be excluded from experience. Strange that he was not aware that between abstract thought and sensory experience there is a continuous spectrum of intermediate stages and that 'imageless thought', so beautifully described by his contemporary, William James, is also an experience of its own kind. Another effect of the traditional dichotomy between 'sensuality' and 'reason'.

Ingarden's approach to the same problem, though motivated by a similar concern, was far more subtle. As mentioned above, he would accept Royce's claim about 'the supersensuous truth of novelty' (and of duration in general), but not about the impossibility of our perceiving it. It is, according to him, *perceived*, though not in the form of sensory perception, but rather intuited in the sense of the phenomenological *Wesensschau*. Ingarden showed convincingly that this is what the Bergsonian intuition really means, even though some passages of Bergson, taken literally, seem to exclude it. Ingarden is rightly critical of those passages and shows that they must not be taken literally if Bergson's epistemology as well as his metaphysics are to remain consistent. In his intuition Bergson certainly claimed to attain something more than his own private and momentary datum, something which, as he believed, can in principle be intuited by all persons who read and honestly try to understand his own introspective analysis. But what else is this 'something' than the universal essence of duration, common, like the Heraclitean *Logos*, to all human percipients? (It would be in line with his own philosophy if Bergson would have claimed with Heraclitus that this dynamic *Logos*, though present in all human minds, is not generally recognized.) "Every of his (i.e. Bergson's) assertions about consciousness", says Ingarden, "is intended, according to his whole basic metaphysical attitude, as an assertion about *a real fact* and claims to have absolute validity (*absolute Geltung*)".[6] Even if we insist on the absolute fluidity of our stream of thought and if we require that our intuition must dynamically coincide or, as Ingarden picturesquely says, 'flow together' (*mitschwimmen*) with its flux, the flux itself must have a definite essence, a definite 'whatness', a definite *structure* or *form* which must be distinguished from its contingent, individual content.[7]

The emphasis on this point is generally lacking in Bergson, or at least

in his explicit pronouncements; his concern about novelty and the uniqueness of *each* moment led him to overlook the meaning of the all-
important word '*each*' which clearly suggests that the intuition of
novelty has a *universal* character and thus cannot be confined to a
particular, passing moment of private life. Since novelty is inseparable
from duration, the same thing is true of the intuition of duration in
general; it must have the same form, the same *whatness*, the same *essence*
not only in various successive moments of Bergson's life, but in different
minds as well, at least in those who understand him. Otherwise the
Bergsonian term '*sympathie intellectuelle*' would be devoid of meaning;
duration, though always different and unique in its concrete content, must
be *structurally* the same in different moments of one person as well as in
different minds. In this sense the intuition of it is an insight into its
*universal essence*.

## NOTES

[1] *E.C.*, p. 8.

[2] *Ibid.*, pp. 7–8.

[3] J. Royce, 'The Reality of the Temporal', *International Journal of Ethics* **XX** (1910)
263–264. Concerning the relations between Royce and Bergson cf. my article 'Time
and Eternity in Royce and Bergson', *Revue Internationale de Philosophie* No. 79–80
(1967), 22–45.

[4] Cf. M. Čapek, 'Eternal Return', in *The Encyclopedia of Philosophy*, Macmillan, 1967,
III, pp. 61–63 with the relevant bibliography.

[5] Roman Ingarden, 'Intuition und Intellekt bei Henri Bergson', in *Jahrbuch für
Philosophie und phänomenologische Forschung* **V** (1921), esp. 398–461.

[6] Ingarden, *op. cit.*, p. 418.

[7] *Ibid.*, p. 416.

# CHAPTER 15

# DURATION AS CONCRETE UNIVERSAL.
## BERGSON AND CROCE

This, of course, does not mean that Bergson is a Platonist; his duration is a *concrete universal* and as such akin to the Aristotelian or Hegelian view. As hinted above, Ingarden is fair enough to recognize that Bergson's polemic was intended only against those 'forms' and 'essences' which were artificially separated from their concrete, dynamic content and whose fictitiously static character is due precisely to their being artificially 'lifted' out of the stream of experience. Ingarden thus entirely agrees with Bergson's criticism of the view which regards the 'essence' of a certain thing as a mere average of its successive, static snapshots; the real 'essence' of 'form' of duration is *inherent in* and *temporally co-extensive with* its dynamic content. He correctly points out that Bergson's criticism of the idea of absolute disorder, the criticism which implies that there is no reality completely devoid of order or structure, requires that even the dynamic processes, that is, duration itself, have a certain 'order', a certain 'structure', a certain recognizable, universal *whatness*, the correct grasping of which is one of the main goals of Bergson's epistemology.[1] This implication of his own thought was overlooked by Bergson when in the fourth chapter of *Creative Evolution* he claimed that the idea of 'becoming in general' is always necessarily inadequate.

What is not determinable is not representable; of 'becoming in general' I have only a verbal knowledge. As the letter $x$ designates a certain unknown quantity, whatever it may be, so my 'becoming in general', always the same, symbolizes here a certain transition of which I have taken some snapshots; of the transition itself it teaches me nothing.[2]

It is clear that the target of Bergson's criticism is precisely *becoming as an abstract universal* which, in virtue of its artificial separation from any concrete experienced flux, is inevitably converted into a static symbol belying its dynamic origin. On the other hand, when in the passage quoted above (Ch. 7) he writes that the perception of melody will remain an inadequate image of duration as long as we first do not efface "the difference among the sounds, then do away with the distinctive features of sound itself, retaining of it only the continuation of what precedes into what follows and the uninterrupted transition, multiplicity without divisibility and succession without separation" in order finally to rediscover basic time, he clearly is trying to grasp duration in the sense of *concrete* universal. For what else is '*le temps fondamental*' which, while inherent in any concretely experienced content, should not be confused with any particular contingent one? Does not this passage justify the distinction (though certainly *not separation*!) between a particular, contingently experienced succession – in this case the succession of sounds – and the dynamic, imageless duration which *transcends* any of its concrete exemplifications precisely in virtue of its universality? Is not this imageless experience of duration what Ingarden calls for – the phenomenologically intuited *essence* (Wesen) of temporality? Using Hegelian terminology, we may then say that the Bergsonian duration "is not merely a *common feature* which has its self-subsistence over what is specific, but it is much rather something self-specifying and self-particularizing, *which remains unperturbedly by itself in its other*".[3] (The last italics mine.) It is obvious that such self-identity of the whatness of duration certainly does not imply any *negation* of duration itself as Bergson feared; on the contrary, only if we insist on the *universal character* of duration, that is, on its *omnipresence* and on its *all-pervasive character*, does Bergson's temporalism acquire the meaning which its author intended to give it.

But we have seen the reaction of both Lovejoy and Ushenko to that passage; it shows that Bergson's fear that any talk about *duration in general* or *becoming in general* would be almost fatally misunderstood was only too well founded. The fact that both these critics interpreted his notion of "succession without separation" – which is just another name for the indivisible continuity of duration, – with the successionless Eleatic continuity is very instructive; they simply mistook the *concrete dynamic universal* for the *abstract universal* which is merely a *static* symbol of the former. It was this static symbol under the name 'homogeneous time' that was the primary object of Bergson's persistent criticism; and when Ushenko and Lovejoy tried to be in this respect more Bergsonian than Bergson they were merely breaking open an already open door. What J. N. Findlay says in his comment on the Hegelian concrete universal applies to the Bergsonian duration; it is "not merely common to the Species and Individuals it informs; it is *differently realized* in each of them".[4] It is because of its immanence in all particular fluxes that the universal duration does not lose its dynamic, successive character as Lovejoy and Ushenko and frequently even Bergson feared.

The affinity of Bergson's and Hegel's thought was noticed not only by James, whose words were already quoted, but also by Benedetto Croce, a critical disciple of Hegel who concluded his book *Cio che è vivo è cio che è morto della filosofia di Hegel* with a brief reference to this kinship.[5] This does not mean that there was any influence of Hegel on Bergson; all available evidence points to the contrary. Bergson and Croce met in person only once – at the International Philosophical Congress at Bologna in 1911. It was then that Croce called Bergson's attention to certain points common to Bergson and Hegel. Bergson was apparently completely unaware of it, objecting that Hegel was an 'intellectualist'; and when Croce suggested that he read *Phenomenologie des Geistes* so he could see the process-like character of Hegel's philosophy, Bergson with a disarming sincerity and to Croce's great amazement admitted that he never had read Hegel! He promised to read him, but Croce, recalling this conversation thirty-eight years later in 1949, doubted that he ever kept this promise.[6] Whitehead's admission in his 'Autobiographical Notes' was equally candid: "I have never been able to read Hegel."[7] On the other hand, it is only fair to stress that neo-Hegelians – Royce excepted – rarely read Bergson attentively, if at all. This sort of a situation – where one great thinker is not acquainted with or not even interested in the thought of another in spite of the kinship of their thought – is far more common than we tend to believe. The relation of Husserl to Bergson is another example. At Bergson's Centenary Congress in 1959, Roman Ingarden recalled that when he reconstructed for Husserl Bergson's description of *durée pure*, Husserl listened attentively and finally exclaimed: "*Das is fast so, als ob ich Bergson wäre.*" ("It looks as if I myself were Bergson.")[8] But this was in 1917 – ten years after Husserl wrote his *Vorlesungen über die Phaenomenologie des inneren Zeitbewusstseins* – and this was apparently the first contact of Husserl with Bergson's thought, and not even a direct one since it was mediated by Ingarden.

But the history of ideas is not our primary concern here. Let us retain in mind the conclusion of this chapter: the Bergsonian duration is a *concrete universal,* inseparable from and inherent in all individual durations. For this reason we must not say that psychological events are *in* time or *in* duration. Such relation between the concrete changing content and the immutable container is meaningful only as long as time is regarded as a homogeneous receptacle independent of its own content; such is precisely the Newtonian and Kantian mathematical time. But such time – *le temps homogène* of Bergson, *il tempo aritmetico* de Croce – is a mere fiction, an *abstract universal,* a static symbol divorced from its dynamic content and artificially reified. In rejecting this reification, Bergson, Royce and Croce are in full agreement. In duration there is no distinction between container and content; psychological events in their ceaseless emergence constitute their own duration.

But two cautions are necessary. First, we have to be on guard against taking the grammatical plural – 'events' – in the conventional arithmetical sense. We have seen that they are not comparable to the externally related arithmetical units; their apparent arithmetical multiplicity results from the artificial, analytical dissection of their indivisible continuity. We cannot then say that they *compose* their duration in a similar sense in which an arithmetical sum is *composed* of the units or a set is constituted by its members. Their diversity is of a *qualitative* kind and we have seen the reasons for terming it *heterogeneous continuity.* But here another caution is in place. This qualitative diversity is basically of an *imageless* type. Let us recall Bergson's analysis of the duration of the perception of melody or his analysis of a 'changelessly' enduring sensory quality. We have seen how in the latter case novelty – and consequently heterogeneity and succession – creeps in in spite of the fact that the sensory quality persists. This explains why it is that Bergson regards the succession of diverse sensory qualities – like the tones in the melody – as a mere *accidental* exemplification of the basic imageless heterogeneity of duration which can be grasped only if we abstract from the *sensory* diversity of its contingent content. (The two main reasons why this point was so seriously misunderstood by both Lovejoy and Ushenko were that (a) they recognized only one kind of multiplicity – that of the arithmetical kind; (b) they failed to see that qualitative diversity may be of a more elusive and subtler kind than the crude differences of sensory qualities.) This contingency of the concrete content of duration was one of the reasons why the illusion of time as an empty receptacle *filled up* by concrete events originated. But the difference between the events 'filling up' time and the succession of events *exemplifying* universal duration is basic. It is as basic as that which separates the dynamic concrete universal from its abstract, static and spatialized counterpart.

It is thus not accurate to characterize Bergson's philosophy as *nominalistic*.[9] Nominalism was and still is regularly connected with psychological and metaphysical atomism – and nothing is more foreign to the letter and the spirit of Bergson's thought. Within such an atomistic framework no universal perception of novelty is possible; for the awareness of the novelty of any atomic event should be gained by its explicit comparison with *all* other atomic events. To use Whitehead's language, we would not be able to know the novelty of any event *without knowing everything else*! Whitehead points out that this objection is based on the failure to distinguish between 'cognizance by adjective' and 'cognizance by relatedness'. The cognizance of novelty definitely belongs to the second category.

The very nature of the relatedness may impose on A some character. But the character is gained through relatedness, not the relatedness through character.[10]

The character of novelty is precisely of this kind; it is constituted by the relatedness of each particular event to the *whole* antecedent context of events. The whole past indivisibly subsists and colors each particular present moment, and it is this coloration, this immanent subsistence of the ever increasing past context which accounts for the freshness and originality of every present 'occasion'. When Whitehead speaks of "becomingness of nature, colored with all the hues of its content"[11], he is merely expressing in different words the fact that duration – including its novelty – is a *concrete universal*. The contingent content of the present may vary and – as in the case of a persisting sensory quality – even remain the same for some time; but what James called "the everlasting coming of novelty into being"[12] characterizes each moment precisely because of its inseparability from its whole antecedent context. Each particular novelty is different from any other since the whole antecedent context is changing continuously 'like a snowball' and it is thus different for each novelty; but all particular novelties agree in that each of them is different from the totality of its own past by the very fact that it transcends it. In this sense, in Whitehead's words, "'creativity' is the universal of universals characterizing ultimate matter of fact".[13]

## NOTES

[1] Cf. Note 7 of the preceding chapter. Bergson's criticism of the idea of disorder is in *C.E.*, pp. 240f.

[2] *C.E.*, pp. 333–334.

[3] Hegel, *Lesser Logic*, § 163; quoted by J. N. Findlay, *Hegel: a Re-examination*, Macmillan, New York, 1958, p. 225.

[4] Findlay, *ibid.*, p. 226.

[5] Cf. Note 9 of Chapter 8 of this part.

[6] Cf. Note 8 of Chapter 8 of this part.

[7] A. N. Whitehead, 'Autobiographical Notes', in *The Philosophy of Alfred North Whitehead* (ed. by Paul Schilpp), Northwestern University, Evanston, 1941, p. 7.

[8] Roman Ingarden, 'L'intuition bergsonienne et le problème phenomenologique de la constitution', in *Bergson et nous. Bulletin de la Société française de Philosophie* **53** numero spécial (1959), pp. 165–166.

[9] The claim made by V. Jankélevitch in his book *Henri Bergson*, Presses Universitaires, Paris, 1959, p. 225. Cf. Bergson's explicit rejection of nominalism in *M.M.*, pp. 150–151.

[10] A. N. Whitehead, *The Principle of Relativity*, Cambridge Univ. Press, 1922, Chapter II 'The Relatedness of Nature', p. 18–19.

[11] *Ibid.*, p. 21.

[12] W. James, *Some Problems of Philosophy*, p. 149.

[13] *Process and Reality*, p. 31. A very similar view is upheld by David Bohm, even though he is apparently not aware that the ultimate character of becoming which he repeatedly stresses imposes certain limitations on "qualitative infinity" and "logical inexhaustibility" of nature. If the "basic reality" is, according to him, "the infinite totality of nature in the process of becoming", then becoming is the highest valid generalization. If such generalization is not valid, being merely "an approximate, conditional and relative truth", then becoming is *not* ultimate. (D. Bohm, *Causality and Change in Modern Physics*, Harper Torchbook, 1961, pp. 164–170.)

CHAPTER 16

# AN OUTLINE OF
# BERGSON'S PHILOSOPHY OF MATHEMATICS

Is it possible to speak of Bergson's philosophy of mathematics except in a negative sense? This is the usual, textbook interpretation of his philosophy and the fact that he insisted so much on the inapplicability of the concept of arithmetical multiplicity not only to psychology, but also, as we shall see, to physics apparently substantiates it. Yet, this is an oversimplification which does not stand a critical and attentive reading of the relevant texts.

There are two central ideas of what may be called Bergson's philosophy of mathematics: first, the inseparability of the concept of number from that of spatial magnitude; second, the view that the act by which the intuition of spatial magnitude is posited is itself a *durational* act, an instance of *durée réelle*. While the first idea is fairly known and was widely criticized, the second was largely ignored, although it was the only one which was compatible with the rest of his philosophy. Let us consider the first one first.

In the second chapter of *Essai sur les données immédiate de la conscience* Bergson tried to show that numerical multiplicity is nothing but the multiplicity of homogeneous units juxtaposed in the homogeneous medium of ideal space. He pointed out first that a mere *enumeration* or listing of members of a certain class is not counting.[1] When we really count instead of mere enumerating, for instance, when we count sheep in the herd, we deliberately disregard individual qualitative differences between them; in considering their number we even disregard their common features which makes them to belong to the same species and treat them as homogeneous units each of which is qualitatively identical to other and each of which still remains distinct from other. What differentiates such qualitatively undistinguishable units must be *a principle of differentiation other than qualitative* and this is precisely space. In other words, it is the juxtaposition of these units in the ideal medium of space which make them diverse and distinguishable despite their qualitative identity.[2]

Bergson's view seems to be radically opposed to Kant who correlated counting with an a priori intuition of time. He wrote in both editions of his *Critique of Pure Reason*:

Thus number is nothing but the unity of the synthesis of the manifold of a homogeneous intuition in general, an unity due to the fact that I generate time itself in the apprehension of the intuition.

And in *Prolegomena*:

Geometry is based on the pure intuition of space. Arithmetic produces its concepts of number through successive addition of units in time.[3]

Thus the duality of arithmetic and geometry is based on the duality of time and space. More specifically, the necessary and synthetic character of arithmetical propositions is due to the fact that the arithmetical operations take place in the ideal medium of time while the similar character of geometrical propositions is due to the underlying a priori intuition of space. Although Kant's view on the relations of arithmetic to time was far free from ambiguities, it was understood in the sense given above, especially when Schopenhauer interpreted him in this way, and when William Rowan Hamilton, himself a Kantian, defined algebra as the "science of pure time in order of progression"[4]. The successive character of counting played a large role in this view. But it was frequently pointed out that *all* intellectual operations – in truth, *all mental processes* – have a similar successive character: why then single out arithmetical operations for such privileged linkage with the "a priori intuition of time"? Furthermore, Bergson showed that an attentive analysis of the process of counting leads to a significantly different conclusion.

This analysis consists of three steps:

(a) In the first place, Bergson shows that no counting can result as long as we remain conscious of one single act of counting without retaining the recollection of preceding acts. Counting thus requires a certain temporal span of consciousness, a mnemic retention of the anterior acts without which we would be forever confined within the narrow limits of the moment in which one single act of counting actually is taking place.

(b) But such mnemic retention of anterior acts can occur in two very different forms and thus it is only a *necessary*, but not a sufficient condition of counting. Thus in listening to successive strokes of the clock, I either count them or I do not. In the first case I not only retain mnemically successive sounds, but I also juxtapose them in the ideal medium of intuited space, that is, I locate them, stripped of their qualitative differences, along a geometrical line, called 'axis of numbers', that is, in 'one-dimensional geometrical space'. The units, though homogeneous, remain nevertheless distinct in virtue of their juxtaposition. In this case, and in this case only, the conditions required for counting are present.

(c) But there is the *third* possibility: instead of either remaining in a pure present, or in juxtaposing the remembered past moments in the ideal one-dimensional space, *I can retain them mnemically without counting.* Bergson illustrates it by his analysis of retrospective counting – the analysis which, in my view, belongs to the subtlest in psychological and epistemological writings:

> Whilst I am writing these lines, the hour strikes on a neighbouring clock, but my inattentive ear does not perceive it until several strokes made themselves heard. Hence I have not counted them; and yet I only have to turn my attention backwards to count up the four strokes which have already sounded and add them to those which I hear. If, then, I question myself carefully on what has just taken place, I perceive that first four sounds had struck my ear and even affected my consciousness, but that the sensations produced by each one of them, instead of being set side by side, had melted into one another in such a way as to give the whole a peculiar quality, to make a kind of musical phrase out of it.... In a word, the number of strokes was perceived as a quality and not as a quantity; it is thus that duration is presented to immediate consciousness, and it retains this form so long as it does not give place to a symbolical representation derived from extension.[5]

We are thus returning to the same notion of 'heterogeneous continuity' or 'succession without separation' of which 'qualitative multiplicity' is just another name. The passage just quoted shows how this type of multiplicity is related to numerical multiplicity. The former is clearly more fundamental; from it and by it the multiplicity of countable units is generated. In the empirical description given above Bergson shows that it was our refocussed attention that dissociated our immediate memory of the original unitary (though *heterogeneous*!) successive whole – a sort of 'musical phrase' into the distinct and countable units. This refocussing clearly had two correlated and complementary effects by which the immediate recollection of the original temporal *Gestalt* was modified; it abstracted from subtle qualitative differences of the four successive sounds which were due to their interpenetration; and in making them thus qualitatively identical, it separated them by ideally *juxtaposing* them. In truth, these two effects are two aspects of one and the same act – the act of juxtaposition; it would be idle to argue which aspect comes first.

But this act of juxtaposition itself is still a *temporal, durational* act. For, as was shown in Chapter 9 of this part, there are no durationless psychological instants; we have seen that not even Russell accepts them, – despite his love of mathematical continuity. It is true that within the *content* of the juxtaposing act its temporality is not disclosed; but this does not make the act itself devoid of duration. As we shall see in Part III, the shortest mental event is contemporary with a long succession of elementary physical events; therefore it cannot be instantaneous. It is an apparent paradox that the act *negating* duration (i.e. positing juxtaposition which is by its own nature the relation between *simultaneous* terms) is itself *durational*. This twofold, paradoxical or 'dialectical' character is even more conspicuous in the act of counting and Bergson is fully aware of it:

In a word, the process by which we count units and make them into a discrete multi-plicity has two sides; on the one hand we assume that they are identical, which is conceivable only on condition that these units are ranged alongside each other in a homogeneous medium; but on the other hand the third unit, for example, when added to the other two, alters the nature, the appearance and, as it were, the rhythm of the whole; without this interpenetration and this, so to speak, qualitative progress, no addition would be possible. *Hence it is through the quality of quantity that we form the idea of quantity without quality.* (Italics added.) [6]

This is the *second* aspect of Bergson's philosophy of mathematics which could be equally well called *psychology* of mathematics since it deals with the psychological dynamism by which the concept of number and arith-metical addition are generated. Arithmetical units and the correlated spatial intervals on the axis of numbers while being completely homoge-neous and external to each other, cease to be so on the metamathematical level; for if they would remain *absolutely* distinct and separated or absolutely homogeneous and undistinguishable, no number greater than one could ever be conceived, no intuition of space *including* two points could ever arise. Two point-numbers are posited on a geometrical line ('one-dimensional space' or 'axis of numbers') not only *outside* each other, but also as being *together, beside* each other; and this *togetherness* resides in the *act* of juxtaposing intuition, not in its content; that is, not on the geometrical, but on the meta-geometrical level. This integrating or, to use Whitehead's term, 'prehensive' character of spatial intuition was correctly stressed by Robert Blanché who, however, did not seem to be aware that it was just another term for 'the quality of quantity' in the last quotation.[7] It is the unity of the act which makes possible a differentiation of the juxtaposed elements which themselves are devoid of any qualitative difference: Their only difference comes from their being 'beside'; but there is no 'beside' without 'together'. This junction of the separative and prehensive character of spatial intuition is due to the fact that this intuition itself is *durational* and as such it both joins and separates; it separates in its content while unifying by its act.

In the passage quoted above the temporal, mnemic character of the intuition of number is made clear: without the mnemic interpenetration of successive units no idea of *ordinal* number could possibly arise. For all arithmetical units being completely homogeneous, there would be no difference between 'first', 'second', and 'third'. Thus René Berthelot, the severe, but very attentive critic of Bergson, was not right when he claimed that Bergson's claim about the correlation of number and spatial intuition applies validly to the *cardinal* numbers only and that his analysis leaves out *ordinal* numbers entirely.[8] But it is true on the whole that Bergson's attention was more attracted to cardinal numbers whose correlation with the underlying spatial intuition is more conspicuous while in ordinal numbers it is obscured and complicated by the more prominent presence of the mnemic element. It is equally true that the attention of both Aristotle and Kant was more turned toward ordinal numbers and this explains their views about the correlation of time and number. A more systematic comparison of what we called Bergson's philosophy of mathematics with the views of both Aristotle and Kant would go beyond the scope of this study.

The affinity of Bergson's 'qualitative multiplicity' with Brouwer's basal intuition of 'two-oneness' or 'many-oneness' had been already pointed out. (Cf. Ch. 11 of this Part.) We shall return to this point briefly at the end of this chapter. Bergson's relation to Husserl has been also discussed. But it is important to add that Aron Gurwitch pointed out the similarity of Bergson's 'qualitative multiplicity' with Husserl's analysis of the apprehension of plurality. Gurwitch's study is one of very few which deal with what we called the *second* part of Bergson's philosophy of mathematics.[9]

As far as the *first* part is concerned – that concerning the correlation of number and the intuition of space – the most known criticism was made by Bertrand Russell. But his sarcastic rendering of Bergson's thought on this subject is nothing but a weird caricature. Thus against all available evidence he charges Bergson with the confusion of numbers with particular collections, completely ignoring Bergson's explicitly made distinction between enumeration and counting. Russell even goes so far as to claim that Bergson does not give *any* reason – whether good or bad – for his view linking spatiality with number[10]; he irresponsibly overlooks the fact that practically the whole second chapter of *Essai* deals with this problem and that in it Bergson *no less than five times*[11] returns to a detailed analysis of the process of counting and the intuition of space involved in it.

But even if we dismiss the effects of plain malice and of inattentive and superficial reading which are clearly visible in Russell's diatribe, it is only fair to admit that there was also a sincere misunderstanding of Bergson's view. Nowhere is Russell's failure to grasp the meaning of this particular view of Bergson more manifest than in his charge that it was due to Bergson's alleged tendency to visualize.[12] In the subsequent discussion he was immediately refuted on this point by Herbert Wildon Carr; but even if the charge were correct, it would be completely irrelevant in the discussion. For although the visual components in the intuition of space are the strongest, one does not have to be a 'strong visualizer' to notice them. Far more elusive are these visual elements present in the idea of number, in particular in larger numbers; one has to be a subtle analyst to detect them. The crucial question is far more general: is spatial intuition an *indispensable* condition of the idea of number or not? Russell clearly misunderstood this question because he confused "intuition of space" with a *crude* visualization, even though Bergson explicitly avoided this confusion. No point was stressed by Bergson more than the *homogeneity* of space, that is, the absence of qualitative sensory differences between its parts. He explicitly assimilated the intuition of space to the 'pure intuition' of Kant.

To the question raised above, Bergson's answer was affirmative, Russell's negative; while René Berthelot conceded that Bergson's view applies only to cardinal numbers. But what is really relevant – besides properly understanding the meaning of Bergson's view – is which answer is substantiated by the very nature of mathematics as well as by its history. Now historically the correlation of the idea of number with the intuition of space is beyond any doubt: let us remember the Pythagorean monads which were both numbers and the points with a bulk; let us remember the

Pythagorean discoveries correlating simple algebraic formulae with geometrical figures. It may be objected that this was characteristic of the
very dawn of mathematics and that the progress of mathematics consisted
in its growing abstractness and its increasing emancipation from intuition.
But this is true only in a carefully qualified sense. While it is true that
the crudely sensory pictures in mathematics continually receded, – for
instance, the just mentioned picture of a point as a tiny sphere – some
subtle visual elements persist even in the most abstract mathematical
thought. This point was obscured by the traditional division of mathematics into supposedly abstract arithmetic and supposedly intuitive,
visualizable geometry; Kant codified this division in his *Critique of Pure
Reason*, although it was dated already in his own time. It has been dated
since the time of Descartes whose discovery of analytical geometry, that
is, of the correlation between the intuitive spatial figures and abstract
algebraical equations, continued the ancient Pythagorean tradition in an
incomparable more complex and more subtle way, free of its original
mystical elements.

Modern arithmetisation of analysis as well as the incorporation of
mathematics into logic seemingly points in a quite opposite direction:
the growing abstractness and the retreat from intuitive elements. Non-
Euclidian geometries in particular showed graphically the imageless
character of modern mathematical thought. There is no question that
all this clearly contradicts any claim about the indispensability of *crude*
visualisation for mathematical thought – the claim that Russell attributed
to Bergson. What Russell and other critics overlook is the presence of
*far more subtle and more elusive elements even in the most abstract
mathematical and logical thought*. Russell thus does not realize that for
instance the *apparently* entirely abstract concept of mathematical continuity is based on the very obstinate habit of mentally magnifying any
spatial interval (even when it is designated as a 'temporal' one) in order
mentally to discern the intermediate points (even when they are called
'instants'); without this 'mental microscope' the dogma of spatio-temporal
continuity loses its convincingness.[13] Equally significant is the all-pervasive
presence of set theory in the foundations of mathematics. Should we recall
again our conclusions of the last chapter of Part I concerning the 'logic
of solid bodies' and restate the arguments showing that 'set' is structurally
isomorphic with material aggregates and that the 'elements' of the set

are in their atomistic nature merely the diaphanous descendants of the material bodies of our macroscopic experience?

Furthermore, as far as the correlation of number and spatiality is concerned, Bergson's view is not as isolated as Russell wanted to suggest. Philosophers, like J. J. Baumann, F. A. Lange, G. Noel, mathematicians like Jules Tannery and Paul Du Bois-Reymond, expressed a similar view.[14] Even more significant is Sophus Lie's axiom (to use Poincaré's nomenclature), according to which *to every point on a straight line corresponds a certain number and vice versa*; i.e. that space is a *Zahlmannigfaltigkeit*.[15] It is difficult to find a more striking confirmation of Bergson's view about the correlation of number and space. It is true that logicists among mathematicians hold a view very opposite to that of Bergson; they claim that it is the *logical* concept of arithmetical series which underlies the geometrical continuum and not *vice versa* as Bergson claimed. This, according to them, is the true meaning of 'arithmetization of the continuum'. But, as mentioned above, this belief is due to the fact that the logicists – including Russell – are little trained in introspective analysis and thus fail to detect more elusive visual elements of spatial intuition in the very ideas of number and of set (class) in general; as Evert W. Beth observed recently (who learned from Piaget to get rid of his own, traditionally rationalistic distrust of psychology) their thinking still depends on the spatial intuition, even when they profess to be independent of it.[16] It is the ideal medium of space which makes the distinction of qualitatively identical units possible; it is the infinite divisibility of the same intuited space which makes the "arithmetization of continuum", that is, its unlimited division into the point-like elements possible. As we shall see in Part III, such belief in the unlimited divisibility of space and time, which itself is an illegitimate extrapolation of our macroscopic experience, decisively affected our mathematical thinking. Thus the modern arithmetization – or rather atomization – of the continuum, despite its subtlety, complexity and apparent abstractness, has its roots in the Pythagorean monads and Democritian atoms. It is against such atomization of continuum that Bergson, Weyl and Brouwer reacted in their different ways.[17]

It may be objected that Brouwer, in correlating number with the intuition of time instead of that of space, holds the Kantian view which is opposite to that of Bergson. It is true that he agrees with Bergson in regarding 'move of time' as the 'initial phenomenon'. As we have seen, Brouwer's own description of this 'initial phenomenon' of 'two-oneness' (or 'two-ity' as he wrote later) has the truly Bergsonian features which is even more significant since there is no evidence of any Bergsonian influence on him. But then does not Brouwer's correlation of number and time place him in the opposition to Bergson, at least as far as the philosophy of mathematics is concerned? This, however, is more apparent than real as an attentive analysis of the following more recent text of Brouwer shows:

Mathematics comes into being when two-ity created by a move of time is *divested of all quality* by the subject, and when *the remaining empty form* of *the common substratum* of all two-ities, as basic intuition of mathematics, is left to an unlimited unfolding, creating new mathematical entities...[18] (Italics added.)

Thus it is not 'the initial phenomenon' of time which is correlated with mathematical entities, but its abstract form divested of all its qualities. But what else is such empty form than 'spatialized time' of Bergson whose temporality is merely verbal because of the homogeneity of its terms? Thus the difference between Bergson and Brouwer even in this respect is more terminological than real: but while Bergson identifies every homogeneous medium with space, there is no such explicit identification in Brouwer.

Now we face the same question which had already emerged in the previous chapter: why homogeneous, mathematical time is still called 'time' despite all of its features which it shares with space? Bergson's statement that "it is through the quality of quantity that we form the idea of quantity without quality" sheds light on this problem.

It is easy to see why the dimension of space that has come to replace time is still called time. It is because our consciousness is there. It infuses living duration into a time dried up as space. Our mind, interpreting mathematical time, retraces the path it has traveled in obtaining it.

Bergson then briefly recalls the process by which the concept of mathematical time has been formed:

From inner duration it had passed to a certain undivided motion which was still closely bound up with it and which had become the model motion, the generator or computer of time; from what there is of pure mobility in this motion, that mobility which is the link between motion and duration, it passed to the trajectory of the motion, which is pure space; dividing the trajectory into equal parts, it passed from the points of division of this trajectory to the corresponding or "simultaneous" points of division of the trajectory of any other motion. The duration of this last motion was thus measured; we have a definite number of simultaneities; this will be the measure of time; it will henceforth be time itself. But this is time only because we can look back at what we have done.[19]

In other words, underlying the spatialized symbol of time, that is, the homogeneous scheme consisting of simultaneous parts devoid of any qualitative differentiation (Brouwer's 'empty form'), is still the temporal connotation reminiscently hovering behind it. It is the reminiscence of real succession, of real duration, which breathes the dynamic meaning into the spatialized scheme. But, on the other hand, the spatializing symbol is continually stifling and distorting the original dynamic connotation as the history of both science and philosophy – including the most recent philosophy – amply testifies. To recover the original meaning behind the static scheme, to rediscover "the quality of quantity by which we form the idea of quantity without quality", is the main, and in a sense, the only meaning of Bergson's philosophy. It is an effort which goes against spontaneous concept formation; in differentiating between the static trajectory and mobility itself, it opens the way to the recovery of the imageless, spaceless duration, forever incomplete and with qualitatively diverse moments. Only in experiencing extreme difficulties which such effort encounters, shall we fully understand the true meaning of Bergson's words: *Philosopher consiste à invertir la direction habituelle du travail de la pensée.*[20]

## NOTES

[1] *T.F.W.*, p. 76.
[2] *T.F.W.*, p. 95.
[3] Norman Kemp Smith, *A Commentary to Kant's 'Critique of Pure Reason'*, Humanities Press, New York, 1962, p. 129.
[4] W. R. Hamilton, *Lectures on Quaternions*, Dublin 1853, p. 2.
[5] *T.F.W.*, pp. 127–128.
[6] *T.F.W.*, p. 123.
[7] Robert Blanché, 'Psychologie de la durée et physique du champ', *Journal de psychologie normale et pathologique* **44** (1951) 420: "Certes, plusieurs instants ne peuvent exister au même moment, et c'est pourquoi un principe de liaison est requis pour faire une durée, mais plusieurs points peuvent-ils davantage exister au même lieu, et un *autre* principe de liaison n'est-il pas nécessaire pour faire une étendue?"
[8] René Berthelot, *Un romantisme utilitaire*, II, p. 181.
[9] Aron Gurwitch, *The Field of Consciousness*, Duquesne University Press, Pittsburgh, 1964, pp. 140ff.
[10] Bertrand Russell, *The Philosophy of Bergson*, Macmillan, London, 1912, pp. 14–15.
[11] *T.F.W.*, pp. 77–79; 86–87; 104–105; 121–123; 127–128.
[12] *Op. cit.*, p. 9.

[13] Cf. M. Čapek, *The Philosophical Impact of Contemporary Physics*, pp. 231–241.

[14] J. J. Baumann, *Die Lehre von Raum, Zeit und Mathematik*, Berlin 1869, II, pp. 668–671; F. A. Lange, *Logische Studien*, Iserlohn 1877, pp. 141ff. The reference to G. Noel is in Bergson, *T.F.W.*, p. 75; that to Paul Du Bois-Reymond *Allgemeine Funktionstheorie* is in Helmholtz-Schlick, *op. cit.*, p. 72; that to J. Tannery's article 'Principes fondamentaux de l'arithmetique' is in Halsted's Introduction to the English translation of Helmholtz's *Counting and Measuring*, Van Nostrand, Princeton, 1931, p. IX.

[15] H. Poincaré, 'Les fondements de la géometrie', *The Monist* VII (1898) 57.

[16] E. W. Beth and J. Piaget, *L'épistémologie mathématique et psychologie, Études de psychologie génetique*, Presses Universitaires de France, Paris, 1961 [Vol. XIV], p. 112. On Beth's change of attitude toward epistemology cf. Jean Piaget, *Genetic Epistemology* (Columbia University Press, 1970), p. 12.

[17] On Brouwer's opposition to the Cantorian atomization of continuum and his view of continuum as "the medium of free becoming" (*ein Medium des freien Werdens*) cf. A. Fränkel, *Einleitung in die Mengenlehre* (Dover, New York, 1946), pp. 238–239; on Brouwer in general cf. Max Black, *The Nature of Mathematics*, Littlefield, Adams & Co., New York, 1959, pp. 186–210. On the relation of Weyl's view of continuum to that of Bergson see Chapter 9 of this part, Note 14.

[18] L. J. E. Brouwer, 'Consciousness, Philosophy and Mathematics,' quoted by E. W. Beth, *The Foundations of Mathematics*, Harper Torchbooks, New York, 1966, p. 618.

[19] *D.S.*, pp. 60–61.

[20] 'Introduction à la métaphysique', in *La pensée et le mouvant*, Paris 1934, p. 241.

Saint Cyr-sur-Loire, 3 juillet 1938

Monsieur,

J'ai bien tardé à vous écrire. Cela tient
d'abord à ce que le volume que vous avez bien
voulu m'envoyer, ayant probablement été expédié
de Prague comme marchandise et non pas par la
poste, a été déposé à la Douane, à Paris, d'où je
n'ai pu le faire retirer qu'après certaines formalités.
Mais cela tient aussi à ce que, malade depuis
longtemps, j'ai traversé dans ces derniers mois
une crise particulièrement grave. J'ai fini par
prendre le dessus, et me voici à la campagne.
Combien j'ai regretté de ne pouvoir, vu mon
ignorance de votre langue, lire votre livre lui-
même! Heureusement, vous en avez donné un
résumé en français; et je tiens à vous dire combien
ce résumé m'a intéressé. Il était impossible de

mieux comprendre l'essentiel de mes vues sur la
durée et sur la matière. En particulier, vous avez
admirablement montré comment, dans quel sens
et dans quelle mesure, la conception de la matière
que j'ai de plus en plus précisée dans mes ouvrages
successifs anticipait sur les conclusions de la
physique d'aujourd'hui. Ce point n'avait guère
été aperçu, pour la raison très simple que mes
vues sur la question, émises à une époque où
l'on considérait comme évident que les éléments
ultimes de la matière doivent être conçus à l'image
du tout, déroutèrent les lecteurs, et furent le plus
souvent laissées de côté comme étant la partie
incompréhensible de mon œuvre. Ils jugeaient
d'ailleurs, probablement, que c'en était une partie
accessoire. Aucun (sauf peut-être, dans une
certaine mesure, le profond mathématicien et

philosophe Whitehead) ne s'est aperçu comme vous
qu'il y avait là pour moi quelque chose d'essentiel, qui
se rattachait étroitement à la théorie de la durée,
et qui était en même temps dans la direction où
la physique s'engagerait tôt ou tard.

Je souhaiterais que votre livre pût être lu en
France. Il faudrait, pour cela, qu'il fût traduit
en français. Mais on aurait peut-être de la peine
à trouver un traducteur compétent et surtout — vu les
difficultés matérielles du temps présent — un éditeur.

Laissez-moi vous adresser, Monsieur, avec mes
compliments et mes remerciements, l'assurance
de mes sentiments bien sympathiques

H. Bergson

# PART III

## BERGSON'S THEORY OF THE PHYSICAL WORLD AND ITS RELATIONS TO CONTEMPORARY PHYSICS

CHAPTER 1

# THE REALITY OF DURATION IN
# THE PHYSICAL WORLD AND ITS IMPLICATIONS

Bergson's first book asserted a sharp dualism of the psychological and the physical world. It was a dualism of a different kind than the dualism of Descartes: instead of the opposition between extended matter and unextended mind, Bergson posited the duality of the timeless, spatial world and the temporal world of psychological events. Had Bergson retained this sharp distinction of the temporal world of mind and the timeless world of matter, he would have run into the same inextricable difficulties as his seventeenth century predecessor. Indeed his difficulties would have been even greater; if it was difficult to conceive of interaction between the Cartesian *res cogitans* and *res extensa*, how could we in any intelligible way conceive of the correlation between the unfolding realm of psychological duration and the completed, timeless, i.e. purely spatial, realm of matter? Thus *Essai sur les données immèdiates de la conscience* was necessarily a provisional stage in the growth of Bergson's thought. But this stage already contained indications of his subsequent development, the main feature of which was that the timelessness of matter was given up without sacrificing the essential duality, or at least *polarity*, of mind and matter.

The question whether the physical world is really timeless cropped up naturally in Bergson's mind as early as in 1889 and traces of it may be found in *Essai* itself. It is true that at least one passage asserts unambiguously that the physical world is by its own nature 'juxtaposition without succession' in contrast to the psychological duration which is its exact opposite.

Thus, within our self, there is succession without mutual externality; outside our self, in pure space, mutual externality without succession.[1]

As we have seen, Bergson himself apparently was influenced by those philosophical implications of classical physics, which, together with the general, ever-present, Eleatic leanings of the traditional philosophy, created the impression that succession does not have a genuine reality outside the human mind. But the enormous difficulties produced by such a bifurcation of reality were too obvious to be ignored. It is, then, only natural that alongside the quoted passage, other less unambiguous statements concerning the status of time in the physical world may be found:

> There is a real space, without duration, in which phenomena appear and disappear simultaneously with our conscious states.... We certainly feel, it is true, that although things do not endure as we do ourselves, nevertheless there must be in them some incomprehensible reason why phenomena seem to succeed one another instead of being unfolded all at once.[2]

But what else is this 'incomprehensible reason' which prevents physical events from being unfolded at once other than their irreducible and unshortenable duration? Do we have any right to call physical space 'durationless' as does Bergson in the first quotation, if we admit that physical phenomena do appear and disappear, being correlated with the successive data of our consciousness? Thus already in this first phase of his thought Bergson admitted, reluctantly and without realizing the full significance of his concession, the reality of time even in the physical world. What was reluctantly conceded in 1889 had become the central part of his theory of matter less than a decade later. In a classic passage from *Creative Evolution*, the reality of physical duration is affirmed with the clarity, conciseness and simplicity of mature thought:

> Yet succession is an undeniable fact, even in the material world. Though our reasoning on isolated systems may imply that their history, past, present and future, might be instantaneously unfurled like a fan, this history, in point of fact, unfolds itself gradually, as if it occupied a duration like our own. If I want to mix a glass of sugar and water, I must, willy-nilly, wait until the sugar melts. This little fact is big with meaning. For here the time I have to wait is not that mathematical time which would apply equally well to the entire history of the material world, even if that history were spread out instantaneously in space. It coincides with my impatience, that is to say, with a certain portion of my own duration, which I cannot protract or contract as I like. It is no longer something *thought*, it is something *lived*. It is no longer a relation, it is an absolute. What else can this mean than that the glass of water, the sugar, and the process of the sugar's melting in the water are abstractions, and that the Whole within which they have been cut out by my senses and understanding progresses, it may be in the manner of consciousness?[3]

Albert Thibaudet compared the significance of this passage with Descartes' famous story of the piece of wax.[4] The comparison is well chosen: both passages illustrate the differences between both thinkers and between the seventeenth and the twentieth century. Descartes illustrates by his piece of wax that the fundamental reality underlying physical phenomena is geometrical space; while Bergson illustrates by his glass of sugar solution his thesis that the fundamental reality even in the physical world is a temporal process. As we shall see, the major features of Bergson's theory of matter, including what has later been called by Whitehead the criticism of "the fallacy of simple location", are present in the passage just quoted.

The reason why Bergson hesitated to recognize the temporality of matter was that to him such recognition implied ascribing basically the same structure to physical time as to psychological duration. We have already pointed out that by his analysis of psychological time Bergson went beyond the area of psychology in claiming that not only the time of our consciousness but *any type of temporality* must have basically the same dynamical organization. It is the same kind of extrapolation which Whitehead later briefly characterized by saying that "the texture of observed experience, as illustrating the philosophic scheme, is such that all related experience must exhibit the same texture".[5] The qualitative content may vary, as it undoubtedly does vary when we pass from introspective data to sensory perception; the texture itself – or what we called the concrete universal of duration – remains the same in its essential features. If this texture implies certain negative implications for psychological duration, we must accept them for physical duration as well, no matter how paradoxical they may at first appear.

Let us recall briefly the five negative propositions by which we, following Bergson in this respect, characterized the structure of psychological duration; and since this structure is the same in *any* duration, these propositions must hold for the duration of the physical world as well.

(1) The dynamic continuity of duration excluded atomism (associationism) in psychology; consequently, it must exclude atomism – at least in its classical form – in physics too. In the same way that the continuity of psychological duration cannot artificially be dissected into a multiplicity of 'mental states', the existence of substantial and mutually external physical particles must be denied. More accurately, their individuality, which is *the individuality of a process rather than that of a thing*, can be meaningfully preserved only by being embedded in the continuity of cosmic interaction. It is immediately obvious that this requires a radical revision of the classical corpuscular-kinetic model of nature.

(2) In the same way that the incomplete character of psychological duration required the reality of genuine novelty, which for Bergson was the basis of his rejection of psychological determinism, this same character of duration implies the rejection of physical determinism in its classic, rigorous, Spinozist-Laplacean form. In other words, each moment of physical duration contains an element of contingency, of indetermination, which resists deduction from its past antecedents.

(3) The reality of novelty implies the diversity of successive phases on both psychological and physical levels. Quantitative, homogeneous time is a fiction in physics as well as in psychology, even though, as we shall see, this fiction is convenient for practical purposes on the macrophysical level.

(4) The negation of the homogeneity of duration implies the negation of its infinite divisibility (mathematical continuity) even in the physical world. In other words, the mathematical, durational instant is *basically* as fictitious in the world of matter as it is in the realm of mind. Again, *the degree of fictitiousness*, if such an expression may be allowed, is less conspicuous in the physical world, since, in the macroscopic – or rather *macrochronic* – perspective, physical processes appear *practically* continuous, that is, divisible *ad infinitum*; even though in reality they have a fine, pulsational or drop-like structure.

(5) Finally, the denial of the homogeneity of time implies the denial of its container-like character in both psychology and physics. There is no such thing as a homogeneous Newtonian receptacle in which the events, whether mental or physical, would take place. In the same way as psychological events *constitute* psychological duration, physical events are fused with the physical becoming. In either case there is no distinction between the qualitative content and the homogeneous, empty container as classical physics and classical epistemology maintained.

After accepting these implications which René Berthelot rightly characterized as *bizarre*[6] (undoubtedly they were so at the beginning of the century!), Bergson faced an additional task: to show why the classical picture of the physical world imposed – and to a considerable extent still imposes – itself upon our mind with such obtrusive convincingness; more specifically, why atomism, strict determinism, the homogeneity, infinite divisibility and container-like nature of time, are notions which we have an almost irresistible tendency to accept even today. In other words, Bergson had to show to what extent the structure of the physical world *approaches* the world of classical physics, without, however, coinciding with it; in other words, that the world of Laplace is an *approximation* or an *ideal limit* of the world of Bergson.

Evidently this is closely related to the question of whether Bergsonism is *monism* or *dualism*. Becoming seems to constitute the nature of the psychological as well as of the physical world; consequently, the term 'dynamic monism' is not inappropriate, though it still may be misleading. It may be misleading because the structure of psychological time, as it is disclosed in the intuition cleared of all accessory and parasitic elements, contains certain features which are not found in the physical processes. This means that certain attributes of duration which are present in its psychological species are absent when we consider the structure of duration in general. Consequently we have to abstract from these accidental features when we pass over from psychological duration to the duration of matter. Thus the philosophy of Bergson is far from being an unqualified form of monism. Not even the adjective 'dynamic' is sufficient to characterize it fully. The duality of the psychological and the physical realms is retained, though not in the classical Cartesian or neo-Cartesian sense of a sharp and irreducible opposition. Process remains the fundamental category; but the structure of process contains some variables which have, so to speak, different values in the physical and psychological events. What are these variables or accidental traits of duration? Is not the whole distinction between accidental and essential features of duration arbitrary and, if not, what is the evidence for maintaining this distinction?

From the standpoint of radical empiricism, whose representative Bergson was before the term was coined by William James, only one answer was possible: the distinction between the accidental and the fundamental features of duration is imposed by experience. Only an appeal to concrete experience will make previous distinctions meaningful; without this appeal the words 'structure,' 'essential attributes,' 'variables,' etc., are nothing but abstract technical terms. The experience to which Bergson refers is nothing but a plain fact which is hardly denied by anybody possessing a minimal acquaintance with elementary physics. But although the fact itself was known for a long time, its full significance and, in particular, the question which it implicitly raised, was largely ignored. What experience was it?

## NOTES

[1] *T.F.W.*, p. 108.
[2] *Ibid.*, p. 110; pp. 209–210. The word 'incompréhensible' is not translated in the English translation of the book.
[3] *C.E.*, pp. 12–13.
[4] Albert Thibaudet, *Le bergsonisme*, I, p. 28.
[5] *Process and Reality*, p. 5.
[6] René Berthelot, *Un romantisme utilitaire*, II, p. 250: "Ainsi la manière dont se sont developpées les idées de Bergson sur la physique peut contribuer à nous faire comprendre les incertitudes de sa pensée, comme la bizarrerie de ses résultats."

CHAPTER 2


# DIFFERENT DEGREES OF TEMPORAL SPAN.
# MICROCOSMOS AS MICROCHRONOS

If we recognize the reality of succession in the physical world, we are admitting that physical becoming and the becoming of our consciousness move on, so to speak, in a parallel way. It is obvious that some physical processes are going on simultaneously with the unfolding and enduring qualities of our psychological duration. For instance, to our perception (including the emotional quality of our impatience) of a piece of sugar which is slowly being dissolved in the glass of water correspond certain processes in the physical world whose features are fairly well known: under the continuing impact of molecules of water the crystalline structure of sugar is loosened and its molecules gradually scattered throughout the whole volume of water. This is the most conspicuous feature of the whole phenomenon; other instances, less conspicuous and discovered later, are equally well explained in terms of the classical atomistic and kinetic framework: the increase of the viscosity of water, lowering of the freezing point and the elevation of the boiling point of the resulting solution, the change of the optical properties (the disappearance of the original transparency), etc. The feature important for our present analysis is that all these processes *take time*; their duration is obviously a species of *duration in general* and shares one of its constant features, that is, its incompleteness, its gradual unfolding. But to what extent can this duration be analogous to the duration of consciousness?

The above description of what is really happening when a piece of sugar is being dissolved in water presupposes the classical, corpuscular-kinetic model of nature, a model whose features are clearly incompatible with those which, according to Bergson, physical duration should have. It presupposes, in particular, the concept of classical time, homogeneous and, consequently, infinitely divisible, whose only indivisible components are durationless instants. Here was the source of Bergson's predicament: the features which, according to him, belong only to *verbal* time, seem to be the attributes of the *physical* time; does this mean that physical time is not a real time at all? But how then explain that physical time seems to possess the same trait of *incompleteness* of psychological duration, as Bergson's example of the piece of sugar dissolving in the water shows? Bergson's answer was that those features which are incompatible with temporality in general merely belong to physical duration *apparently* (or rather approximately) and he tries to explain the source of this appearance.

Let us consider the visual sensation of redness. It cannot be denied that as a psychological quality it remains absolutely indivisible without any subdivision into 'earlier' and 'later' phases. It is a simple and undivided – though, as explained before, not an atomic – act. But by simple experiments it can be shown that this indivisible act is not instantaneous in the mathematical sense; it has a certain duration and, what is even more surprising, it is simultaneous with *an enormous number of successive events in the physical world.* To one second of our sensation of redness corresponds, in the realm of matter, the succession of about four hundred billions of electromagnetic vibrations. It is true that some visual sensations may be shorter, some of them lasting only for one twentieth of a second; this still does not eliminate the fantastic ratio of their duration to the 'duration of things'. The epistemological paradox remains: to the single-ness and qualitative undifferentiatedness of our sensory qualities corresponds the enormous complexity of the physical stimulus; to the temporal unity and indivisibility of a single mental act corresponds an enormously long history of distinctly successive physical events. To understand Bergson's solution of this paradox is a key to the understanding of his theory of matter and its thoroughly unconventional character.

This fact has been widely known since the discovery of spectral analysis. The fact that physical events have enormously shorter periods of duration than psychological sensory qualities was considered in the classical period to be another indication of the infinite divisibility of physical time. If objective time is mathematically continuous, why should there be any lower limit to the duration of physical events? As for the limited divisibility of psychological time and the existence of the specious present, it was believed that these limits concern only the *qualitative content of time*, not time itself; the true present is a mathematical present, the psychological present was significantly called 'specious'. (Similarly, the indivisibility of the Democritean atoms did not apply to their atomic volumes; hence the distinction between the 'physical' and 'geometrical' divisibility, and it is only the latter which was believed to have no limit.) Furthermore, time and again it was also believed that the indivisibility of sensory qualities might be only spurious and that the ultimate elements of the stream of consciousness might have a one-to-one correlation to the elementary physical events. The epistemological untenability of this attitude has already been shown in Part II. It cannot be meaningfully maintained that psychological qualities which are experienced as simple acts or events may at the same time be *really* many. Their physical correlates may be and in fact *are* many; but this is a different assertion from claiming that psychological events themselves *appear* to be simple while they are *in reality* many. To be psychologically real means to be experienced; to be psychologically *indivisible* means *to be experienced as such*, that is, in a single, psychological act. If this were an illusion, this illusion would have to be explained; however, if we claim that what appears to be one is in truth a multiplicity, the question arises: *where* is the locus of this illusory unitary appearance? Not in consciousness, because it consists *ex definitione* of the postulated much shorter sub-sensations. Thus the illusion is not explained, but simply denied; hence, there would be no reason to speak about the 'illusion of consciousness'. The contradiction is obvious.

This disproportion between the duration of sensation and the duration of the elementary events composing the corresponding stimulus may be illustrated in many other ways – and perhaps even more convincingly. We must not forget that the frequency of the red light is the lowest; at the opposite end of the spectrum the frequency of the violet rays is about double, while the duration of the corresponding color sensation remains the same. The frequency still increases and the duration of the elementary vibrations correspondingly decreases outside of the visible portion of the spectrum, for the ultra-violet, X-rays and gamma rays. No human sensation corresponds to the ultra-violet rays, though, as mentioned above, in Part I, some sort of sensation may be present in certain species of insect. Even shorter than gamma-rays are the wave-lengths of cosmic radiation, whose frequency is correspondingly higher and can be expressed by a number having 22 zeros. This means that a single vibration would be of the order of one sextrilionth of a second. But even this extremely small interval does not represent the limit of the divisibility of time in the physical world. According to wave mechanics, the vibratory character seems to belong to matter itself: what seems to be a corpuscular entity is constituted by a string of an enormous number of successive events whose duration is no longer than $10^{-23}$ sec. This is not too far from the hypothetical value of Lévi's chronon.[1] But it is sufficient to remain on the ground of classical physics in order to become aware of how extremely short the duration of microphysical events is. If we consider, for instance, the number of times one molecule clashes with others within one second according to the kinetic theory of gases, we can appreciate the more how far below the temporal threshold of our consciousness the corresponding temporal intervals separating two successive changes lie. For this reason the *microcosmos* can be appropriately characterized as *microchronos*.

It is evident that the *degree of divisibility* of physical duration greatly surpasses the degree of divisibility of psychological duration. As stated above the minimum of psychological duration cannot be shorter than one thousandth of a second. But how exceedingly long a history is represented by the same interval in the physical world! In order to make us understand more concretely the complexity of the physical history which is condensed into one single psychological pulsation of redness, Bergson computes the time necessary for counting the number of corresponding successive physical events: with the greatest economy possible, that is, assuming that each act of counting would last no more than the Jamesian minimum, i.e. 0.002 sec, it would take 250 centuries.[2] From the human point of view and for all practical purposes the number of elementary physical events is infinite and their individual duration is practically indistinguishable from 'the zero-duration' of mathematical instants. It is precisely this character of physical duration which explains why the belief in homogeneous, mathematically continuous time imposed itself so obtrusively on our minds. Not only is the idea of the infinite divisibility of time suggested by an obtrusive analogy of the geometrical line; in addition, the whole experience of classical physics seemingly substantiated our belief that there are no smallest intervals of time. In the same way that the unlimited divisibility of space was more and more confirmed by optical microscopes which disclosed within apparently indivisible spots of our visual field literally new worlds, 'time microscopes', realized by means of ultrarapid cinematography, revealed new events within moments which to us appear as absolutely indivisible.[3] But the situation is different in contemporary physics, especially since the arrival of quantum theory and wave mechanics.

In my previous book I pointed out how and why our belief in mathematically continuous time has been weakened by recent discoveries without affecting its practical and approximate value.[4] Almost two decades before Poincaré coined the term *'l'atome du temps'*,[5] Bergson denied the mathematical divisibility of time and the existence of durationless instants on epistemological and metaphysical grounds; to him the essence of temporality *excludes* the existence of mathematical instants. It is not necessary to repeat all the reasoning which led Bergson to this conclusion; let us only remember that the relation of succession is impossible without the rudimentary element of heterogeneity, that is, without the qualitative incommensurability of the past and the present; that this qualitative difference is only another word for the prolongation of the past into the present. Thus even physical duration contains a rudiment of memory, that is, the 'condensation of the past in the present'; consequently, even physical moments, instead of being durationless instants, are pulsations, although these pulsations are incomparably shorter than the pulsation of consciousness. The existence of temporally thick pulsations is one constant feature which may be found everywhere temporal processes are found; while the length of the pulsation, or, in other words, the rhythm of duration, i.e. the degree of condensation of the past and the present, is variable. Its different values account for the differences in temporal structure in various strata of nature. What was called in traditional language the difference between 'mind' and 'matter' is reducible, according to Bergson, to the difference in the *rhythm* or 'tension' of durations.[6]

The existence of different temporal rhythms is also suggested by certain facts of psychology. In a certain sense these facts may appear even more convincing, as their respective durations are directly observable while it can be always pointed out that the duration of physical events is only *inferred*. The facts to which we are going to refer show plainly that even within one individual consciousness the rate of duration, or, in other words, the temporal span of the specious present, varies according to circumstances. What is called by Bergson 'the condensation of the past into the present' is definitely less intense in states of relaxed attention and reduced activity of mind, and it seems to acquire the lowest degree in dreams and related states. It would be beyond the scope of this book to explain how the author of *Matter and Memory* and *Mind-Energy* applies his doctrine of the different tensions of duration to the various facts of psychology, in particular to the presence of various 'levels of consciousness', each of which is characterized by a certain volume of specious present; how the states of maximum attention and of the greatest intellectual concentration are those where the span of consciousness is widest while the states bordering on unconsciousness and characterized by the greatest passivity are those in which the synthesis of the past and present is loosest.

The modifications of the temporal consciousness in certain dreams and psychasthenic states are thus accounted for: what is a simple, indivisible act for the waking consciousness becomes, in virtue of the decreased tension of duration, a succession of shorter psychological pulsations; hence the amazing rapidity of certain dreams and hypermnesic states of mind.[7] The term 'rapidity' is, of course, only relative as the term 'extreme slowness' would be equally adequate; the choice of a term depends on whether the frequency of the elementary events or their individual temporal volumes are considered. Thus what is experienced in one minute by a normal consciousness may be experienced as an interval of one hour by the consciousness of a psychasthenic, which can thus be properly called a 'slower' one; however, considering the fact that the frequency of elementary, psychological pulsations is greater in the dreamy and psychasthenic states, the term 'greater rapidity' is not inadequate. The choice of the term clearly depends on that of the frame of reference to be adopted.

We shall not consider here to what extent this theory successfully explains the specific facts to which it is applied, for instance dreams, the illusion of paramnesia ('*l'illusion du déjà vu*') and certain characteristics of the intellectual effort.[8] One thing is certain: that without the fundamental idea of different degrees of durational tension all these specific theories of Bergson remain thoroughly unintelligible and that, without grasping it, all fruitful and fair criticism, and possible suggested amendments, remain impossible. What is important for our present purpose is that the conclusion suggested by the psychological facts described above is basically the same as that imposed by the comparison of the temporal span of consciousness with that of elementary physical events: that the *volume of specious present* or *the rhythm of duration* is a variable feature of the structure of time, a variable whose particular 'values' are different not only for various strata of physical nature, but for different 'levels of mind' as well. As we shall see, not only the difference between 'mind' and 'matter', but also their connection and interaction are accountable, according to Bergson, by the variability of the temporal span. In any case, the durationless instant does not exist in psychology or physics.

NOTES

[1] The value of chronon ('atom of time') was estimated by Robert Lévi as being $4.48 \cdot 10^{-24}$ sec. Cf. his article 'La théorie de l'action universelle et discontinue', *Journal de Physique et le Radium* **VIII** (1927) 182f. The number $2.4 \cdot 10^{23}$ indicates the frequency associated with the mass of neutron according to the basic formula of wave mechanics $v = mc^2/h$.

[2] *M.M.*, p. 202.

[3] Cf. for instance A. Magnan, *Premiers essais de cinématographie ultra-rapide* and *Cinématographie jusqu'a 12,000 vues par seconde* (both published by Hermann, Paris, in the collection *Actualités scientifiques et industrielles*, Vol. XXXV and XLVI, 1932. By means of the Kerr cell the exposure time is now reduced to 1/140 000 000 sec!

[4] Cf. M. Čapek, *The Philosophical Impact of Contemporary Physics*, Van Nostrand, Princeton, 1964, esp. Ch. XIII and the relevant bibliography on p. 242.

[5] H. Poincaré, *Mathematics and Science: Last Essays* (transl. by John W. Bolduc), Dover, New York, 1963, p. 86.

[6] *M.M.*, p. 58; pp. 197–220.

[7] Cf. *M.M.*, p. 148; *M.E.*, the article 'Dream'.

[8] *M. E.*, the articles 'Memory of the Present and False Recollection' and 'Intellectual Effort'.

# CHAPTER 3

# TWO FUNDAMENTAL QUESTIONS

Bergson himself was aware that the idea of different rhythms or different "tensions" of duration, or as he sometimes called it, "different degrees of elasticity of duration," is difficult for us to comprehend precisely because of our natural tendency to attribute homogeneity and independence to time with respect to its concrete content. Not only was this belief strengthened by the three centuries of differential calculus and its successful applications to physics; not only was it strengthened by our perennial tendency to symbolize time by a geometrical line whose unlimited divisibility was naturally conferred to time itself, but nothing in physics prior to 1900 remotely suggested that this belief was merely an extrapolation of our limited macroscopic – or rather *macrochronic* – experience. But once we realize that this belief was an illusion – undoubtedly biologically well-founded and useful – the conditions are created for its removal.

But what if our instinctive repugnance to admit the multiplicity of durations of a different temporal span is something more than a mere psychological inhibition? What if it is a fundamental postulate which even Bergson himself was unable to avoid? This is precisely the objection raised by René Berthelot, one of the most systematic and severe, but – unlike Russell – the most attentive, critics of Bergson. According to him, Bergson implicitly admits homogeneous, all-containing time when he speaks of the temporal minimum of consciousness lasting 0.002 sec.[1] What else is this but a reference to a homogeneous, quantitative, measurable time whose tiny *measurable* segment is *filled* by the *minimum sensible* mentioned above? If this objection were correct – and it is unquestionably plausible – then Bergson would be guilty of an intrinsic inconsistency similar to that of the atomistic theory of time, which surreptitiously reintroduced the concept of instant which it overtly denied. Indeed, if the *minimum sensible* lasts 0.002 sec. what else is meant than that it is bounded by *two* instants separated by the interval of homogeneous time of the length specified above?

But the contradiction is only apparent, as becomes obvious when the incriminating passage of Bergson is more carefully rephrased. According to Bergson, every duration, in virtue of its intrinsic heterogeneity, is, strictly speaking, not measurable at all. On this point there is a remarkable agreement between him, Poincaré and Russell, although in the latter two it is expressed in a different language. They both pointed out that time intervals cannot be measured, since, being successive, they cannot be superposed. One cannot displace a unit-interval of time through time in the same way as we can displace a unit of length through space. The superposition of lengths and, more generally, of geometrical figures, and their quantitative comparisons is possible in space, not in time. In Bertrand Russell's words, 'the axiom of free mobility' which makes possible the free displacement of geometrical figures in Euclidian space, and, more generally, in the spaces of constant curvature, does not have its counterpart for time. Poincaré concluded that the equality of successive intervals cannot be intuited precisely because of the impossibility of superposing them; thus it can only be *defined*.[2] The language of Poincaré and Russell sounds quite different from that of Bergson; but what else are all three thinkers trying to convey except the impossibility of applying the quantitative treatment, applicable only to static, geometrical relations, to genuine successions?

In truth, even Bergson's language when attentively analyzed is less different from that of Poincaré and Russell than it at first glance appears. In his very first book he pointed out that what is measurable in motion is not its mobility (*mobilité*) – what is genuinely temporal – but only its *geometrical* aspect, more specifically, the static trajectory which motion, so to speak, leaves 'behind' itself in space. It was the confusion of motion with its trace left in space which generated the paradoxes of Zeno. To measure motion and, more generally, time is thus, in Bergson's words, nothing but "counting simultaneities".[3] Today we would speak of the spatiotemporal coincidences which are by their own nature *instantaneous*, that is, devoid of duration. Duration itself, separating these coincidences, escapes our measurement. We can, it is true, increase the number of such simultaneities indefinitely to narrow the interval of duration separating them; but the interval itself is merely narrowed without ever vanishing.

Even the differential of time, d*t*, while supposedly 'smaller than any conceivable finite interval', retains its character of an interval separating two successive boundary-instants which are to be 'infinitely near' each other without, however, coinciding. For this reason mechanics does not define duration itself, but only the *equality* of two durations: "Two intervals of time are equal when two identical bodies, in identical conditions at the beginning of each of these intervals and subject to the same actions and influences of every kind, have traversed the same space at the end of these intervals." Bergson then observes:

> In other words, we are to note the exact moment at which the motion begins, i.e. the coincidence of an external change with one of our psychic states; we are to note the moment at which the motions ends, that is to say, another simultaneity; finally, we are to measure the space traversed, the only thing, in fact, which is really measurable. Hence there is no question here of duration, but only of space and simultaneities. To announce that something will take place at the end of a time *t* is to declare that consciousness will note between now and then a number *t* of simultaneities of a certain kind.[4]

But – as we asked before – how can Bergson then speak of any duration in a quantitative way? How can he say that the minimum sensation lasts for 0.002 sec.? It was difficult – in truth impossible – to give a satisfactory answer to such a question in Bergson's first book, that is, as long as he believed in the *absence* of duration in the physical world. But in *Matter and Memory* this original view was given up; the reality of physical duration was admitted. This duration, however, consists of a succession of events which in our macroscopic – or rather macrochronic – perspectives are of negligible duration and may thus be very approximately replaced by durationless instants; consequently, the physical duration can be practically replaced by the homogeneous, infinitely divisible "time-length" (*temps-longueur*) in which it is possible to establish quantitative relations between its intervals. Thus when we assert that the elementary *minimum sensible* has a quantitatively definite 'length', we merely say that it is *contemporary* with a certain series of physical events which being *almost* countless and, each of them *almost* without duration, are in their totality *almost* equivalent to a certain segment of homogeneous and measurable time. By this lengthy and rather tedious formulation any seeming contradiction is avoided. What Bergson presupposes is not, contrary to René Berthelot's claim, homogeneous, infinitely divisible time, but time *almost* divisible *ad infinitum*. The whole difference between classical and modern physics can be expressed by such "almosts" of this and similar kind. Bergson's real assumption is the *contemporaneity* of psychological and physical duration, or, more generally, of durations of different degrees of tension. Such 'simultaneity of fluxes' or simultaneity of temporally extensive moments is explicitly conceded by Bergson.[5] He only denied the simultaneity of durationless instants. As we shall see, it is this simultaneity of fluxes which, together with the quasi-instantaneous character of physical events, constitutes the dynamic basis of spatiality.

The answer just given to the basic objection of René Berthelot will provide us with a basis for answering the *second* serious objection. It seems that Bergson's rejection of the mathematical continuity of physical time implies the *atomic structure of duration* in the physical world. What else can the denial of infinite divisibility of time mean except the assertion that no duration is divisible *ad infinitum*? And if it is not infinitely divisible, does it not mean that we have to substitute the finite atomic intervals of time for the classical durationless instants? This would bring Bergson's theory of physical duration into the close neighborhood of the *chronon* theory of time which under the pressure of the discoveries of quantum physics began to emerge in 1913. As already mentioned, it was Poincaré who in 1913 first considered seriously the idea of 'quantum of time'. Six years later A. N. Whitehead, mainly on epistemological grounds, expressed doubts about the physical existence of mathematical instants, and one year later in *The Concept of Nature* admitted more explicitly the possibility of 'quanta of time'.[6] Similar views were expressed subsequently by a number of physicists from J. J. Thomson in 1926 to Arthur March in 1960; the list of them – which certainly can be extended – can be found in my previous book.[7]

But a closer look will show that the only feature which is common to Bergson's 'pulsational' theory of duration and the chronon theory is of a *negative* kind: they both reject the mathematical continuity of physical time. In the form in which it was formulated the chronon theory is both self-contradictory and incompatible with Bergson's view. It is self-contradictory since it implies a surreptitious return to the concept of instant which it purportedly denies. To it the objection of René Berthelot applies fully and is unanswerable: if the duration of 'chronon' is of the order of $10^{-24}$ sec, what else can this mean but the assertion that within the interval of that magnitude separating two successive *instants* there are no instants? Thus the idea to be eliminated – durationless instant – reappears in the concept of the instantaneous boundaries of the chronon itself. The contradiction is obvious. While in most of the atomistic theories of time the reintroduction of the concept of instant was only covert, it was explicitly acknowledged by Ludwig Silberstein in his own theory of "discrete space-time." In referring to several authors advocating the chronon theory, Silberstein wrote:

The reader will notice that while all these authors have in mind 'time quanta', i.e. little bits of time, but finite ones, I propose more radically a discrete manifold of time *instants*, of no extension, and similarly for space.[8]

But such honesty and explicitness hardly saves Silberstein's theory from its intrinsic difficulties. He obviously retains the durationless instants, but instead of assuming that they form a dense mathematical continuum, he assumes that they form a discrete manifold, the number of instants per second being, according to him, no more than of the order of $10^{-18}$ sec. This means, if it means anything at all, that two successive durationless instants are separated by – what? By a temporal vacuum, a sort of 'gap in time'? But then they would coincide and we could not speak of their separation, not even of their distinction, at all. In order to remain distinct they should be separated by *temporal* intervals of the postulated length – $10^{-18}$ sec. This means that Silberstein must concede, like other time-atomists, the reality of 'little bits of time', of 'time quanta', the very idea which he rejects in the above passage. The basic contradiction thus remains.

But it is equally obvious that the chronon theory is incompatible with Bergson's theory of duration. Needless to repeat what was said on this point in Part II. Time, whether psychological or physical, cannot be built out of mutually external parts, whether these parts are temporal segments or point-like instants. The doctrine of mathematically continuous time composed of instants was replaced by the theory of contiguous, finite, temporal segments; in either case the mutual externality of the component parts was retained. The inadequacy of the chronon theory is basically due to the underlying, unconscious tendency to symbolize time by a geometrical line. It overlooks the fact that it is equally inadequate to symbolize the moment of real duration, either by an unextended geometrical point or a linear segment of a finite length. Neither symbolism does justice to both the continuity and the heterogeneity of duration; neither symbolism can adequately convey the immanence of the anterior phase in its successor and the qualitative diversity of both. Yet, without both these features no duration is possible. The fact that the chronon theory covertly reintroduces the concept of durationless instant shows that it is hardly more than a half-hearted, isolated, *ad hoc* correction, a hastily improvised patchwork. While it remains an important symptom of the inadequacy of the classical concept of *mathematically continuous* time and of the correlated concept of mathematical instant, it is hardly anything more than a symptom.[9]

## NOTES

[1] *Un romantisme utilitaire*, II, p. 227.
[2] B. Russell, *An Essay on the Foundations of Geometry*, p. 176; H. Poincaré, *The Value of Science*, Chapter II 'The Measure of Time' (*F.S.*, pp. 223–234, esp. p. 224.)
[3] *T.F.W.*, p. 108.
[4] *Ibid.*, pp. 115–116.
[5] *D.S.*, pp. 52–53.
[6] *The Concept of Nature*, p. 162.
[7] Cf. Note 4 of the previous Chapter. To this list the following references can be added: B. Russell who spoke of "the corpuscles of time" as early as in his *Mysticism and Logic* (p. 124 of Pelican Books ed.) and of "quantized geodetic lines" in *Analysis of Matter* (p. 304); Georges Matisse, *Interprétations philosophiques des relations d'incertitude*, Hermann, Paris, 1936; Edmund Whittaker, *From Euclid to Eddington*, Cambridge Univ. Press, 1949, p. 41; G. J. Whitrow, *The Natural Philosophy of Time*, Thomas Nelson & Sons, London, 1961, pp. 153–157.
[8] Ludwig Silberstein, *Discrete Space Time*, University of Toronto Studies, Physics Series, 1936, p. 127.
[9] H. Margenau, *The Nature of Physical Reality*, McGraw-Hill, New York, 1950, p. 156. Cf. also R. B. Lindsay-H. Margenau, *Foundations of Physics*, Dover, New York, 1957, pp. 76–78.

CHAPTER 4

# THE REJECTION OF THE CARTESIAN DOGMA OF
# THE COMPLETELY EXTENSIONLESS MIND

It has already been pointed out that the original sharp dualism of the *Essai*, which opposed the temporal realm of consciousness to a physical world completely devoid of duration, was attenuated in *Matter and Memory* and *Creative Evolution* by conceding the reality of the duration of matter itself. But so far we have had no opportunity to stress that the original sharp opposition between the 'mental' and the 'physical' was further attenuated by Bergson's modification of his original view about the utterly inextensive character of the mental data. "Neither is space so foreign to our nature as we imagine, nor is matter as completely extended in space as our senses and intellect represent it."[1] Our following exposition will be largely a detailed commentary on this quotation which concisely formulates the way in which the original sharp dualism of Bergson's first book was modified. Instead of a spaceless, psychological duration, he posited a mental realm which, though intrinsically temporal and foreign to the static space of classical physics, participated *in some degree* in spatiality in general. Instead of the durationless space of matter he posited extension which, though far less heterogeneous than psychological duration, still in some degree participates in duration in general. Such a theory clearly implies the assumption of *different degrees of spatiality* and of *different degrees of durational tension* or rhythms. While the meaning of 'different rhythms of duration' has already been explained, it remains to be shown how they are related to the different degrees of spatiality. We can say now, anticipating future discussion, that different degrees of durational tension and different degrees of spatiality are not only closely correlated, but that they are different aspects of one and the same thing.

Thus Bergson's denial of a static space, completely divorced from becoming, did not mean an unqualified negation of spatiality in general. Today the difference between the Newtonian classical space and spatiality in general is known to everybody who is only superficially acquainted with the more recent physics and non-Euclidian geometry; it is equally well known that certain species of spatiality, especially those used in modern cosmologies, are *not* static since they are, in virtue of the changing spatial constant, so to speak incorporated into becoming. Bergson was naturally unaware of such a generalization of the concept of space; its anticipation by one of his contemporary countrymen, Calinon[2], was too bold and solitary to attract his attention. It was again a concrete psychological observation which led him to distinguish between the classical space and spatiality in general, or in his own terminology, between *l'espace* and *l'extension*. In agreement with William James and James Ward he did not believe that our sensations are originally extensionless and that only gradually in the process of our mental development the medium of space is constructed in which these sensations are then localized.

Such a view, upheld by the British associationists, is not borne out by psychological experience. An attentive observation, undistorted by the Cartesian metaphysical prejudices, discloses that *every sensation* is by its own nature *extensive*; this is true not only of the sensations of sight and touch, but all other sensations as well, though less prominently. As James wrote in his *Principles of Psychology*:

Now my first thesis is that this element, discernible in each and every sensation, though more developed in some than in others, is the original sensation of space, out of which all the exact knowledge about space that we afterwards come to have is woven by processes of discrimination, association, and selection. 'Extensity,' as Mr. Ward calls it, on this view, becomes an element in each sensation just as intensity is. The latter every one will admit to be a distinguishable though not separable ingredient of the sensible quality. In like manner extensity, being an entirely peculiar kind of feeling indescribable except in terms of itself, and inseparable in actual experience from some sensational quality which it must accompany, can itself receive no other name than that of *sensational element*.[3]

Referring to this passage Bergson observed:

Contemporary psychology is more and more impressed with the idea that all our sensations are in some degree extensive. It is maintained, not without an appearance of reason, that there is no sensation without 'extensity' or without a 'feeling of volume'. English idealism sought to reserve to tactile perception a monopoly of the extended, the other senses dealing with space only in so far as they remind us of the data of touch. A more attentive psychology reveals to us, on the contrary, and no doubt will hereafter reveal still more clearly, the need of regarding all sensations as primarily extensive, their extensity fading and disappearing before the higher usefulness of tactile, and also, no doubt, visual extensity.[4]

And on the same page:

> But the truth is that space is no more without us than within us, and that it does not belong to a privileged group of sensations. *All* sensations partake of extensity; all are more or less deeply rooted in it; and the difficulties of ordinary realism arise from the fact that, the kinship of the sensations one with another having been extracted and placed apart under the form of an indefinite and empty space, we no longer see either how these sensations partake of extensity or how they can correspond with each other.

But the acceptance of this view has far-reaching consequences. It seems directly to contradict Bergson's affirmation of the non-spatial character of psychological duration on which all his criticism of the fallacy of spatialization is based. If all sensations are extensive, our imagination in virtue of the sensory nature of images will also become tinged with spatiality, and with it a considerable portion of our psychological self, whose spacelessness had been so vigorously stressed in Bergson's first book. This is what Bergson admits when he says "space is no more without us than within us".

Yet, the discrepancy is less serious than it appears. The whole opposition between the *Essai* and *Matière et Mémoire* is only the difference between the sweeping simplicity of the original view and its more balanced and more mature formulation. It is true that Bergson should have been more careful in his use of terms in the last quotation; it cannot be denied that the term 'space' in the last quotation is out of place and either 'spatiality' or 'extension' should be used instead. Extensiveness or 'voluminousness' of sensory qualities has neither the homogeneity nor infinite divisibility of geometrical space.

As James pointed out, it is difficult to speak even of the *dimensions* of psychological extension or voluminousness: "Its dimensions are so vague that in it there is no question as yet of surface as opposed to depth; 'volume' being the best short name for the sensation in question."[5] In most instances, Bergson's language makes this distinction between what we today call *geometrical* and *representational* space; it is the latter which is called by him 'extension', the former being called simply 'space' ('*l'espace*'). Bergson's opposition to the doctrine of inextensive sensations may be traced to his first book in which he criticized the psychological theories of Bain, Lotze and Wundt on this point[6]; in truth, his claim in the same book that the associationist psychology, though inapplicable to 'the deep self' (*le moi profond*), is at least approximately applicable to its superficial strata communicating with the external world, definitely implies that a certain degree of spatiality or extensiveness is not altogether foreign to our mind.

If, in proportion as we get away from the deeper strata of the self, our conscious states tend more and more to assume the form of a numerical multiplicity, and to spread out in a homogeneous space, it is just because these conscious states tend to become more and more lifeless, more and more impersonal. Hence we need not be surprised if only those ideas which least belong to us can be adequately expressed in words: only to these, as we shall see, does the associationist theory apply.[7]

Bergson was cautious enough not to say that conscious states *assume* the form of numerical multiplicity, but only *tend* to assume it; but since, according to him, numerical multiplicity and homogeneous space are correlated, this implies that in some sense the sensory 'superficial' strata of our self are, so to speak, *nearer* to the space of the physical world than the deep self. Only the latter, the locus of pure becoming, *durée réelle*, is altogether foreign to space. Thus even his first book indicates that the duality of the temporal, mental realm and the spatial world of matter is less sharp than it appears. But this was at the time nothing more than a mere hint at a theory whose systematic formulation and elaboration had to wait until the publication of *Matter and Memory*.

Nevertheless it was a precious hint and Bergson merely had to draw out all the consequences implied in it. Sensations and sensory images, though more or less external to each other, nevertheless share the successive character of 'the stream of consciousness'; thus they represent the intermediate zone between the extensionless or *practically* extensionless temporality of imageless and non-sensory thought and the *almost* homogeneous and *almost* mathematically continuous space of the external world. (As we shall see, even physical space is not identical with the geometrical and homogeneous space of classical mechanics.) This intermediate position of sensory data between our inner extensionless dynamic core and physical space accounts for the mixed character of our external perception, which is "no more within us than without us." Perception is not a mere internal phantom, a sort of veridical hallucination which happens to be in agreement with the object which it represents, even though it is completely external to it; nor does it altogether coincide with its object as naive realists (and modern neorealists) believe. It participates in the general subjectivity of all mental data, but by its extensiveness it also participates in a certain degree in the objectivity of physical space. But all this sounds intolerably vague as long as Bergson's theory of different degrees of spatiality is not fully explained.

Before we venture further let us sum up the results so far obtained. The main thesis of his first book was qualified rather than thoroughly modified or rejected; in *Matter and Memory* the nature of mind remains as foreign to the classical, homogeneous, static and mathematically continuous space as it was in the first book. But in addition, the inapplicability of classical, static space to physical reality is now stressed, since even matter itself is conceived in terms of duration while still retaining the extensive character disclosed in our sense perception. We shall see that the most concise and also most accurate term by which Bergson's concept of matter is described is 'extensive becoming.' Thus the over-all view of reality in *Matter and Memory* is far more complex than in the *Essai*. Instead of the original dualism of spaceless becoming and timeless matter, there is one single dynamic reality stratified, so to speak, into at least three or four different regions: (a) 'deep self' (*le moi profond*) corresponding to the pure duration of the first book, foreign to space and extension, the locus of imageless thought, of the deepest feelings and volitions; (b) sensory perception and imagery, corresponding to the 'superficial self' (*le moi social*) of the first book, extensive and multiple in its nature, even though its extension is different from geometrical space and whose multiplicity, while being more pronounced than the qualitative multiplicity of the innermost duration, still remains distinct from the multiplicity of arithmetical units; (c) physical reality itself where the tendency toward numerical multiplicity is far more pronounced and whose extension is so much closer to the homogeneous space of Euclid and Newton that for centuries it was either identified with it (Descartes) or lodged in it (atomism and classical physics in general); (d) geometrical space itself, static, homogeneous, and continuous (i.e. divisible *ad infinitum*), to which no concrete reality exactly corresponds, even though the macroscopic space approximates it closely enough for practical purposes. Only this ideal limit is completely devoid of succession; all other strata participate in duration, even though the rhythms of their durations, corresponding to different temporal spans and different degrees of spatiality, vary from one region to another.

## NOTES

[1] *E.C.*, p. 222.
[2] A. Calinon, 'Les espaces géométriques', *Revue philosophique* **XXVII**, 595.
[3] W. James, *The Principles of Psychology*, II, pp. 135–136.
[4] *M.M.*, pp. 213–214.
[5] James, *op. cit.*, p. 136.
[6] *T.F.W.*, pp. 93–94.
[7] *T.F.W.*, p. 136.

CHAPTER 5

# THE CORRELATION OF
# DIFFERENT TEMPORAL RHYTHMS
# WITH DIFFERENT DEGREES OF EXTENSION

While Bergson was definitely influenced by James in his view about the intrinsically extensive character of *all* sensations – a view which was clearly implied in Kant's *Transcendental Aesthetic* – he went beyond James and Kant in his effort to *correlate* psychological extension with the general temporality of consciousness. One of the fundamental questions of *Matter and Memory* – the question overlooked by almost all critics and commentators, both friendly and hostile – was as follows: is it possible to *relate* the extensive character of our sensory perception of the physical world to the basic temporal structure of our consciousness as it was described and analyzed in the *Essai*? Bergson's answer was surprising: what we call extension is merely another aspect of 'diluted' or 'extended' duration, i.e. of duration with a restricted temporal span. More specifically, *different degrees of spatiality* correspond to *different degrees of durational tension*; the acceleration of the temporal rhythm (i.e. the restriction of the temporal span) generates *ipso facto* extension itself. *"L'extension apparaît seulement, disons-nous, comme une tension qui s'interrompt."*[1] This sentence and all other sentences of a similar nature occurring in *Creative Evolution* certainly had a mysterious and even mystical ring which baffled equally those who disagreed and those who were generally sympathetic to Bergson's philosophy of duration. The characterizations of matter as 'the inversion of *élan vital*' or as '*réalité qui se défait*' were generally considered as picturesque similes without any definite content; sometimes they even appeared as simple *jeux des mots*, a mere juxtaposition of the rhymed words: *faire–défaire*, *tension–extension*. While the adversaries of Bergson hurried to attack what they regarded as the weakest and the least tenable part of the doctrine, the disciples and sympathizers of Bergson either remained silent or expressed politely their embarrassment or their reservations.[2] We are approaching the most difficult, most elusive as well as the least known and least understood part of Bergson's thought which nevertheless was, according to Bergson's own words, an essential part of what he himself called his "theory of duration".[3]

For this reason, nothing would be more unfair than to dismiss the above expressions of *Creative Evolution* too hastily. It remained generally unnoticed that they summed up, unfortunately in a needlessly metaphorical style, the precise reasonings of *Matter and Memory* and *Introduction to Metaphysics*. Only in the context of these two works can the meaning of the enigmatic passages in *Creative Evolution* become clear. Today when a large number of contemporary thinkers spend an enormous amount of effort and time in trying to give an intelligible meaning to the most abstruse passages of Husserl and Heidegger, and to the most obscure aphorisms of the early Wittgenstein, it is only fair to adopt toward the difficult passages in Bergson's writings the same attitude of "initial logical charity" which John N. Findlay recommended to the study of Hegel.[4] Such an attitude toward what Bergson called "the ideal genesis of matter" will prove itself far more rewarding than a hasty reading of the relevant texts suggests (and also more rewarding than the same attitude adopted to some authors mentioned above).

The explanatory elucidation of the correlation between duration and extension – the elucidation which even Whitehead accused Bergson of not having given[5] – is to be found in the following crucial passages of *Creative Evolution*, provided that they are read with care and that they are placed into the proper context of his two previous works, mentioned above:

Let us then concentrate attention on that which we have that is at the same time the most removed from externality and the least penetrated with intellectuality. Let us seek, in the depths of our experience, the point where we feel ourselves most intimately within our own life. It is into pure duration that we then plunge back, a duration in which the past, always moving on, is swelling unceasingly with a present that is absolutely new. But, at the same time, we feel the spring of our will strained to its utmost limit. We must, by a strong recoil of our personality on itself, gather up our past which is slipping away, in order to thrust it, compact and undivided, into a present which it will create by entering. Rare indeed are the moments when we are self-possessed to this extent: it is then that our actions are truly free. And even at these moments we do not completely possess ourselves. Our feeling of duration, I should say the actual coinciding of ourself with itself, admits of degrees....

Now let us relax the strain, let us interrupt the effort to crowd as much as possible of the past into the present. If the relaxation were complete, there would no longer be either memory or will – which amounts to saying that, in fact, we never do fall into this absolute passivity, any more than we can make ourselves absolutely free. But, in the limit, we get a glimpse of an existence made of a present which recommences unceasingly – devoid of real duration, nothing but the instantaneous which dies and is born again endlessly. Is the existence of matter of this nature? *Not altogether, for analysis resolves it into elementary vibrations, the shortest of which are of very slight duration, almost vanishing, but not nothing. It may be presumed, nevertheless, that physical existence inclines in this second direction, as psychical existence in the first.* (Italics added.) [6]

This is obviously a view of matter inspired by Leibniz' famous statement which, via Ravaisson, influenced Bergson: "*Omne enim corpus est mens momentanea sive carens recordatione.*" [7]

Much more recently, A. N. Whitehead, in one of his last essays, formulated a nearly identical view. We shall return to it later. The passage quoted (in Arthur Mitchell's translation, which probably can be improved) refers to two different mental attitudes or two "levels of thought": one characteristic of intellectual effort and of non-sensory thought which is constituted by a greater mnemic span or, in Bergson's words, by a greater "condensation of the past in the present"; the second, more or less passive, the level of sensory images and impressions, characterized by a narrower temporal span and by either the absence or at least a reduction of volitional effort. The passage clearly implies that although the active and passive attitudes are mutually antagonistic, there is a gradual transition between them, that is, the whole spectrum of intermediate attitudes between the maximum effort and its absence. This view is expressed more fully in various sections of *Matière et mémoire* and *L'Energie spirituelle*, in particular in his essay '*L'Effort intellectuel*'.

It is true that the antagonism between abstract non-sensory thought and concrete imagery has been known since the time of Plato, in truth since the time of Parmenides; but it has never been studied with a greater subtlety and care than for the last three quarters of a century. William James' unsurpassed analysis of imageless thought in his *The Principles of Psychology* established in the stream of thought the distinction between 'the places of rest' ('substantive parts') and 'the places of flight' ('transitive parts') which corresponds to the distinction between the sensory nucleus of thought and its non-sensory fringe, consisting of the 'feelings of relations' or 'the feelings of tendency'. [8] This dichotomy corresponds very approximately to the two antagonistic mental attitudes described by Bergson. (In truth, I am inclined to believe that James' view, which in its essential features was already expounded in 1884, did influence Bergson's thought even though Bergson himself was not clearly aware of it.)[9] Imageless thought was studied further by Alfred Binet and Albert Burloud in France, by the Würzburg school and Gestalt psychologists in Germany[10]; in the English speaking countries the interest in such studies, after their temporary eclipse by the influence of behaviorism and the sensualistically oriented neopositivism, was revived again, particularly by Brand Blanshard in his now classic *The Nature of Thought*.[11] For our present purpose suffice it to say that Bergson's description is based on well recognized psychological facts; what is novel in his approach is that he correlated the dichotomy between imageless active and sensory passive thought *with the differences in the temporal span*. Following his first book there was the tendency in Bergson's thought to equate the intensity of effort with the degree to which the past is "condensed" in the present: the wider the mnemic span, the greater the effort, the wider the field of choice and thus the greater indetermination. This makes intelligible Bergson's claim that consciousness in reducing its temporal span moves in the direction of the far less condensed, "diluted" duration of the physical world.

But what still remains obscure is how *extension*, which is the most conspicuous character of matter, can arise by a mere reduction of the temporal span. The distrust which Bergson's view of the physical world aroused was due precisely to the impression that matter seemed to be conceived of in exclusively temporal terms, as a mere *succession of events*. Is not such a 'temporalisation of matter' as arbitrary and artificial as the 'spatialization of time' which Bergson so effectively criticized? The weight of this objection is greatly diminished in light of the fact that Bergson, as we pointed out in the last chapter, gave up the Cartesian myth of the entirely extensionless mind; thus the character of extension, being inherent in mind itself, does not arise *ex nihilo*, it only becomes more pronounced when the mnemic span is reduced:

The more we succeed in making ourselves conscious of our progress in pure duration, the more we feel different parts of our being enter into each other, and our whole personality concentrates itself in a point, or rather a sharp edge, pressed against the future and cutting into it unceasingly. It is in this that life and action are free. But suppose we let ourselves go and, instead of acting, dream. At once the self is scattered; our past, which till then was gathered together into the indivisible impulsion it communicated to us, is broken up into a thousand recollections *made external to one another*. They give up interpenetrating in the degree that they become fixed. Our personality thus descends in the direction of space. It coasts around it continually in sensation. We will not dwell here on a point we have studied elsewhere. Let us merely recall that *extension admits of degrees*, that all sensation is extensive in a certain measure, and that the idea of unextended sensations, artificially localized in space, is a mere view of the mind, suggested by an unconscious metaphysic much more than by psychological observation. No doubt we make only the first steps in the direction of the extended, even when we let ourselves go as much as we can. But suppose for a moment that matter consists in this very movement pushed further, and that physics is simply psychics inverted. (Italics added.)[12]

As regards space, we must, by an effort *sui generis* follow the progression or rather the regression of the extraspatial degrading itself into spatiality. When we make ourselves self-conscious in the highest possible degree and then let ourselves fall back little by little, we get the feeling of extension: we have an extension of the self into recollections that are fixed and external to one another, in place of the tension it possessed as an indivisible active will. But this is only a beginning. Our consciousness, sketching the movement, shows its direction and reveals to us the possibility of continuing it to the end; but consciousness itself does not go so far. Now, on the other hand, if we consider matter, which seems to us at first coincident with space, we find that the more our attention is fixed on it, the more the parts which we said were laid side by side enter into each other, each of them undergoing the action of the whole, which is consequently somehow present in it. Thus, although matter stretches itself out in the direction of space, it does not completely attain it; whence we may conclude that it only carries very much further the movement that consciousness is able to sketch within us in its nascent state.[13]

What else can this mean but that matter *extends* itself in space without being *extended* therein, and that in regarding matter as decomposable into isolated systems, in attributing to it quite distinct elements, which change in relation to each other without changing in themselves (which are "displaced," shall we say, without being "altered"), in short, in conferring on matter the properties of pure space, we are transporting ourselves to the terminal point of the movement of which matter simply indicates the direction?[14]

Our task is now to restate these key passages in a language as rigorous and as systematic as possible and to avoid metaphors as much as the structure of our language will allow it. As previously mentioned, the central idea is that of the *variable temporal span*, that is, of the variable synthesis of the immediate past within the temporally extensive present. It should be needless to recall that this is an empirical fact that cannot be disputed even though the estimations of the limits of this span vary. (According to James, the maximum is 12 sec, the minimum about 0.002 sec.) When the rhythm of consciousness is accelerated or when the mnemic span is reduced, this can mean only one thing: that the unity of the original denser present is differentiated into a succession of shorter presents. In saying this, we must not forget that the term 'unity' in the previous sentence should not be taken in its usual arithmetical sense; as repeatedly stated before, no temporal moment is ever barely 'one' in the sense in which a pebble or a bead is. Nor is the succession of new shorter presents ever 'many' in the same arithmetical sense in which a string of beads is. It is thus more accurate to say that *the virtual multiplicity of the original denser present* tends to become *the actual multiplicity* of its sub-moments without ever becoming it entirely. Thus what in a more concentrated duration is a 'recent past' merging with the rearward edge of the denser ('specious') present, in a 'diluted' duration becomes a more *remote* past which subsists *outside* of the shorter present rather than *within* it, even though an attenuated mnemic link still persists as in every duration. This phenomenon occurs in hypnogogic states, in certain dreams and in certain psychasthenic states and it accounts for a subjective lengthening of the intervals of time. The psychological significance of the hallucinatory 'swelling of time' produced by opium intoxication and reported by Thomas de Quincey was recognized as early as Herbert Spencer's *Principles of Psychology*.[15] Spencer also conjectured that the specious present varies widely from one zoological species to another. Historically famous instances of the pronounced, subjective lengthening of time in psychological literature were the dreams reported by Lavalette and A. Maury, to name at least two.[16]

It is thus clear that what was described above is nothing but *the tendency toward exteriorization of successive phases*. Bergson refers to this fact when he writes that "the indivisible impulse of the past" is "broken up into a thousand recollections made external to one another"; or when James describes how the imageless transitive parts of the stream of thought, by a slackening of our attention, dissolve into "substantive parts", that is, into sensory images.[17] On the level of sensory images and perceptions the separation of successive states is sufficiently pronounced to justify the application of the atomistic associationistic psychology *à la* Hume or Condillac. But even here such application is only approximative, and it is interesting that Bergson, in order to illustrate his thesis of the mutual immanence of the temporal phases of every process, selected as apparently the least favorable instance of it, a visual perception of an unchanging, motionless object viewed from the same angle and from the same distance.[18] A more attentive introspective analysis discloses that the discontinuity of sensory qualities is far from being absolute and that Gestalt relations exist even on this traditional hunting ground of the atomistic psychology. But *the tendency toward juxtaposition, toward exteriorization*, is undoubtedly more pronounced and it is precisely this tendency which, *further accentuated*, constitutes the spatiality of the physical world. As we shall see, this tendency toward exteriorization always remains a *tendency* only; it never goes to its uttermost limit – a completely mutual externality of the juxtaposed terms.

This is also the reason why, when we say that in a diluted duration 'the volume of the present' is narrower, we must be on guard against all parasitic associations clustered around this word. We must especially be on guard against the geometrical connotation of the word 'volume'. Recall all previous warnings against the atomistic theory of time. Unlike spatial segments, temporal 'segments' have no point-like extremities. It has repeatedly been stressed that the imageless qualitative difference between two successive moments both *separates* and *links* them; for the mnemic link which joins the present to its immediate ancestor is precisely the act by which the novelty of the present is constituted; for it is an immediate recollection of the antecedent moment which makes the present different from it and it is the emergence of this qualitative difference which is *the very essence* of novelty. The whole paradoxical nature of duration consists of this relation *which separates as well as unites* in the act which is both a retention of the past and the emergence of the present. It would be otiose to repeat what has been already said about this basic 'two-oneness' of duration. What is important in this context is that while the dynamic linkage between successive moments can be made looser, it can never be interrupted completely; and this is the reason for Bergson's claim that while matter *tends* toward a complete juxtaposition of its parts, it never attains it: in other words, that 'it *extends* itself in space without being *extended* therein'.

If we bear this in mind, it becomes understandable that the reduced tension of duration brings up another effect: *the reduction of novelty itself*. For, since the novelty of the present is due to the quality which differentiates it from its antecedent context and is thus inseparable from the mnemic link which joins it to the past, an attenuation of this link means a reduction of the qualitative difference between two subsequent moments; the novelty of the present is less pronounced and the successive phases will tend to be *more similar* to each other. Thus together with the reduction of the temporal span and the tendency toward exteriorization, there are two other concomitant features characterizing 'extended' or 'diluted' duration; the tendencies toward *homogeneity* and toward *determinism*.

A present moment, being, by virtue of its lesser degree of novelty, less differentiated from its ancestor, will yield itself more easily to the deductive effort which will derive it from its antecedent. For any consistent rigorous form of determinism, such as that of Democritus, Spinoza and Laplace, implies a *complete* negation of novelty. Its successful application is possible only when the irreducible difference between successive phases – call it an element of contingency, of novelty, of indetermination – can be neglected for practical purposes. This, as we shall see, is the case with physical duration. A complete equivalence of cause and effect, i.e. the mutual deducibility of one from another regardless of 'the direction of time', was the ideal of classical deterministic explanation. The principle *causa aequat effectum* graphically shows how closely the homogeneity of successive phases and strict determinism are correlated. Émile Meyerson, in a series of now classic works, showed how this type of explanation was based on an assumed 'identity in time', that is, on the simultaneous elimination of both *novelty* and *heterogeneity in time*. It is hardly surprising that Bergson was among the first who recognized the significance of Meyerson's epistemological analyses.[19]

Thus there are *four features* of extended duration: the tendency toward the exteriorization of successive phases, the narrowing of the present moment, the tendency towards homogeneity and strict determination, that is, toward the qualitative identity of successive moments. It is needless to stress the fact that not only the last two features, but *all four* are closely related, being only different and complementary aspects of the same dynamic process which Bergson calls a *relaxation* (*relêchement*) *or extension* of time. It is clear that various expressions of Bergson, like, for instance, 'regression of the extraspatial degrading itself into spatiality' and other of similar kind have only a metaphorical significance and that visual associations which they may evoke are nothing but virtual traps for our imagination. All such expressions, as long as they are isolated from their broader and more comprehensive context, are nothing but simple assemblages of words devoid of any coherent meaning and possessing only a certain emotional color. This vague emotional tinge is about the only thing which the majority of Bergson's adversaries grasped and against which they reacted so violently. Even sympathetic readers found the passages quoted above unintelligible and concluded, in Bergson's own words, that this was only an accessory part of his doctrine.[20] The *very opposite* is true: it is an *essential* part without which the whole structure of *Matter and Memory* and *Creative Evolution* will remain hopelessly obscure.

## NOTES

[1] *C.E.*, p. 267.

[2] William James even in his enthusiastic letter to Bergson about *Creative Evolution* (June 13, 1907) admitted that Bergson's theory of matter was obscure to him. The same admission by James Ward in his letter to James of April 12, 1909. (Cf. Ralph B. Perry, *Thought and Character of William James*, II, pp. 618–620; 652.) Vladimir Jankélevitch, without doubt one of the best specialists in Bergson's thought, regarded the passages in Ch. IV of *Matter and Memory* as "*les plus obscures et les plus embarrassantes*" and the corresponding passages of *Creative Evolution* as "paradoxical". Cf. the first edition of his book *Bergson*, Alcon, Paris, 1931, pp. 158, 240.

[3] Bergson's letter to the author July 3, 1938.

[4] J. N. Findlay, *op. cit.*, p. 26.

[5] A. N. Whitehead, *The Function of Reason*, The Princeton Univ. Press, 1929, p. 23.

[6] *C.E.*, pp. 218–220.

[7] G. W. Leibniz, *Theoria motus abstracti seu rationes motuum universales, a sensu et phenomenis independentes*. Cf. *Philosophische Schriften* (ed. by E. J. Gerhardt), IV, p.221

[8] W. James, *op. cit.*, Chapter IX, especially, pp. 243–256.

[9] Cf. my articles 'Process and Personality in Bergson's Thought', *The Philosophical Forum* XVII (1959–1960), pp. 25–42; 'La signification actuelle de la philosophie de James', *Revue de métaphysique et de morale* 67 (1962), esp. pp. 308–313. The basic ideas of the Chapter IX of *The Principles of Psychology*, 'The Stream of Thought' are contained in his article 'On some omissions of introspective psychology', *Mind* IX (1884) 1–16.

[10] Cf. A. Binet, 'La pensée sans images', *Revue philosophique*, vol. LV (1903); A. Burloud, *La pensée conceptuelle*, Paris 1927; G. Humphrey, *Thinking. An Introduction to Experimental Psychology*, Science Editions, John Wiley & Sons, New York, 1963; Jean M. Mandler-George Andler, *Thinking: From Association to Gestalt*, John Wiley & Sons, New York, 1964.

[11] *The Nature of Thought*, George Allen & Unwin, London, 1948, in particular Chapter VII 'The idea as Image'.

[12] *C.E.*, pp. 220–221.

[13] *Ibid*, p. 227.

[14] *Ibid.*, p. 223.

[15] H. Spencer, *The Principles of Psychology*, 3rd Ed., I, pp. 216–217.

[16] A. Maury, *Le sommeil et les rêves*, Paris, 1878, pp. 161–162; H. Taine, *De l'intelligence*, 16th Ed., I, pp. 400–404.

[17] W. James, *op. cit.*, I, p. 244.

[18] *C.E.*, p. 4 (Already quoted in Chapter 8, Part II.)

[19] Cf. Part II, Chapter 5. Note 8.

[20] Cf. Note 3.

# CHAPTER 6

# JUXTAPOSITION AS THE IDEAL LIMIT OF
# DISTENDED DURATION

The extreme theoretical limit of the process of distention of duration, as it was described above, would be, properly speaking, a *complete suspension of time* or, rather, its complete transformation into a homogeneous and static space. For by virtue of the increasingly restricted temporal span the successive phases of duration would become more and more *external* to each other until their complete mutual exclusion would become equivalent to the complete externality of the juxtaposed terms. The present moment would shrink to a mathematical instant which, being without duration, would lose its concrete character of novelty and thus would be qualitatively identical to the past. The past itself, lacking any qualitative differentiation with respect to the present, would lose its constitutive character of pastness; it would be a purely verbal 'past' which, instead of *preceding* the present, would *coexist* with it, since the essence of succession consists in the qualitative differentiation between the anterior and subsequent moments. This qualitative differentiation depends, as we have seen, on the fact of *elementary memory*, that is, on the elementary survival of the past in the present. But there is no such survival within a durationless instant; *mens momentanea* lacks *recordatio*. By the same token, the present deprived of novelty, and thus being qualitatively identical with the past, *would not follow it*, since its consecutive character would be purely verbal. Thus in such an obviously impossible limit case, the succession of heterogeneous phases would pass over into the juxtaposition of an infinite number of mathematical, qualitatively identical instants whose more appropriate name would be 'points.' This would be the timeless geometrical world of Spinoza and Laplace in which the future is not only necessary, but literally *pre-exists*, or rather *co-exists*, alongside the so-called 'present' and the so-called 'past'. It would be an entity in all respects similar to classical space, that is, to the mathematical continuum of points without any qualitative differentiation and thus without succession.

As early as his first book Bergson insisted not only that space is a homogeneous medium – which is hardly controversial – but also, conversely, that every homogeneous medium is space.[1] Although this converse claim was nothing but an application of the principle of *identity of indiscernibles*, it appeared rather surprising because of the generally prevailing belief that time is a homogeneous medium as well. It would be otiose to return to the reasons which led Bergson to insist on the fundamental heterogeneity of real time; the whole of Part II deals with it.

Even more paradoxical is the first negative consequence of the view stated above: that classical space conceived as a three-dimensional continuum of juxtaposed homogeneous points *does not possess any concrete physical reality*. This claim appeared especially shocking and paradoxical at the time when Bergson first formulated it in 1896. For although the first faint signs of the coming upheaval in physics were then already discernible, nothing indicated how far reaching this upheaval would be nor how deeply the traditional concepts, including that of space, would be affected. For not only classical Newtonian physics, but *every physics prior to 1900*, insisted either explicitly or implicitly on the physical reality of Euclidian space. In this respect there was no significant difference between the *plenists* who, like Aristotle, the Stoics, Descartes and most of the modern aether theories, insisted on the impossibility of the void, and the atomists of all ages who lodged matter *in* space while leaving the interstices of the void between its parts. In both cases space was regarded as an *objectively existing receptacle of everything physically real* and the question as to whether this container was filled by matter continuously or discontinuously was of secondary importance.

In truth, there was even a strong tendency to regard space not only as a reality among other realities, but as the most real of all, as something closely related to *ens realissimum*. In 1671 Henry More listed all twenty attributes which are common to space and the Supreme Being, such as unity, simplicity, immobility, eternity, immensity, completeness, uncreatedness, etc.[2] Space was regarded by Gassendi as the basis of the divine omnipresence; by Newton as *sensorium Dei*[3]; by Spinoza as one of the divine attributes. This tradition of the divinization of space was traced by Alexander Koyré to the late medieval philosophy[4]; in truth, its roots were contained in the view of Archytas of Tarentum – who, according to F. M. Cornford[5], *invented* the idea of space – that "perhaps place (ὁ τόπος) is the first of all beings, since everything that exists is in a place and cannot exist without a place".[6] It was thus hardly accidental that, as Koyré observed, the void of godless atomism became, for Henry More, God's own extension, "the very condition of His action in the world."[7] Modern space-mysticism or aether mysticism can be found not only among poets such as Hölderlin and Maeterlinck, but even among serious physicists like Oliver Lodge and Hermann Weyl.[8] Although in materialistic metaphysics, space is generally stripped of such a mystical connotation, it still enjoys the highest metaphysical rank; it still remains a sort of *ens realissimum*, being together with matter the only thing which possesses a genuine and underivative reality. But while Bergson's denial of the privileged ontological status of space appeared fantastic in 1896, it appears much less so now in the light of modern relativistic physics.

## NOTES

[1] *T.F.W.*, pp. 98.

[2] *Enchiridium Metaphysicum*, London 1671, VIII.

[3] Cf. M. Čapek, 'Was Gassendi a Predecessor of Newton?', in *Proceedings of the XIth International Congress for History of Science, Ithaca 1962*, Paris 1964, pp. 705–709.

[4] A. Koyré, 'Le vide et l'espace infini au XIVe siècle', in: *Études d'histoire de la pensée philosophique*, Paris 1961, pp. 33–84.

[5] F. M. Cornford, 'The Invention of Space', in *Essays in Honor of Gilbert Murray*, Allen & Unwin, London 1936, pp. 215–235.

[6] Quoted by Max Jammer, *Concepts of Space*, Harvard Univ. Press, 1957, p. 8.

[7] A. Koyré, *From the Closed World to the Infinite Universe*, Johns Hopkins Univ. Press, 1957, p. 154.

[8] Cf. Oliver Lodge, *Ether and Reality*, Ch. X 'Life and Mind and their Use of Ether', with the characteristic quotations from Newton, Larmor and Maxwell; H. Weyl, *Was ist Materie?*, Berlin 1924 who speaks of aether in religious terms and quotes Hölderlin (p. 76); finally Maurice Maeterlinck, *La vie de l'espace*, Paris, 1928.

# CHAPTER 7

# THE NEGATION OF INSTANTANEOUS SPACE IN THE RELATIVISTIC PHYSICS

In various expositions of the special theory of relativity, whether they are technical, popular or semi-popular, we do not frequently find the expression which I placed as the title of this chapter. We do not read about the 'negation of instantaneous space'. All we read is 'the negation of absolute simultaneity'. In truth, we read, if not more frequently at least as frequently, about the 'relativization of the simultaneity of distant events'. In such a formulation the true significance of the revision to which the classical concept of space was subjected is semantically obscured. The most interesting and philosophically most significant implications of such revision are then inevitably *not* grasped. This is just another illustration of the fact that radical conceptual revisions require radical revision of the language; otherwise we are pouring a new wine into the old vessels with the familiar result. Though in this case the famous Biblical simile should be slightly modified: the old vessels do *not* break apart, since nothing is more stubbornly resistant than "ordinary language," whose resistance measures the inertia of the mental habits of which it is a depository. In any case the 'new wine' is inevitably spoiled. For if one thing should be clear, it is that the new concepts of physics are *not* expressible in the language of *homo faber*, in the language fashioned by the macroscopic *milieu*.

Although the term 'space-time' itself was first used only in this century with the advent of relativistic physics, its conceptual counterpart was implicitly present in the very structure of classical physics in the way in which classical space and time were related. The fact that time in rational mechanics played the role of independent variable led Lagrange to use the word "*la géometrie à quatre dimensions*",[1] which anticipated Minkowski's term "the four-dimensional world". Nevertheless the similarity of words is dangerously deceptive; as we shall see, nothing will bring the meaning of the *relativistic* space-time into a clearer focus than its comparison, and its *contrast*, with its *classical* counterpart. The precise meaning of classical space-time as a mode of union between its spatial and temporal components may be graphically illustrated by a three-dimensional model in which the time-dimension is represented by a straight horizontal line while parallel vertical planes, all perpendicular to the 'time-axis', represent successive instantaneous spaces, each of them containing an instantaneous configuration of material particles and thus symbolizing 'the state of the world at a given instant'. It is true that such a model has one dimension less than the real four-dimensional manifold, but classical physics assumed that with the exception of this single feature, the diagram correctly represented the relation between space and time. The history of the physical universe is thus represented by the continuous succession of instantaneous spaces, each of them representing the state of the world at a given instant. In other words, space at any moment is nothing but an instantaneous cross-section, an instantaneous cut across the four-dimensional world process. Each such cut is nothing but a three-dimensional layer containing all the events which have the same date in the stream of the Newtonian absolute time; in other words, all the events which are *absolutely simultaneous*. In the words of Hermann Weyl: "All simultaneous world-points form a three dimensional *stratum*, all world-points of equal location a one-dimensional fiber." [2] This is only a modern formulation of Gassendi's statement about the correlation of absolute simultaneity and the absoluteness of space: "*Et, ut quodlibet Temporis momentum idem est in omnibus locis; ita quaelibet Loci portio omnibus Temporibus subest*".[3]

From this it is evident that the concepts of absolute simultaneity and of instantaneous space are not only closely related but *basically identical*; each instantaneous space is nothing but a class of the events absolutely simultaneous with each other at that instant. Or, in other words: the terms 'the class of the points constituting a three-dimensional layer at an instant' and 'the class of all the events absolutely simultaneous at the same instant' are *altogether synonymous*. Newton said the same thing when he wrote that "each indivisible moment of duration is everywhere" ("*unumquodque durationis indivisibile momentum ubique*")[4]; there is an objective cosmic *Now* and this Now is *everywhere*, since each Now-moment stretches transversally in infinity, across the whole universe. The extent of this stretching is precisely instantaneous space itself,

*Momentenraum* in Carnap's terminology.[5] The essential idea, the importance of which was realized only later in contrast to the revolutionary ideas of the relativity theory, is that at a particular moment an instantaneous cross-section in the four-dimensial world-process *can* be obtained; in other words, that by this operation space can *always* completely and unambiguously be separated from time. In the unambiguity of this operation lies the objectivity of 'Everywhere-Now', that is, the objectivity of absolute simultaneity. All instantaneous, successive spaces are integrated into one changeless, Newtonian space which is both the absolute frame of reference (by means of which the true or absolute motions can be distinguished from the relative and apparent ones), and also the objective substratum of all absolutely simultaneous events.

How successive instantaneous spaces are strung together into one changeless space was a question in which classical physics took little interest. Few philosophers took a real interest in it. Yet, the question becomes meaningful as soon as it is realized that each individual space in virtue of its instantaneous nature is *foreign* to duration, and consequently no change and no motion can take place in it. Change and motion can take place only in an *enduring* space; but can such an enduring space be obtained by juxtaposing durationless spaces even if this juxtaposition is called 'succession'? This obviously is Zeno's problem in a slightly different form; the only difference is that while Zeno faced the problem of constructing motion out of motionless positions, classical physics tried to construct enduring space out of instantaneous spaces; or, in other words, the spatio-temporal becoming out of becomingless 'instantaneous states of the world'. Descartes, following in this respect the Arabian atomists, and realizing that no dynamical link joining the passageless and mutually external instants could be found in these instants themselves, felt this difficulty acutely. He was consistent when he looked for such a link *outside* the instants themselves. Like the Arabian atomists he found the connecting link between instants in the continuous divine creation by which the perpetually perishing world is perpetually re-created.[6] It was another instance of a *deus ex machina* solution; another appeal to God – in this case not to intervene *in* the world machine, but to keep the world machine *moving in time*.

An altogether different approach would be to regard the four-dimensional world process as a *primary given reality* and not as something to be reconstructed or to be explained from the processless entities. In such a view instantaneous spaces are nothing but *fictitious cuts* across the four-dimensional becoming, mere artificial static snapshots to which nothing physically real corresponds. This view was implied in Bergson's criticism of "the cinematographical mechanism of thought"[7] by which the continuity of temporal processes is artificially dissected in static 'states'; and it was *explicitly* contained in his claim, discussed in the previous chapter, according to which geometrical space is never physically realized. But who could have paid any attention to such views in 1896 or even in 1907 when the idea of static space as the absolute frame of reference and the objective substrate of absolute simultaneity was still generally accepted? Physicists, who were mostly unaware of metaphysical subtleties and even of logical difficulties, continued to accept confidently what may be called the *'stratified structure'* of spatio-temporal world-history, each three-dimensional stratum representing one instantaneous space constituted by the class of objectively simultaneous world-points. If they adopt a less dogmatic attitude today, it is because the pressure of new facts forced them to do so.

There is no place here for surveying all the facts and experiments which have led to the relativity theory. Let us only recall that the crisis of the traditional belief in the objective reality of 'Everywhere-Now' had its source in the difficulties which physicists faced at the end of the nineteenth century when they tried to construct a coherent mechanistic model of aether. For aether, postulated as an all-pervading substrate of the electromagnetic and possibly gravitational interactions, was in the eyes of the physicists of that time a concrete, physical embodiment of the Newtonian absolute space. In such an all-pervading medium the electromagnetic vibrations supposedly take place of which the luminous vibrations are only a small fraction.

The material atoms themselves, according to William Thomson's speculations, may be nothing but vortex-configurations within the continuity of such a medium. Despite the enormous difficulties which the construction of such a model faced – the aether would require an extremely small density which would suggest the properties of an extremely rarefied gas, and an extreme elasticity characteristic of solid substances (since only in solid substances can the transverse vibrations occur – it was still hoped that it could retain at least *the basic kinematic features* of the absolute frame of reference by means of which 'the real' motions would be differentiated from the merely 'apparent' and relative ones.

It is true that from the time of Newton it was obvious that all the inertial systems were dynamically equivalent and thus no mechanical observations in them can reveal to us the reality of absolute motion. Yet it was still hoped that they are not *optically and electromagnetically* equivalent and that sufficiently refined observations of the electromagnetic phenomena would reveal to us an 'aether wind' by means of which the true, absolute motion of the Earth through absolute space would be eventually disclosed. This was what Michelson's interferometer experiment, and after it the Trouton-Noble experiment with a vertically suspended condenser, were devised for, and repeatedly performed, without any positive result. Nor could the negative result of such experiments be explained by the 'aether-drag' hypothesis which was incompatible with the observed aberration of light. Nor could it be explained by the Fitzgerald-Lorentz contraction understood in its original realistic sense. This situation led Henri Poincaré to declare as early as 1899 that absolute motion is undetectible in principle.[8]

To conclude from this that absolute motions simply *do not exist since the absolute frame of reference does not exist either* was but one small methodological step, but an enormously difficult one psychologically and philosophically. It was made by Einstein, and amounted to the daring claim that there is no such thing as an 'Everywhere-Now'. This was a direct challenge to the traditional belief in the objective reality of absolute space as an all-containing receptacle which stretches transversally at each moment across the whole universe. In Whitehead's words, there is no such thing as "nature at an instant"; in Eddington's words, there are "no world-wide instants".[9] The logically correlated ideas of absolute space and absolute "Now" stand and fall together; one cannot survive without another. It is against this that our Newtonian subconscious so violently reacts. This resistance shows itself in certain peculiarities of language. We frequently read that absolute space is 'unobservable' or that absolute motion cannot be 'detected'; or that the simultaneity of distant events is merely 'relative'. All such expressions are misleading; they tend to obscure the tremendous *ontological* significance of the special principle of relativity. It is then hardly surprising to hear, for instance, Aloys Müller in 1912 say that '*ratio cognoscendi*' should not be confused with '*ratio essendi*', and that absolute space, though undetectible, is still as real as the equally unobservable Kantian 'thing-in-itself'.[10] The same claim was repeated by such a prominent philosopher as Hans Driesch in 1930 and by a severe critic of Bergson, René Berthelot as late as 1934.[11] There was no such confusion in Einstein's mind; he wrote explicitly in his *Autobiographical Notes*:

*There is no such thing as simultaneity of distant events*; consequently *there is also no such thing as immediate action at a distance* in the sense of Newtonian mechanics. (Italics added.)[12]

From our point of view a converse statement is equally true: *because* there is no such thing as an immediate action at a distance, there is no simultaneity of distant events. If we remember that, as d'Abro correctly observed, a purely spatial distance joining two simultaneous point-events in the Newtonian world is nothing but *a world-line of a particle moving with an infinite velocity,*[13] *and because there are no infinite (i.e. instantaneous) velocities in nature* (the velocity $c$ being the limit velocity), there is no absolute simultaneity, no absolute 'Everywhere-Now.' In other words, since instantaneous space itself is nothing but a dense network of 'instantaneous interactions' and since such instantaneous interactions do not exist, the negative conclusion is inevitable: *there is no such thing as space separable from time; there is no such thing as a purely spacial distance.* There are only *spatio-temporal* distances. Instantaneous spaces are mere artificial cuts in the continuity of the spatio-temporal, or, as Bergson says, "extensive" becoming. In Whitehead's concise summary, since the advent of relativity, "the spatial relations must now stretch across time".[14]

From such a point of view there is a semantically slightly misleading note in Einstein's quotation above. His reasoning is clearly a *modus ponens*:

If there is no simultaneity of distant events, then there is no instantaneous interaction in nature.
There is no simultaneity of distant events.
Therefore there is no instantaneous interaction in nature.

But it is quite clear that the antecedent and the consequent in the major premise are *logically equivalent* and thus can be exchanged. In truth, the converse proposition ("If there are no instantaneous interactions in nature, there is no absolute simultaneity of distant events.") is psychologically *more revealing* and corresponds better to the natural order. It is the limiting character of the velocity of light, implied by its constancy in all

inertial frames, which excludes the possibility of instantaneous interactions and thus removes the possibility of absolute simultaneity. But
then our formulation should be far more radical than Einstein's. The very
adjective 'distant' in the major premise should be dropped entirely, since
it always retains the connotation of 'purely spatial distance'. When such
connotation creeps into the minor premise, Einstein's reasoning acquires
the following self-contradictory character: "There is no simultaneity of
spatially distant (i.e. simultaneous) events...." In other words, the very
language in which the minor premise was formulated is a depository of
the Newtonian conceptual framework! This accounts for the fact that
the true, radical meaning of the conclusion may be and frequently *is*
obscured. Its meaning becomes clear only in the following rephrasing:
*there are no mere spatial distances at all.* Yet, although this conclusion
differs considerably in its psychological impact from Einstein's above
conclusion, it differs from it only semantically. For, once again, the
concepts of 'purely spatial distance' and 'instantaneous interaction' are
altogether synonymous.

It may be objected that while it is impossible to separate space from
time, it is equally impossible to separate time from space. Yet, the
situation is not altogether symmetrical. In the widely popularized formulation that both spatial and temporal relations are equally relativized
in the special theory of relativity, a number of important facts are
obscured. The main misunderstanding is due to a lazy semantic confusion.
It is gratuitously assumed that the relativization of simultaneity is a
relativization of time, while it should far more correctly be called a
*relativization of juxtaposition.* What the textbooks of relativistic physics
call 'space-like connections of events' are neither connections nor are they
'space-like' in the classical sense. They are really *relations of causal
independence,* while their alleged 'space-like' character, defined in the
sense of the juxtaposition of the simultaneous terms, varies from one
observer to another. The term 'space' thus being fully relativized loses
its original connotation as objective, Newtonian space.

Thus philosophically the most significant result of the special theory
of relativity is that *while there is no juxtaposition of events which would
be a juxtaposition for all observers, there are certain types of succession
which remain successions in all the frames of reference.* These types of
succession are represented by the causal series, i.e. by the world-lines,

including also the world-lines of photons. In other words, unlike juxta-position (i.e. 'simultaneity of distant events'), the irreversibility of the world-lines retains an *absolute* significance, possessing a genuine and objective reality independent of the conventional choice of the system of reference.

There is no room here to go into the details of what is an elementary consequence of Minkowski's formula for the constancy of the world-interval. Its significance was pointed out as early as 1911 by Langevin and stressed by A. A. Robb in 1914.[15] I myself dealt with it several times at different places, and I concluded that *the objective status of becoming was strengthened rather than weakened by the special theory of relativity*.[16] This is also shown in that the past and future in the relativistic physics are separated even *more effectively* than in the physics of Newton. While prior to Einstein this separation was made by the infinitely thin, world-wide Present, spreading instantaneously in three dimensions, after Einstein, it is due to the four-dimensional double cone of Elsewhere, or Elsewhen, separating unambiguously and objectively (i.e. for *all* observers) the causal ancestry of each event from its posterity.

All this follows from the relativistic space-time diagram, provided it is properly and attentively interpreted. The same diagram also sheds a new light on one of the basic theses of Bergson's philosophy of the physical world: that *juxtaposition* is merely an *ideal*, i.e. *unrealized limit* of suc-cession. This is precisely equivalent to the basic relativistic affirmation that purely spatial distances, the so-called Now-lines, are mere ideal, never realized limit-cases of concrete spatio-temporal actions. Both in the Newtonian and Einsteinian physics, the faster the causal action, the larger the ratio of the spatial and temporal component of space-time. But while there is a definite upper limit for the velocities in the relativistic world, there was no such limit in the world of Newton; in principle *all* values of velocity between 0 and ∞ were equally admissible. Thus for a considerable time, indeed until the time of Olaf Römer, if we do not take into account the prophetic insights of the ancient atomists, the velocity of light was believed to be infinite, that is, its propagation instantaneous. 'Now' and 'Seen-now' were regarded as synonymous terms, and they still remain so to our common sense as soon as we forget what we learned in a high school course of physics. In truth, the cosmic 'Now' is nothing but an enormous extrapolation of our limited macroscopic experience of 'Seen now'. From the fact that the propagation of light does not need an appreciable time in our daily surrounding it was naturally, but wrongly, inferred that this was true even for larger regions of time-space. (I purposely shall use the term 'time-space' from now on to avoid the misleading static connotation lurking behind the more conventional term 'space-time'.) Thus the Euclidian straight line stretching instantaneously *in infinitum* in both directions was an enormous extrapolation of the *practically* instantaneous 'visual line' or 'optical ray' joining the objects of our biological surroundings. In this way the concept of distance timelessly joining two bodies was generated; and it is by the same token that two bodies, or any number of them, are posited as existing simultaneously or, what is the same, placed in the all embracing receptacle of space.

Nothing is more natural, and more justified, on the level of our sensory perception. But the belief generated by the pressures of our limited environment held firm even after our experience was widely extended. Even after the discovery of the finite velocity of light, man continued to believe that beneath the time-consuming luminous messages from the celestial bodies there is the 'unseen Now', that is, the network of instantaneous relations constituting absolute space and physically concretized in the luminous or electromagnetic aether. Furthermore, gravitation was still regarded as being propagated 'instantaneously', that is, with infinite velocity, its actions thus being embodiments of purely geometrical distances stretching instantaneously across the whole infinite universe. But even when both the electromagnetic aether and the infinite velocity of gravitation departed from physics with the advent of relativity, the idea of the geometrical container, the old Democritean void, timelessly underlying the concrete and time-consuming physical interactions, survives in our subconscious. It survives because the effects of enormously long ancestral experience cannot be overcome at once; it survives in spite of the explicit relativistic warnings that such a belief surreptitiously reintroduces discredited concepts – absolute simultaneity and an absolute frame of reference. As long as the surreptitious influence of the idea remains unchecked, no genuine insight into the nature of relativistic time-space remains possible. As long as it survives, Bergson's and Whitehead's claim that instantaneous space is an artificial cut across extensive becoming remains forever unintelligible.

Let us consider again the relativistic space-time diagram. The causal front-cone symbolizing the potential future posterity of the event has a limited opening according to the relativity theory, which is nothing but an expression of the fact that there is a definite, upper limit to the velocity of causal actions. No action whatever radiating from Here-Now can ever reach the region of Elsewhere; *a fortiori*, no physical action can spread *perpendicularly* to the 'axis of time'. Such impossible actions would represent the world-lines moving with infinite velocity. As the diagram shows, they would constitute both *a realization of instantaneous space as well as a suspension of time*. For the time-component of a purely spatial line would vanish; the time-interval separating the points on a geometrical line is *zero by definition*. Mathematically the same result is obtained when in the Lorentz transformation we substitute $c = \infty$; in other words, if we assume that the velocity of light is infinite. Then the Lorentz equations pass over into the classical Galileo transformation. In other words, the instantaneous space of Galileo and Newton is an *ideal, never realized limit* of the Einsteinian dynamic time-space. It is, so to speak, the relativistic time-space *suspended in its process of becoming*; or, in Bergson's anticipatory phrase, an artificial, static snapshot of the only true reality – 'extensive becoming'.[17]

It is true that even the physicists of the classical period faced the obtrusive reality of becoming. They had to fit it in their system. The only way they could do so was to have the instantaneous spaces strung along in a dense mathematical continuum perpendicular to the 'axis of time'. In this way the separation of space from time, vaguely suggested by our common sense, was completed. Their method had its roots in Zeno's ancient attempt to reconstruct motion out of motionless points, time out of instants; but, unlike Zeno, they failed to see that this method, consistently applied, makes time, motion, and becoming impossible. At best, when it is tempered by a compromise with our irreducibly temporal experience, it produces merely a clumsy imitation of concrete processes. This clumsiness was the main target of all Bergson's books. He lived long enough to see that contemporary physics began to move in the direction he generally anticipated.

This, however, leads us to the complex problem of the relation of Bergson's thought to that of Einstein. We shall see that this problem is complex for two reasons. First, Bergson's criticism of *certain interpretations* of the relativity theory obscured to the reading public not only the implicit agreement between his philosophy and the relativistic physics, but also those agreements which he explicitly pointed out. Second, his justified criticism of certain interpretations of the relativity theory not only obscured to Bergson the implicit agreements to which we referred above, but also led him to his indefensible criticism of the relativization of simultaneity, indefensible even from his own point of view! We shall analyze in due time how this confusion originated. For our present purpose, let us retain the result of the foregoing discussion. Unlike the physics of Galileo and Newton, the relativistic physics insists that the angular opening of the frontward causal cone (as well as the rearward causal cone) is limited by the fact that the light velocity $c$ cannot be surpassed. In the limiting case, its complete opening (or 'flattening', corresponding to the angle $\pi$ in a two-dimensional diagram or to $2\pi$ in a three-dimensional model of space-time) would correspond to the physical realization of the Newtonian instantaneous space, of the 'Everywhere-Now' which our imagination is so reluctant to give up. Yet, such a limit is never physically realized. The spatial-temporal connection never passes over into a purely spatial connection. This is nothing but Bergson's conclusion formulated in relativistic terms: juxtaposition is a merely ideal, fictitious limit of succession or, as he says, of "distended duration".

## NOTES

[1] Joseph Louis Lagrange, *Oeuvres*, Paris, 1867–1892, Vol. IX, p. 357.

[2] H. Weyl, *Philosophy of Mathematics and Natural Science*, Princeton Univ. Press, 1949, p. 95.

[3] Pierre Gassendi, *Syntagma philosophicum*, in *Opera omnia*, Florence, 1727, I, 198.

[4] *Newtoni Opera* (ed. Horsley), III, p. 172.

[5] R. Carnap, 'Über die Abhängigkeit der Eigenschaften des Raumes von denen der Zeit', *Kant-Studien* 30 (1925), 339–340.

[6] On this problem cf. Jean Wahl, *Du rôle de l'idée de l'instant dans la philosophie de Descartes*, 2nd Ed. Paris 1953; A. N. Whitehead, *The concept of nature*, Cambridge Univ. Press, 1920, p. 71; M. Čapek, *The Philosophical Impact of Contemporary Physics*, Van Nostrand, Princeton, 1961, pp. 49–51, 162.

[7] Cf. the very title of Chapter IV of *C.E.*: 'The Cinematographical Mechanism of Thought and the Mechanistic Illusion'.

[8] E. Whittaker, *A History of the Theories of Aether and Electricity*, Thomas Nelson & Sons, London, 1953, II, p. 30.

[9] A. E. Eddington, *The Nature of the Physical World*, Cambridge Univ. Press, 1933, pp. 42–47; A. N. Whitehead, *Science and the Modern World*, Macmillan, New York, 1926, p. 172.

[10] A. Müller, *Das Problem des absoluten Raumes und seine Beziehung zum allgemeinen*

*Raumproblem*, Braunschweig, 1912, p. 42f.

[11] H. Driesch, *Relativitätstheorie und Weltanschauung*, 2nd Ed., Leipzig, 1930, p. 22; R. Berthelot, *Bulletin de la Société française de philosophie* **34** No. 5 (1934), pp. 172–183.

[12] *Albert Einstein, Philosopher-Scientist*, p. 61.

[13] A. d'Abro, *Bergson ou Einstein?*, Paris, 1927, pp. 304–305.

[14] A. N. Whitehead, *An Enquiry Concerning the Principles of Natural Knowledge*, 2nd Ed., Cambridge Univ. Press, 1925, p. 6.

[15] P. Langevin, 'Le temps, l'espace et causalité dans la physique moderne', *Bulletin de la Société française de philosophie* (Séance du 19 octobre 1911), p. 37; A. A. Robb, *Geometry of Space and Time*, Cambridge University Press, 1936, p. 22.

[16] Cf. *The Philosophical Impact of Contemporary Physics*, Ch. XI, XII, XIII, XVII; 'Relativity and the Status of Space', *The Review of Metaphysics* **IX** (1955), pp. 169–189; 'The Myth of Frozen Passage: the Status of Becoming in the Physical World', *Boston Studies in Philosophy of Science*, II, Humanities Press, New York, 1965, pp. 441–465; 'Time in Relativity Theory: Arguments for a Philosophy of Becoming', in *The Voices of Time*, New York, 1966, pp. 434–454.

[17] The term used by Bergson in *C.E.*, p. 345, But "the extensivity of the becoming" in Arthur Mitchell's translation is not a very fortunate rendering of *"le devenir extensif"* of the original.

CHAPTER 8


BERGSON AND EINSTEIN.

THE PHYSICAL WORLD AS EXTENSIVE BECOMING


This leads us to consider in a more detailed way the relation of Bergson's thought to relativistic physics; more specifically, Bergson's comments on Einstein's theory made in his book of 1923. An extensive critical commentary on this book and of the critical reactions which it aroused would require a detailed and documented monograph beyond the scope of the present study. In truth, such a monograph has already been written by A. d'Abro in 1927. Although written from a completely hostile point of view, it will help us to focus our attention on the main objections against Bergson's interpretation of the relativity theory. That it is a proper *interpretation* of the theory and not the theory itself about which he was concerned, was stated explicitly by Bergson himself:

In short, there is nothing to change in the mathematical expression of the theory of relativity. But physics would render a service to philosophy by giving up certain ways of speaking which led the philosopher into error, and which risk fooling the physicist himself regarding the metaphysical implications of his views.[1]

Thus we have to bear in mind that, unlike some other critics of the relativity theory, in particular Hans Driesch, Jacques Maritain and Bergson's critic, René Berthelot, Bergson does accept the fact of the constant velocity of light and its mathematical expression, the Lorentz transformation. He does accept Einstein's rejection of the absolute frame of reference and consequently he rejects the existence of the motionless aether. He fully realizes that only when we completely and unambiguously reject the existence of the absolute frame of reference can complete relativity of motion and the full dynamic equivalence of all inertial systems be upheld:

Let us now return to the three assertions with which we set out: 1) $S'$ shifts with respect to $S$, 2) light has the same speed in both systems, 3) $S$ is stationed in a motionless ether. It is clear that two of these express facts, and the third, a hypothesis: we now have no more than the two facts. But, in that case, the first will no longer be formulated in the same way. We stated that $S'$ shifts with respect to $S$; why did we not just as readily declare $S$ to be shifting with respect to $S'$? Simply because $S$ was judged to be sharing the absolute immobility of the ether. *But there is no longer any ether, no longer absolute stability anywhere.* (Italics added.)[2]

Whether Bergson was consistent when he retained absolute simultaneity and the separability of space from time after rejecting so clearly and unambiguously the reality of the absolute frame of reference is another question which, unfortunately, must be answered negatively as the subsequent discussion will show. On the other hand, this should not blind us to Bergson's other valid and penetrating comments on relativity. The main defect of nearly all discussions of Bergson's book *Durée and simultanéité* was due to the fact that his interpretation of the relativity theory was either enthusiastically accepted or passionately rejected. Certainly a more subtle and more attentive approach is required. Only thus are we able to disentangle the valid elements in his interpretation from those which are not and which, as I shall try to show, are incompatible not only with the relativity theory correctly understood, but also with Bergson's thought itself.

Let us first consider those valid aspects of Bergson's comment which cannot be questioned by any consistent relativist. The titles of the first two chapters of *Durée et simultanéité* are 'La demi-relativité' and 'La relativité complète' respectively. In these two chapters he shows very clearly the distinction between two kinds of relativity: that which was admitted by classical Newtonian mechanics, and the new mechanics of Einstein. Only when we keep the difference between these two kinds of relativity in a sharp focus can certain misunderstandings be avoided. In classical mechanics there were two distinct categories of motions: (a) absolute, i.e. those which were taking place with respect to the absolute space of Newton; (b) relative, which were taking place with respect to other bodies only. Thus a body can be either in absolute motion and at relative rest or at absolute rest and in relative motion. For instance, as long as the sun was believed to be at rest in absolute space, as Newton still believed, it was also moving relatively with respect to the earth, the planets and their satellites. On the other hand an inhabitant of the earth while being at relative rest with respect to the earth, was supposed to be *always* in absolute motion. When the proper motion of the sun was discovered, the situation became only more complicated without being changed basically; the absolute frame of reference was simply shifted but not eliminated. Eventually, it was placed in the hypothetical aether which – apart from its local, tiny, vibratory displacements – was motionless as a whole. It was only when the search for the evidence of absolute motion by Michelson and others repeatedly failed that the distinction between absolute and relative motions had to be dropped and with it inevitably also the concept of absolute frame of reference, whether it was represented by the motionless aether, by Carl Neumann's hypothetical 'Body $\alpha$'[3] or simply by Newton's instantaneous space. In other words, *all* the motions are now relative. The 'half-relativity' of classical mechanics was superseded by the *complete* relativity of the Einsteinian mechanics.

Bergson shows that in various expositions of relativity, the traditional idea of absolute frame of reference very often still unconsciously lingers in the mind of the authors. This is merely a special instance of a far more general phenomenon among contemporary physicists: classical concepts quite often survive underneath conscious rejections and verbal denials. And there is hardly any other idea which is more firmly entrenched in our intellectual subconscious than the idea of an all embracing container-like space in which, and with respect to which, bodies move. Only when we succeed in eliminating from our subconscious the obstinately persisting idea of absolute space, can we grasp fully the basic claim of the special relativity theory that there is no privileged frame of reference and that *all* inertial systems are dynamically equivalent. Then and only then we can realize that *the perfect reciprocity of appearances* is a logical consequence of such equivalences. About it no possible disagreement can exist among those who understand the special theory. From such a point of view, if two systems move with respect to each other with constant speed the contraction of lengths and dilatation of time must occur in *both* of them. Since the claim that time is 'slowed down' in the system $S$ with respect to $S'$ and also is being 'slowed down' in $S'$ with respect to $S$ is clearly selfcontradictory, the only possible interpretation respecting both the law of contradiction and the principle of reciprocity of appearances is to regard the dilatation of time (as well as the contraction of the lengths) as merely *referential*, and not as something taking place in a physical, ontological sense *within* the system observed. In Max Born's words, "the contraction is only a consequence of our way of regarding things and not a change of physical reality."[4]

The main difference between the contraction understood in the original Lorentz' sense and the contraction in the sense of Einstein is as follows: Lorentz contraction was assumed to be real since it was supposed to take place only in the bodies moving *with respect to absolute space*; similarly the dilatation of time was supposedly occurring only in the clocks in *absolute* motion. Neither of them was believed to occur in the bodies and clocks being at absolute rest. The situation changed radically after the elimination of the absolute frame of reference, that is, after the denial of both absolute rest and absolute motion. The contraction of the lengths and dilatation of time became both *reciprocal* and *referential*. To speak of such changes as *really* taking place would mean to slip back to the Lorentz-Newtonian idea of an absolute frame of reference and, in addition, to get entangled in a contradiction. For two clocks cannot be, as Herbert Dingle recently stressed again, such that *each* is slower than the other![5] To avoid this contradiction many writers of textbooks on relativity made a highly unsatisfactory compromise: while nominally speaking of the equivalence of *all* inertial systems, they unwittingly made one of them a privileged one in claiming that the appearances taking place in all other systems, moving with respect to it, *do not occur* in it! This privileged system in the discussion of "the paradox of twins" was our Earth. Against this unconscious anti-relativism, we might even say unconscious geocentrism, Bergson vigorously protested. The insistence on the reciprocity of appearances is one of the most persistent themes of *Durée et simultanéité*.

A concrete analogy with which he illustrates the reciprocity of appearances is especially well chosen. Two persons of equal height when separated by distance in space will appear to each other reduced in size. The greater the distance in space, the greater will be an apparent decrease of height. This is an illusion of perspective which is completely reciprocal; for neither of them regards the other as a dwarf because since very early childhood they are both used to interpreting the reduction of stature as a mere effect of perspective. A similar situation exists in the relativistic contractions of lengths and the dilatations of time. But there are two important differences: these two effects are not the effects of distance, but the effect of *differences in velocities*. The greater this difference is, the more pronounced is the apparent contraction of length and the dilatation of time; but like the effects of the perspective of distance, they are both *referential* and *reciprocal*. Bergson then concludes:

> The plurality of times which I obtain in this way does not exclude the unity of real time; on the contrary, it presupposed it. In the same way the apparent reduction of James' height on different canvasses representing his body at different distances indicates that the real James retains the same height.[6]

What Bergson calls here "the unity of real time" is nothing but *the equality of local or proper times*. On this point even d'Abro must concede grudgingly that Bergson is right.[7] What we then call 'lengthened' or 'dilated' times are nothing but proper or local times observed from different frames of reference. The proper time, separating two events within its own frame of reference, remains unaffected; it *appears* to be dilated only to an external observer who moves with a certain velocity with respect to the events observed.

Against this, the 'hyper-relativists' like d'Abro insist that the distinction between 'apparent' and 'real' in the special theory of relativity is without foundation and that 'the proper time' is not in any sense more real than any other, so-called 'apparent' times. The same view was upheld by Max Born who apparently did not realize that this contradicted his previously quoted statement that the dilatation of time should not be understood in a naively realistic way.[8] This Born said in answering the objections of some anti-relativists who claimed the dilatation of time and contraction of lengths, being physically uncaused, violate the principle of causality; but since the contraction of lengths – and the same is true of the dilatation of time – is not a change of physical reality, 'it does not come within the scope of cause and effect'. But if these effects are not physically real, how can they be better characterized than by the adjective 'apparent' – the very term which Born rejects? D'Abro is equally inconsistent. On the one hand he uses nearly the same simile as Bergson to illustrate an apparent lengthening of temporal intervals. Instead of comparing it with the effects of perspective, he compares it to the apparent distortion of the object reflected in a concave mirror. At the same time, however, he claims that the question, which duration out of the whole set of durations of a certain event is real, is as meaningless as to ask which color of an object is real.[9] For, evidently, the color of an object depends on the illumination, on the angle of observation, on the transparency of the medium, on the motion of the observer; all different resulting colors are *equally* real.

Such inconsistencies and vacillations become intelligible only if we understand the underlying motive from which they spring. It is the fear that the distinction between 'apparent' and 'real' is contrary to the principle of relativity, since it surreptitiously reintroduces the absolute system of reference. Thus if we claim that the 'proper' or 'local' time of the atomic vibration is its *true* time, we select, by the same token, one particular system, that is, the system in which the atom, or rather its nucleus, is at rest. But is this system more privileged than any other? D'Abro and others obviously fear that any such claim would lead us back to the pre-relativistic mechanics.

Unfortunately – or rather fortunately – such fear is completely unfounded. Even a relativist does not question the obvious fact that *every observer is at rest with respect to himself* without equating his own immobility within his system with the absolute motionlessness of Newtonian space. More accurately, omitting any confusing reference to a conscious observer, we could say that *every* system is motionless with respect to itself; and this unique and privileged relation of 'being at rest with respect to itself' was correctly recognized by Bergson as an *"absolutist element of relativism"*. It is, after all, nothing but the physical expression of the law of contradiction, for to 'move with respect to itself' is obviously impossible. From this the privileged character of the 'proper' or 'local' time as well as 'proper length' naturally follows.[10] The local duration thus must be regarded as *the causal nucleus* of all other *apparently* dilated durations, and as such cannot be compared – as d'Abro does – with various colors of an object. For obvious reasons: these varied colors are different wave patterns *reflected* from the object and therefore *external* to the object itself while the proper duration is *inherent* in the very object, or rather in the very event, itself. It *appears* to be lengthened in any other frame of reference by the effect of the "perspective of velocity" in a similar way as a physical object appears distorted and increased in size in a concave mirror placed outside of it. In choosing this second metaphor, d'Abro, without realizing it, came very close to Bergson's view.[11]

But Bergson's correct insight into the equality of proper durations was unfortunately obscured by his unwarranted identification of 'apparent' and 'unobservable'. He claimed that both the dilatation of time and the contraction of lengths are *unobservable in principle* and as such they can never be verified empirically. He reached this conclusion by the following spurious reasoning: every observer perceives only his own local time in which no dilatation occurs. By means of Lorentz's formula he computes that his own duration must appear dilated in any other system which moves with respect to his own with a certain velocity. But as soon as he himself enters any other system this dilatation will vanish, even though he will conclude that it must now take place in the original frame of reference. Thus the relativistic modifications of time and length are always 'phantasmatic', being never experienced by any concrete observer, but only 'attributed' to an external observer who moves with respect to the 'attributing' observer.[12] The illusion thus arises out of the fact that a real observer, in mentally identifying himself with such an external observer, *imagines* himself to perceive the modifications of space and time which nevertheless remain imaginary and disappear as soon as the first observer is effectively placed in the other system.

What is true in the foregoing paragraph had already been pointed out: no observer can experience the dilatation of his local time. But what was *invalid* in it clearly escaped Bergson: his assumption that no observer can perceive anything going on *outside* his own system. Therefore he cannot perceive any other duration, any other length but his own. A strange, almost solipsistic assumption! By the same logic, the relativistic increase of mass would be unobservable; for the observer could perceive only the masses associated with his own system – the masses which remain constant; thus the experiments of Kauffmann, Guye and Bucherer, made long before Bergson wrote his book, verifying the increase of the mass of the electron, would have been impossible! Bergson was apparently trapped by the spell of the words "within his own system" ("*à l'intérieur de son système*") which he took too literally, unwittingly making of each system a sort of self-contained monadic unit. How strange a view for the thinker who always insisted on cosmic continuity and interaction and who truly anticipated Whitehead's attack on the fallacy of simple location! Furthermore, not only, as d'Abro correctly pointed out, was the dilatation of time verified indirectly when the relativistic theorem for the addition of velocities explained quantitatively the reduced velocity of light in moving water far more simply than the artificial 'aether-drag' hypothesis; but in 1941 a direct verification took place when B. Rossi and D. B. Hall found that the rate of disintegration of the mesons is apparently slowed down

by their relative motion with respect to the observer at the rate anticipated by the relativistic formula.[13]

The second limitation of Bergson's comment was his belief that the
dilatation of time in the general theory of relativity has the same apparent,
referential and reciprocal character as in the special theory. The failure
to distinguish between the special and the general theory vitiated his
criticism of Paul Langevin's thought-experiment which became famous
under the name 'clock paradox' or 'the paradox of the twins'. Langevin
presented it first in 1911 and it may be summarized as follows.[14] Suppose
that an observer is placed inside of a projectile which moves with a velocity
very close to the velocity of light. Let us assume that it is only by 1/20.000
less than the limit velocity $c$. Let us further assume that after moving
away from the Earth for a year, the projectile will reverse its velocity and
start coming back to the Earth. In this way the total duration of the trip
*for the observer in the projectile* will be two years. According to Newton
we would expect that his twin brother on the Earth would be two years
older at the time of his brother's return. Not so, according to Langevin.
The interval of two years, lived by a traveller, will correspond to *two
centuries* on the Earth. In other words, the twin brother who stayed on
the Earth, at the moment of the return of the traveller would be dead
long ago. The reason for the discrepancy between the traveller's time and
the time of the earthly observer is that the former would be effectively
lengthened or dilated with respect to the latter. It would be an instance
of the relativistic dilatation of time.

Langevin's thought experiment was criticized by both Bergson and
Lovejoy[15] on the ground that it is incompatible with the reciprocity of
appearances required by the special theory of relativity. Since there is
no privileged frame of reference, the separation of the space traveller and
the stationary earthly observer could be equally well described by adopting
the projectile itself as the system of reference. Then it would be the earth
which would move away from the spaceship with the same velocity as
in the first description, though with an opposite sign. By applying the
same reasoning to the earthly observer as Langevin applied to the traveller
in the projectile we would arrive at the conclusion that while two centuries
elapsed in the space-ship, only two years elapsed on the earth! Since two
clocks cannot both go slower and faster than each other, Bergson concluded that the dilatation of time is only referential, an effect of 'the

perspective of velocity', and not an effective lengthening of time. According to Bergson, Langevin's reasoning is a perfect example of "half-relativity" ("*demi-relativité*"), which is nothing but an imaginative hang-over from the Newtonian – he might have even said *geocentric*! – physics.

And Bergson was right as long as we remain within the framework of the special theory of relativity. But in the confused discussion which followed the publication of *Durée et simultanéité* this essential point was overlooked both by Bergson and his opponents. Neither Bergson nor Lovejoy pointed out with sufficient emphasis that the idea of time-retarding journey is incompatible with the special theory of relativity which deals with the inertial motions only. Yet, in Langevin's speculations an enormous acceleration is introduced at the moment when the projectile reverses its path: not only should there be first an enormous negative acceleration by which the speed of the projectile would be reduced to zero, but then there should follow an equally enormous acceleration in the opposite direction. It is this acceleration which places the '*voyage au boulet*' clearly beyond the purview of the special theory. Thus only within the framework of the general theory can Langevin's spectacular space trip be correctly analyzed.

This was recognized as early as 1918 by Einstein, then by Thirring, Whitehead and Reichenbach; most recently it was stressed again by David Bohm.[16] According to the principle of equivalence, the phenomena of acceleration are equivalent to the effects of gravitational field; and such field *slows down effectively the proper or local time*.[17] Then there is no longer the symmetry of appearances and the argument of Bergson and Lovejoy loses its force. We may then conclude that while the time-retarding journey is meaningless within the special theory, it is a physically admissible possibility within the general theory of relativity.

Yet, this crucial difference between the special and general theory was not stressed by the physicists who tried to answer Bergson's objections. D'Abro makes appeal in his answer to the Lorentz transformation of the special theory of relativity. His polemic is often ill-tempered and ill-informed, otherwise he would not have so grotesquely misunderstood Bergson's criticism of Einstein as an attempt to rehabilitate the Newtonian concept of absolute motion.[18] André Metz, while correctly pointing out that the reciprocity of appearances holds only within the special theory and thus does not apply to Langevin's space trip, still claims that the effective dilatation of the time – to wit the *proper* time – of the traveller begins *before* the acceleration at the moment his return takes place, while asserting, on the next page, a complete reciprocity of appearances between the stationary observer and the traveller who continues his space trip *without return*.[19] This inconsistent oscillation between the standpoint of the general and that of the special theory could hardly have a convincing effect on Bergson, whose last answer to André Metz showed the signs of impatience.[20] Nor was Bergson convinced by the letter of Jean Becquerel who, curiously enough, in his answer to Bergson's objections used exclusively the conceptual framework of the special theory. This is even more curious, since Becquerel in a book which appeared before that of Bergson correctly insisted on the *effective* slowing down of the proper time in the gravitational field.[21] He need have only stressed the equivalence of such a field with the effect of acceleration to show the immanent, non-referential character of the dilatation of time in Langevin's projectile.

It is true that even so it can be pointed out that, kinematically speaking, acceleration is relative too. In other words, an equivalent description of the space trip can be given by placing the stationary observer in the spaceship itself from which the earth would move away and toward which – after being first decelerated and then accelerated in an opposite direction – it would return. This is what Bergson stressed in his answer to Becquerel. Unquestionably, this rejoinder would be correct in a universe *in which only the earth and the space-ship existed* or in a universe populated exclusively by the inertial systems. But this is impossible. As Whitehead pointed out, "the traveller has an equal right to say that the earth started with great velocity on a voyage in space, and suddenly stopped and came back". But then he should add that "*curiously enough, just as the earth stopped at its furthest distance, a star* [he could have said as well: the rest of the universe or at least our galaxy] *came up to him and then retreated*".[22] The essential importance of this reference to the cosmic – or at least to the galactic – gravitational field was omitted in the confused discussion between Bergson and his opponents. We may then conclude that if Bergson did not grasp the physical feasibility of a 'time-retarding journey', it was not entirely his fault. We must bear in mind that Langevin formulated his thought-experiment *prior* to the formulation of the general principle of relativity and thus inevitably phrased it in an inadequate language, whose incompatibility with the special theory Bergson rightly pointed out.

We may also add that Bergson had no reasons to fear that the possibility of a 'time-retarding journey' would be in conflict with his own version of temporalism, and if he ever had such fears, they were quite superfluous. No matter how strange it may sound, an attentive and rigorous analysis of the apparent paradox of the twins will show clearly that topological, i.e. the non-metrical unity of time underlying the diversity of the relativistic times, is not affected. As far as the status of time in the special theory is concerned, there is no problem about it; about the equality of proper or local times there is a rare agreement even between d'Abro and Bergson. As the latter says in *Durée et simultanéité*:

> Not only do the multiple times of the relativity theory not destroy the unity of real time, but they even imply and uphold it... Without this unique and experienced duration, without this time common to all mathematical times, what would it mean to say that they are *contemporary*, that they are contained in the same interval?[23]

Bergson's error was that he failed to see the implications of this passage for the general theory, in particular for the apparent paradox of the time-retarding journey. In other words, he failed to see that the unity of time immanent to all frames of reference should be understood in a *non-metrical* sense; more specifically, that the same stretch of duration can underlie discordant metrical temporal series. By considering more closely this so-called 'time retarding journey' let us explain what this 'topological unity of duration' means.

In the first place, we must get rid of semantic confusions which the misleading terms easily suggest. Strictly speaking, there is no such thing as 'time retardation' or 'time lengthening' at all. Such kinematic and geometrical terms are borrowed from our visual experience of the bodies moving in space. When a car is moving faster than another, the slower one will be overtaken, 'left behind'; the faster one will be 'ahead'. Similarly when one geometrical interval is lengthened with respect to another one, their extremities will not coincide. Nothing of this sort, no "dislocation in time," happens in Langevin's thought experiment. Both twins are obviously living *in the same moment of time* when they are being separated; they are later again *at the same moment of time* when they are reunited.[24] (More accurately, when the space traveller visits the grave of his brother after his return.) We said 'at the same moment of time'; but of *which* time do we speak here? Certainly *not* of *metrical* time, since the time of the spaceship and that of the earth are metrically different. Yet, these two different metrical times, since they are *bounded by the same successive moments*, are necessarily *contemporary*; in other words, they express two complementary aspects of *one and the same* stretch of universal duration which underlies them both, even though it is measured differently in each of them. If we designate the moment of separation $A$ and the moment of return $B$, then the succession of $B$ after $A$ remains a succession in *all* systems because of the causal dependence of $B$ on $A$. Such succession is thus a *topological invariant* not affected by the effects of relative motion nor by the dynamic effects of acceleration. What is modified is the *rhythms* of the local times, that is, the local time units whose different degree of dilatation in different gravitational fields account for different measuring of time in two systems. But these metrical differences do not affect the irreversibility (in Langevin's sense) of the underlying common duration.

It is strange that Bergson apparently failed to realize that the idea of different metrical rhythms suggested by the general theory not only does not conflict with his thought, but even agrees with the spirit of *Matter and Memory*. This is what Z. Zawirski pointed out.[25] But it is clear to any attentive reader of this book of Bergson, especially of those passages in which he described the variability of temporal span. Thus very short intervals of public time may be experienced by a dreaming person or by a person under various drugs as subjectively very long. Yet, this subjectively long psychological interval remains *contemporary* with the subjectively much shorter interval of a person awake with the normal span of his psychological present. The analogy with the relativistic paradox of differently aging persons is obvious. We have only to substitute for different psychological 'clocks' of two persons, one of whom is awake and another dreaming, the different 'biological clocks' of the space traveller and his brother staying on earth. Then the apparent relativistic paradox loses much of its mysterious appearance. After his awakening a dreamer will find himself *at the same moment of the public time* as a person awake. Similarly, after his return a space travelling person will find himself *at the same moment of cosmic duration* as his earthly brother. This analogy between the relativistic and introspective time was noticed recently by David Bohm and Herbert Dingle.[26] In either case, the measuring unit, whether it is the span of a mental present or an oscillation of the pendulum, is affected.

Bergson failed to grasp the implications of his thought because, in defending correctly the unity of cosmic duration underlying metrically discordant times, he confused this unity with oneness in the metrical sense. This is especially surprising since the rejection of homogeneous, quantitative time and the insistence on the heterogeneity of duration is, especially in *Matter and Memory*, one of the most persistent themes of his thought. But in *Duration and Simultaneity* Bergson surprisingly turned into a defender of Newtonian time. He did not do it in an explicit and conscious fashion. The discrepancy between his rejection of homogeneous and quantitative time and the explicit affirmation of the empty, homogeneous time of Newton would have been too obvious. Yet, the reassertion of the Newtonian time is implied by his unambiguous defense of the objective simultaneity and of the separability of space from time.

We claim that a single time and an extension independent of duration continue to exist in Einstein's theory considered in its pure state; they remain what they have always been for common sense.[27]

If this were true, the instantaneous space which at each moment of time could be carved out of the four-dimensional becoming would be a substrate of all absolutely simultaneous events; we would thus return to the physics of Newton and to the common-sense belief in the objectivity of the cosmic Now, spread instantaneously across the whole universe. Needless to say that the existence of what Eddington called "the world-wide instant", stretched out perpendicularly to 'the axis of time' is completely incompatible with the relativity theory. But it is far more significant that it is incompatible not only with both the spirit and the letter of the whole of Bergson's thought, but even with his explicit statements made in *Durée et simultanéité*.

(a) We pointed out in Chapter 6 of this part that, according to Bergson, geometrical, instantaneous space is a mere ideal limit of concrete physical extension, a limit which is nowhere physically realized. He repeatedly insisted on the fictitious, unreal character of the mathematical instants without apparently realizing that this excluded the existence of three-dimensional 'world-wide instants', in other words, of instantaneous spaces as well. In the fourth chapter of *Creative Evolution* he explicitly stressed that what is real is *extensive becoming* in which successive "instantaneous states" are mere artificial and static snapshots due to "the cinematographic mechanism" of our mechanistically oriented thought. But in *Duration and Simultaneity* he speaks about the possibility of "instantaneous vision" by which our imagination – unlike the luminous signals unrestricted in its flight – can always draw the true "Now-line" and thus unambiguously achieve a separation of space from time. But what else is his "extension independent of duration" than an artificial, instantaneous cut across the universal becoming, a static snapshot whose artificiality and illegitimacy was so strongly denounced in all his books?

(b) Moreover, the affirmation of 'extension independent of duration' is equally incompatible with several unambiguous assertions of Bergson's book on relativity. We have seen that 'instantaneous space' and 'the class of simultaneous events' are synonymous terms within the Newtonian framework. Yet, the absolute space of Newton is rejected unambiguously by Bergson. Let us remember his explicit negation of the absolute frame of reference since "there is no longer any ether, no longer absolute stability anywhere".[28] Not only this: while he admits the simultaneity of two temporal intervals ("*la simultanéité de deux flux*"), he explicitly rejects the simultaneity of two instants:

Now from the simultaneity of two flows, we would never pass to that of two instants, if we remained within pure duration, for every duration is thick; real time has no instants... The instant is what would terminate a duration if the latter came to a halt... Real time cannot therefore supply the instant; the latter is born of the mathematical point, that is to say, of space.[29]

What else is 'instantaneous space independent of duration' than a three-dimensional layer of simultaneous events? The contradiction is obvious; one cannot reject one while retaining the other.

Instead of contradiction we should speak rather of a certain *distraction* of Bergson's thought which becomes understandable when we analyze its psychological origin. The most significant part of Bergson's comment on the relativity theory is his criticism of the static misinterpretations of Minkowski's space-time. It was natural for him, who was never tired of criticizing various forms of the spatialization of time, to reject the idea of four-dimensional hyper-space of which time was only one dimension comparable to the remaining three. He recalls how his opposition to the view of time as "the fourth dimension" of space goes back to his very first book.[30] But this static misinterpretation of the Minkowski's fusion apparently concealed from Bergson its true dynamic significance. It was this opposition which drove him back to an unconscious acceptance of the Newtonian idea of space separable from time – contrary to his explicit espousal of the constancy of the luminous velocity, of the Lorentz transformation and the relativity of motion. His error was to believe that there is no middle ground between the Newtonian idea of metrically unique time and a complete denial of time implied in the static misinterpretation of space-time. He apparently did not realize that the relativistic space-time is really *time-space* in the sense explained in our Chapter 7; that in it the future is separated from the past even more effectively than in the Newtonian scheme; that the future, unreal for my present Here-Now, is *intrinsically unobservable*, that is, *physically unreal* for any other conceivable frame of reference, no matter how far in the contemporary region from my present Here-Now it is located. He apparently did not realize that since the advent of relativity, according to Whitehead's profound remark already quoted, "spatial distance must stretch across time", in other words, that there are no *purely spatial distances at all*. Thus he failed to see that the relativistic space-time, correctly interpreted, far from implying the elimination of becoming, reintroduces it into the physical world; that the relativistic time-space in its structure resembles his own 'extensive becoming', since in both of them instantaneous cuts are equally artificial and equally unreal.

Yet, the name of Whitehead should remind us that we should not be excessively severe to Bergson. For Whitehead's thoughts on relativity show similar hesitations. On one side, in *Science and the Modern World*, Whitehead correctly stresses that the relativization of simultaneity was "a heavy blow at the scientific materialism, which presupposes a definite present instant at which all matter is simultaneously real. In the modern theory there is no such a unique present instant" (p. 172). Yet, in his other writings he explicitly tried to save the intuitive, traditional concept of simultaneity! This shows that the idea of a huge, instantaneous 'Now', spread transversally through the whole universe, is well entrenched in the human mind, even in the minds of those who have, at least intermittently, a full insight into its inadequacy. As early as in his *Concept of Nature* Whitehead made the following significant remark:

The difficulty as to discordant time-systems is partly solved by distinguishing between what I call the creative advance of nature, which is not properly serial at all, and any

one time series. We habitually muddle together this creative advance, which we experience and know as the perpetual transition of nature into novelty, with the single-
time series which we naturally employ for measurement. The various time-series each
measure some aspect of the creative advance, and the whole bundle of them express
all the properties of this advance which are measurable. The reason why we have not
previously noted this difference of time-series is the very small difference of properties
between any two such series.[31]

In another passage Whitehead explicitly linked his idea of the "creative
advance of nature" to Bergson's view of time, and it is natural that
Bergson in *Durée et simultanéité* noted with satisfaction this agreement.[32]
Whitehead's view that various metrically discordant times are, so to
speak, various manifestations of the single, non-metrical time is very
similar to Bergson's claim that the diversity of the relativistic times
presupposes "the universal time" underlying all of them, since all of
them, while metrically different, remain *contemporary*. Yet, although both
Whitehead and Bergson drew a valid distinction between 'contemporaneity' and 'co-instantaneity' (in Bergson's terms: between "the simultaneity of fluxes" and "the simultaneity of instants"), insisting that only
the first one is meaningful, they apparently forgot this distinction when
they upheld the absoluteness of simultaneity. Thus they, too, muddled
together 'creative advance of nature' with one single temporal series – the
very confusion against which Whitehead warned! It was the confusion
of the topological invariance with the metrical invariance; for if anything
is certain in the relativistic view of the world, it is the fact that the
simultaneity of distant events – unlike the separation of the causal past
from the causal future – is *not* topologically invariant. (The expression
'creative advance into novelty' is merely a poetic Whitehead's phrase for
the absolute separation between the causal past and the causal future,
invariant in all the frames of reference.)

It is in the very nature of any traditional, deeply ingrained intellectual
habit that even when its inadequacy is realized, it still persists in our
subconscious and crops up again in moments of relaxed attention. In this
particular case it is extremely easy to confuse the unambiguous separation
of the past from the future, on which both Bergson and Whitehead
insisted, with the Newtonian "Now" stretching transversally across the
four-dimensional world process. But this is due to the fact that the concept
of 'Elsewhere', that is, of the four-dimensional region of contemporary
independence separating the absolute past from the absolute future, is so

utterly unfamiliar to us and has never been even remotely foreshadowed in our previous intellectual history. How strongly the traditional idea of the Newtonian universal 'Now-cuts' is entrenched in our mind is indicated by the language used in numerous textbooks on relativity and cosmology. After insisting on the inseparability of space and time in the part dealing with the special theory, the authors return to the pre-relativistic language when they deal with the general theory and with cosmology. They speak of the curvature of *space*, of the infinity or the finite extent of *space* – and not of *space-time*. Yet, any separate statement about space, whether Euclidian or Riemannian, is pregnant with the misleading idea of the universe as a whole, existing *at a certain instant* and consisting of the *simultaneously* existing parts. It completely disregards the fact that on the extragalactic and even on the galactic scale the width of the contemporary region (Eddington's 'Elsewhere', Whitehead's 'Co-presence'), that is, the indeterminacy of 'Now', stretches for millions and even billions of years!

We can summarize the previous discussion in the following way: d'Abro was right in pointing out that Bergson erred in assuming the unobservability of the dilatation of time and the contraction of the lengths. But Bergson was right when he insisted on their *referential* and *reciprocal* character. As d'Abro himself conceded, Bergson was right about the equality of proper times *within the special theory*. Yet he was wrong when he claimed that such equality must also exist in the accelerated systems; hence his misguided rejection of the possibility of Langevin's "time-retarding journey". On the other hand, Bergson was right when he claimed that the non-reciprocity of appearances is incompatible with the special theory; and it was in the language of the special theory in which Langevin's thought-experiment was phrased in 1911. D'Abro was right when he pointed out that Bergson was inconsistent when he, while accepting the curved space of the general theory, denied its implication – the influence of the gravitational field on our spatio-temporal measurements. But he was grievously wrong when he misunderstood Bergson's defense of *"mobilité absolue"* – the term which to Bergson was synonymous with *durée* – as an attempt to save the Newtonian concept of absolute motion. Finally, he was wrong on the most valid aspect of Bergson's comment— the rejection of the static interpretation of space-time – when he claimed that our experience of passage is as subjective as the sensations of color and sound.[33]

It is clear that to separate what is living from what is dead in Bergson's interpretation of the relativity theory is a complex task. Such a task cannot be achieved when either a passionately hostile or uncritically enthusiastic attitude is adopted. Bergson's general insights are far more significant than his errors and occasional vacillations which, as I tried to show, most frequently resulted from not always fully grasping the implications of his own thought. Such vacillations prevented him from seeing more clearly the similarity of his own "extensive becoming" and the structure of the relativist time-space undistorted by neo-Eleatic misinterpretations.

## NOTES

[1] *D.S.*, p. 185.

[2] *Ibid.*, pp. 32–33.

[3] Carl Neumann, *Über die Principien der Galilei-Newtonschen Theorie*, Leipzig, 1970, p. 15.

[4] Max Born, *Die Relativitätstheorie Einsteins und die ihre physikalische Grundlagen*, Berlin, 1921, p. 189.

[5] H. Dingle, Introduction to the English translation of *D.S.*, p. XVI.

[6] *D.S.*, pp. 73–74.

[7] D'Abro, *Bergson ou Einstein?* Paris, 1927, p. 138.

[8] Max Born, *op. cit.*, pp. 189–190.

[9] D'Abro, *op. cit.*, p. 75; 14.

[10] Cf. my article 'La théorie bergsonienne de la matière et la physique moderne', *Revue philosophique* CLXIII (1953), 28–59, especially pp. 43–44.

[11] This apparent character of the dilatation of time in the special theory of relativity was stressed among others besides Max Born also by Jean Becquerel, *Le principe de relativité et la théorie de la gravitation*, Paris, 1922; René Dugas, *A History of Mechanics* (transl. by J. R. Maddox), Central Book Co., New York, 1955, pp. 495f.; J. L. Synge, *Relativity: the Special Theory*, Interscience Publishing Co., New York, 1956, pp. 119–120.

[12] *D.S.*, pp. 109–113.

[13] D'Abro, *op. cit.*, pp. 117–118.

[14] Cf. Note 15 of the previous chapter; also Langevin's article 'L'évolution de l'espace et du temps', *Revue de métaphysique et de morale* XIX (1911) 455–456.

[15] *D.S.*, pp. 74–78; A. O. Lovejoy, 'The Paradox of the Time-Retarding Journey', *The Philosophical Review* 40 (1931), 48–68; 152–167.

[16] A. Einstein, 'Dialog uber die Einwände gegen die Relativitätstheorie', *Naturwissenschaften* VI (1918) 697–702; H. Thirring, *Naturwissenschaften* IX (1921) 209; A. N. Whitehead, 'The Problem of Simultaneity', in *Aristotelian Society* (Suppl.) III (1923) 34–41; H. Reichenbach, *The Philosophy of Space and Time* (transl. by Maria Reichenbach and John Freund), New York, Dover, 1956, pp. 192–194; David Bohm, *The Special Theory of Relativity*, W. A. Benjamin, New York, 1965, Chapter XXX.

[17] J. Becquerel, *op. cit.*, p. 240.

[18] D'Abro, *op. cit.*, pp. 43–53. D'Abro's erroneous charge was due to the fact that he

confused Bergson's term "mobility" – which is synonymous with "qualitative change" or "duration" – with "motion" in the sense of spatial displacement. While Bergson insisted on the absolute character of mobility (*mobilité*) in the sense of duration or change, he never denied the relativity of motion in the sense of spatial displacement. On the contrary, he accepted Einstein's radical relativization of motion without reservations. Cf. Note 2 of this chapter.

[19] A. Metz, *La relativité*, Paris, 1923, pp. 65–75.

[20] The polemic between Bergson and André Metz took place in *Revue de philosophie* **31** (1924); first Metz's article 'Le temps d'Einstein et la philosophie, A propos de la nouvelle édition de l'ouvrage de M. Bergson *Durée et simultanéité*', pp. 56–88; Bergson's reply "Les temps fictifs et le temps réel", pp. 241–260; Metz's rejoinder on pp. 437–439; and Bergson's final remark on p. 440. The editors of the review did not publish the whole text of the final rejoinder by André Metz, probably because it was even more impatient than that of Bergson; it contained a reference to Einstein's private letter to A. Metz in which he charged Bergson for not recognizing the absolute character of the spatio-temporal coincidences (July 2, 1924). Bergson's intention was clearly very opposite – to *save* the absolute character of the spatio-temporal coincidences; but this good intention was marred by his failure to differentiate between the simultaneity of the *isotopic* and *heterotopic* events. In rightly insisting that the simultaneity of the events occurring at the same place can never be dislocated in a succession except in the imagination of an external observer, he maintained the same view about the simultaneity of distant (heterotopic) events.

[21] *D.S.*, Appendix I (pp. 163–172). Cf. Note 17 of this chapter.

[22] A. N. Whitehead, *article quoted* in Note 16, pp. 34–35.

[23] *D.S.*, p. 118. (Professor Jacobson's translation differs slightly from my own, given in my book (p. 207).)

[24] This is what I stressed in my discussion with Professor O. Costa de Beauregard at Congrès Bergson in 1959. Cf. *Bulletin de la Société française de philosophie* **54** (Paris, 1960), p. 81.

[25] Z. Zawirski, *L'évolution de la notion du temps*, Cracow, 1934, pp. 305–306.

[26] D. Bohm, *op. cit.*, p. 172; H. Dingle, Introduction to the English translation of *D.S.*, p. XXXVI.

[27] *D.S.*, p. 30.

[28] Cf. Note 2 of this chapter.

[29] *D.S.*, p. 52–53.

[30] *D.S.* p. 59; *T.F.W.*, p. 109 (about homogeneous time as "a fourth dimension of space"). Bergson's criticism of the static interpretation of the relativistic space-time (Chapter VI of *D.S.*) is again seriously vitiated by his insistence on the separability of space from time.

[31] *The Concept of Nature*, p. 178.

[32] *Ibid.*, p. 54; *D.S.*, p. 62.

[33] Cf. Note 18 of this Chapter; also Note 9. In his other book *The Evolution of Scientific Thought from Newton to Einstein*, Dover, New York, 1950, d'Abro says explicitly: "it is therefore necessary to postulate that our consciousness rises along the world line of our body through space-time, discovering the events on its course" (p. 206).

CHAPTER 9

# LIMITATIONS AND USEFULNESS OF
# THE CORPUSCULAR MODELS

Since the space of Newton and Euclid was an integral part of the classical model of reality, its denial by Bergson – and by modern physics – seriously affects this model itself. More accurately, the claim that the Newtonian-Euclidian void is a mere ideal and a physically unrealized limit, does affect other constituent parts of the same model. We shall see how all other components of the corpuscular-kinetic scheme of nature – matter, motion, energy and causality itself – were profoundly modified and how these modifications largely depended on the revisions to which the concepts of space and time have been subjected. We shall see how, in the light of Bergson's analysis, the above mentioned classical concepts while being, strictly speaking, *distortions* or *simplifications* of the physical reality, retain their practical usefulness on the macroscopic – and macrochronic – level; that their very artificiality guarantees their usefulness on the human level. This has been illustrated already by several examples. *For practical purposes*, that is, from the standpoint of the human 'specious present', time is infinitely divisible, since even if the microphysical events have the duration of $10^{-24}$ sec, they can be *practically* regarded as instantaneous. *For practical purposes*, the physical interactions on the scale of the human body are instantaneous, and in this respect undistinguishable from the relations in the geometrical, instantaneous space. *For practical purposes*, the velocity of light in comparison to the velocities perceptible by man is infinite, which means that for *practical purposes* the separation of space from time is only negligibly inaccurate and the simultaneity of events remains *practically* absolute in our daily surrounding. This was the reason why the effects of the finite velocity of light were observed for the first time on a much larger scale of magnitude, i.e. in the observation of the eclipses of the satellites of Jupiter. But even after this discovery nobody suspected how deeply further investigations in this direction would shatter the confident extrapolations of our limited experience. We shall see also that for *practical purposes* the separation of matter in mutually external bodies is very approximately valid, not only on the macroscopic, but even on the molecular level.

Let us briefly recall the main features of the classical picture of the physical world: in the Euclidian, that is, homogeneous, infinite and mathematically continuous space the material particles of constant mass move according to the laws of Newton's mechanics. In this concise statement a reference is made to *all* basic concepts of classical physics – space, time, matter and motion – whose features and mutual relations I summarized in my previous book.[1] This is what I called "the classical corpuscular-kinetic model of nature" and, although it has never been realized completely and consistently, it remained the ideal to the realization of which physics from 1600 to 1900 definitely tended. Even the collateral trends – dynamism, energetism and plenism – were based on the same fundamental concepts; for even the fluid-theories of matter, which insisted on the reality of the physical *plenum*, covertly reintroduced – in the form of 'the atoms of aether' – the discontinuity of matter to which they were opposed.

Among the four basic concepts two of them stood out conspicuously: *matter* and the *void*, both indestructible and uncreatable because of their immutability. In this way the static aspects of the physical reality were stressed. It is true that the obtrusive reality of change could not be completely eliminated, since it was clearly present in the form of *displacement*. Since displacement was nothing but a change of position which affected neither the vehicle of motion (i.e. matter) nor its container (i.e. space) it appeared somehow less real than matter and space themselves. Two other facts further tended to obscure the reality of succession. The laws of motion themselves assumed the form of the *conservation laws* of certain quantities which – whether it was mass, energy, momentum or angular momentum – tended by their very constancy to obscure the successive character of motion itself. Furthermore, the very principle of determinism, of which the conservation laws were only more definite expressions, in reducing the world history to a huge, cosmic tautology, tended to eliminate succession altogether. It is then hardly surprising that the status of time in such a model appeared even more shadowy than that of motion. The mechanistic view of nature from its very beginning, that is, from the epoch of ancient atomism, tended toward a *relational theory of time* in which time was regarded as depending on motion or, in Epicurus's words, as "an accident of accidents", which was, so to speak, *twice* removed from the solid, fundamental reality of matter and space.

We may then conclude that, although change and succession were not altogether eliminated in classical physics, there was a definite tendency to do it. There was a similar tendency to reduce all diversity of nature to mere *differences of configuration*, that is, to one remaining difference, that of 'full' and 'empty'. It is true that from Descartes to Lord Kelvin there were persistent efforts to overcome even this distinction. Émile Meyerson, in his now classic and richly documented works, showed that the ideal limit of these two trends was 'acosmism', i.e. the dissolution of all diversity in space and time in the Parmenidean One, and for this reason these trends could not have been successful in such an empirical science as physics.[2] Thus even the world of classical physics showed features which differentiated it from a completely homogeneous and completely static space. Yet, the fact that such space served as an all-embracing container of the whole physical content, and that the physical content itself was interpreted in terms of homogeneous matter, shows unmistakably how close classical physics came to the unrealizable ideal of the Eleatic philosophy. The fact that outstanding contemporary scientists – to name only Sir James Jeans among physicists and Kurt Gödel among mathematicians – looked with admiration at such philosophers as F. H. Bradley and John E. McTaggart is highly significant and hardly accidental.[3]

Historians of philosophy are generally aware that atomism was an attempt to reconcile experience with the native Eleatism of the human mind. In other words, to admit plurality and change in its most 'innocuous' form: plurality in the form of *diversity of position*, change in the form of *change of position*. Since the main feature of 'Bergsonian physics' is to restore the objectivity of qualitative becoming, its critical attitude toward atomism and kineticism is not surprising. This was the reason why at the beginning of Part III we emphasized that Bergson's philosophy of physics shares its anti-atomistic attitude with Bergson's psychology. This rejection of atomism must now be explained more fully.

In the physical world as it is disclosed in our sensory perception and in the refined perception of the physicist, the independence of its constituent parts is very pronounced. It is so pronounced that our senses perceive it as an aggregate of mutually impenetrable bodies which can influence each other only from outside, i.e. by pressure and impact. The kinetic-atomistic conception of science is thus only a detailed and systematic elaboration of this original naive sensory perception. We resort to this atomistic conception when the testimony of our senses is hazy or less definite. It is certain that our original experience with liquid substances does not itself yield *immediately* to the atomistic-kinetic interpretation. Hence the original plausibility of the Aristotelian physics, which viewed liquidity together with other secondary qualities as inherent features of external reality. What was required first was a creation of the concept of particle deprived of any other qualities than the impenetrability, shape and inertia. This had been done prior to Aristotle, although Aristotle's authority obscured the significance of this achievement for more than twenty centuries. The eventual triumph of the Democritian conception of reality coincided with the foundation of classical physics. Some recent psychological investigations by Jean Piaget showed how early and spontaneously the human mind tends towards kinetic and atomistic interpretation of macroscopic experience, since macroscopic experience itself suggests it.[4] It is also evident that both main components of the classical corpuscular-kinetic model of reality – space and matter – are conceived of as made of parts *external to each other* in the most absolute sense. The points of the homogeneous space are mutually as external as the material particles.

Two reasons inspired Bergson to reject this additive conception of the physical reality. First, according to him, even the physical world is by its own essence a durational process and as such it cannot be made of mutually external parts. Consequently neither the concept of space, nor that of matter atomistically conceived, can adequately represent it. The second reason was epistemological and followed directly from his biological theory of knowledge:

*A priori* and apart from any hypothesis on the nature of the matter, it is evident that the materiality of a body does not stop at the point at which we touch it: a body is present wherever its influence is felt; its attractive force, to speak only of that, is exerted on the sun, on the planets, perhaps on the entire universe. The more physics advances, the more it effaces the individuality of bodies and even of the particles into which the scientific imagination began by decomposing them: bodies and corpuscles tend to dissolve into a universal interaction. Our perceptions give us the plan of our eventual action on things much more than that of things themselves. The outlines we find in objects simply mark what we can attain and modify in them. The lines we see traced through matter are just the paths on which we are called to move. Outlines and paths have declared themselves in the measure and proportion that consciousness has prepared for action on unorganized matter – that is to say, in the measure and proportion that intelligence has been formed. It is doubtful whether animals built on a different plan – a mollusk or an insect, for instance – cut matter along the same articulations. It is not indeed necessary that they should separate it into bodies at all. In order to follow the indications of instinct, there is no need to perceive *objects*, it is enough to distinguish *properties*. Intelligence, on the contrary, even in its humblest form, already aims at getting matter to act on matter. If on one side matter lends itself to a division into active and passive bodies, or more simply into coexistent and distinct fragments, it is from this side that intelligence will regard it; and the more it busies itself with dividing, the more it will spread out in space, in the form of extension adjoining extension, a *matter that undoubtedly itself has a tendency to spatiality, but whose parts are yet in a state of reciprocal implication and interpenetration.* (Last italics added.) [5]

The passage is obviously inspired by Bergson's biological theory of knowledge as it has already been explained in Part I. The biologically selective character of our sensory perception is clearly illustrated by a concrete example. Physical reality itself is an undivided flux whose constituents, in virtue of their mutual immanence, are not *parts* in the usual sense of the word. But, though undivided, this flux is *not* uniform and undifferentiated; its various 'parts' (*sit venia verbo*) are not equally interesting to human and, more generally, to animal organism. Only where the physical reality is, so to speak, sufficiently condensed and capable of affecting an organism or at least its special sensitive parts called sensory organs, does the phenomenon of physical body arise. Conversely it may be said that those portions of the physical reality which act on our senses and are perceived as physical bodies, are at the same time the areas accessible to our practical activity, modifiable by the influence of our body or of our tools.

It is thus clear that those areas of the physical reality which affect our senses produce the appearance of physical bodies which mark off the sphere of our manual or, more generally, instrumental influence: the boundaries of such "spheres of influence" are perceived by us as the contours of physical bodies. In this way arises our sensory perception which perceives the physical reality as an aggregate of solid bodies resting or moving in empty space. Classical physics was only an elaboration and systematization of this tendency, cleared of the haziness and inconsistencies still present at the level of naive perception.

But the concrete physical reality does not yield itself to such a neat division into 'full' and 'empty'. Even superficial observation shows that the difference between the body and its surrounding space, in other words, between matter and void, is never as clear-cut as may appear at the first glance. In the first place, the volume of a body is never full in an absolute sense; this is shown by trivial experiences concerning the porous nature of all bodies, even those which appear the most compact. The interstices and gaps within the material structure are usually too small to be perceived by our sight or touch. In virtue of their smallness they do not affect the cohesive forces ensuring the apparent solidity of the body, which thus appears to us as a *solid* and *manipulatable whole.* The compactness of solid bodies is thus an illusion from the theoretical point of view, but this illusion has a *practical usefulness* and *justification.* How little a portion of the apparently compact body is really 'full' was disclosed by the discovery of crystalline structure and especially by the discovery of the complexity of the atom: only one ten-thousandth of what has been considered the radius of the atom is 'filled up' by the nucleus in which practically the whole mass of the atom is concentrated. As early as 1914 Jean Perrin concluded that "more than we realize matter is prodigiously lacunal and discontinuous".[6]

It would be equally incorrect to assert that the space surrounding the bodies is absolutely empty. Even when we disregard the presence of the subtle gaseous medium in which solid bodies are usually bathed and which, as in the case of the air, is not registered by our sight and touch, the surrounding space is not as physically empty as the concept of absolute void would require.

As the above quotation of Bergson showed, the surroundings of every body is pervaded by its own gravitational field which, according to the equivalence principle, constitutes *the very essence* of its materiality. Besides, it is surrounded by perpetual electromagnetic disturbances which guarantee its visibility. On the other hand, the body itself is pervaded by the gravitational and electromagnetic influences from outside, and these influences represent the threads which, no matter how subtle and tenuous, link the body in question with the rest of the universe.

It is true that for the bodies of our daily experience their gravitational links with other bodies than the Earth are generally negligible, though not always so. The explanation of the tides by the combined action of the sun and the moon is so familiar that it is hardly worth mentioning.

But even the mutual gravitational action between the bodies of our daily experience is detectible, as Cavendish showed in his classical experiment. Yet, the gravitational thread thus discovered is so exceedingly tenuous that it would be a serious lack of biological economy if our perception would register it. Similarly, the mass contained in the waves of light which are spreading from, for instance, my fountain pen, is, as Einstein's relation $m = E/c^2$ indicates, so insignificant that my sight and touch commit only a negligible inaccuracy if they confine all perceptible mass of the pen in a narrow quasi-cylindrical volume and when they represent the surrounding space as completely devoid of its mass. *Practically*, our senses are correct. But ontologically speaking, their implicit assertion is, if not false, at least grossly simplified and incomplete. The useful illusion of a compact and isolated body is a result of the biologically selective character of our sensory perception which carves out of the continuity of the physical flux the contours of the visible and touchable body. In doing this, it eliminates all biologically useless links by which it is embedded in the universal cosmic interaction, and also it ignores all biologically useless discontinuities which constitute its prodigiously complex and porous structure.

When we observe that a thing *is* there where it *acts*, we shall be led to say (as Faraday was) that all the atoms interpenetrate and that each of them fills the world. On such a hypothesis, the atom or, more generally, the material point, becomes simply a view of the mind, a view which we come to take when we continue far enough the work (wholly relative to our faculty of acting) by which we subdivide matter into bodies. Yet, it is undeniable that matter lends itself to this subdivision, and that, in supposing it breakable into parts external to one another, we are constructing a science sufficiently representative of the real. It is undeniable that if there be no entirely isolated system, yet science finds means of cutting up the universe into systems relatively independent of each other *and commits no appreciable error in doing so*. (Italics added.)[7]

To sum up: space can be empty only to our senses or with respect to our action. 'Empty' in this sense means nothing but 'penetrability without noticeable resistance' or optical transparency without noticeable absorption, but *never an absolute void*. Similarly, 'full' means the absence of visible pores in the body and its manipulability as a single, cohesive whole.

Some objections naturally rise in our mind. Every physicist will admit that the distinction between empty and full space as established by our perception is far from being accurate. But can we accuse that distinction which is established by the exact science itself of the same inaccuracy? If it is true that the perceptible molar body is only apparently compact and that its apparent compactness hides from our senses the gaps and interstices within its structure, did not this fact impose upon the scientists a necessity to postulate the elementary, material particles whose compactness is absolute and unbroken by any kind of pores? Are not porous bodies of our macroscopic experience formed by the aggregates of these elementary compact particles? Every physicist also admits that the real physical space is not empty, as it is pervaded by all sorts of physical actions. But there was always the strong tendency to interpret this *actio in distans* by means of corpuscular models. Let us briefly recall Newton's emanation theory of light, Le Sage's kinetic theory of gravitation, numerous corpuscular models of aether from Huygens to Maxwell. All these attempts had one thing in common: the dynamical action through space, whether it was the interstellar space or the microscopic space separating atoms, was reduced to some kind of material, and, ultimately, corpuscular model. And if the interaction of the aethereal atoms had to be explained, it was always possible to postulate sub-aether whose moving particles were to explain what the model of aether left unanswered. The danger of a possible infinite regress implied in such kinds of explanation did not discourage even the most outstanding thinkers, as the case of Leibniz and Huygens showed.[8]

If this view is accepted, then the attribute of compactness, if it does not belong to the molar bodies, or even not to the molecules and atoms, *does* belong to the atoms of aether or sub-aether. The absolute emptiness, though not belonging to the interplanetary space, *does* belong to the space separating the basic particles. The fundamental Democritian distinction between the plenum and the void is thus fully preserved.

But only within the framework of classical physics. Once matter itself, in the general theory of relativity, was dissolved into a local curvature of the heterogeneous, non-Euclidian time-space, the whole distinction between 'full' and 'empty', on which the concept of well defined and sharply localizable body is based, lost its original meaning. In truth, this distinction was already challenged by the special theory of relativity which fused together two distinct classical concepts – matter and energy. This was the meaning of the equation $E=mc^2$. The philosophically most significant consequence of this conceptual fusion was the impossibility of considering any mass apart from its energetic content and context. In classical physics the inertial mass was an immutable *vehicle* of motion which was in no way affected by its own displacement. Consequently, it was independent of its velocity and *ipso facto* of its kinetic energy. This was in accordance with the classical dogma according to which the inertial mass of matter represented its immutable, substantial core, what Newton called 'residing force' (*vis insita*), whose presence marked it off sharply from its surrounding void. This immutability is now lost. The inertial mass is clearly a *relational property*, since, being a function of its own velocity, it depends on its relation to something *external* to itself, in this case on the frame of reference in which its effects are observed. An "apparent" increase of the mass of the electron when its relative velocity is increased is far less apparent than the physicists originally believed. It is clearly due to a *very real* reaction of the surrounding energetic field, and since this reaction was different in different frames of reference, the mass itself lost its invariance.

The second example is equally well known. The total mass of the aggregate is not equal to the arithmetic sum of the masses of its components. It is affected, in general decreased, by their energetic binding. Thus the mass of the alpha particle, instead of being the sum of the masses of two protons and two neutrons, is measurably less. There could hardly be a more striking illustration of the failure of the atomistic ideal according to which the elementary building blocks of the universe remain eternally the same in all their combinations. Instead of being independent, they are affected by their own energetic context. In truth, the resulting 'mass defect' can under certain conditions wipe out their 'rest mass' entirely. This was observed for the first time in 1932 after the discovery of positive electrons. Their very short 'life' is due to their quick 'dematerialization'. More specifically, a positive electron is *fused* with an electron of the opposite charge in such a way that *the total mass* of the whole couple is converted in the electromagnetic radiation.

It is true that the vanished mass of the couple is preserved in the mass of the "created" quantum of radiation according to Einstein's same equation. In a similar way the law of conservation of charge is formally not violated since the total charge of the vanished couple is zero – the same as the zero charge of the radiation. But this formal preservation of the conservation laws is small consolation for those who would like to uphold the intuitive corpuscular models. For the concept of mass contained either in the kinetic energy of an accelerated particle or in the quantum of "created" light is so abstract and so devoid of the original, intuitive content that the concomitant disappearance of particles would still be regarded as a complete annihilation by every consistent atomist from Democritus to Boltzmann. Similarly, the discovery of the opposite process, the materialization of radiation into a couple of oppositely charged electrons would be regarded by classical atomists as a true *creatio ex nihilo*. These facts clearly showed how far the 'particles' of the contemporary physics lack even the two most fundamental classical properties: indestructibility and uncreatibility.

What is even more significant, the phenomena of materialization and dematerialization of the microphysical particles are far more general than was suspected in 1932. The term 'perishing' or 'short-lived' particles is now common in microphysics. If the 'life-time' of the positive electron appeared fantastically short, $10^{-8}$ sec, what should we say about the neutral $\pi$-meson which lasts for only $10^{-14}$ sec or of particles like $\omega$, $\varrho$, $\eta$-mesons which last only $10^{-23}$ sec?[9] In truth, there is apparently not a single fundamental particle which would be everlasting in the Lucretian sense. Even such massive particles like neutrons "decay" into protons in 750 sec with the concomitant emission of negative electrons and neutrinos. Also, as Niels Bohr pointed out as early as 1943, this decay cannot be characterized in a classical sense as a disintegration of an aggregate into its pre-existing components. $\beta$-electrons are *created* during the process of emission in the same way as the photons and neutrinos are.[10] Thus the phenomena of materialization and dematerialization of 'particles' are not basically different from the emission and absorption of quanta. In either case there is no ejection of the pre-existing particles from the atoms or its nucleus. Nor is there any persistence of the particles in the atoms or in the nucleus which they invade. The traditional term 'particle' clearly designates nothing but temporary modification of the spatio-temporal medium from which various 'corpuscles' arise and into which they return. To ignore this fundamental link between 'particles' and their surrounding was the basic fallacy of atomism – "the fallacy of simple location" as Whitehead called it.

Paradoxically closest to the everlasting atoms of Lucretius seem to be the photons from very distant nebulae which, without being eternal, still 'persist' through no less than several billions of years. But what kind of persistence is it? Certainly not the persistence of a well defined, sharply localizable corpuscular entity describing a rectilinear path between the nebula and the Earth; otherwise we would completely ignore the fundamental difference between the corpuscular theory of Newton and the modern theory of photons. In other words, we would overlook entirely the *undulatory nature* of radiation; and it is precisely the vibratory pattern that survives across such enormous intervals of time-space, although, as the fact of the red-shift shows, *not* without modification. But then is it not far more appropriate to speak of a *string of events* between the emission or 'creation' of the photon and its absorption or 'annihilation' rather than of a particle persisting through time? The surprising discovery of the undulatory character of matter by Louis de Broglie shows that the so-called material particles should be treated in a similar way. Not only are they lacking permanence, but even within their 'life-time' they lack the undifferentiated duration of the classical atoms. The electron itself consists of vibrations, and since it is clearly hopeless to interpret these vibrations in an intuitive fashion as the periodic motions in a hypothetical material medium – let us not forget the failures of all mechanistic models of an aether and of any subaether – it is correct to say with Gaston Bachelard that "there are temporal events at the very foundation of its [electron's] existence".[11] The same is true of any other corpuscles. Thus not only are some 'particles' so evanescent that they deserve the name 'events', but even more steady particles are in reality constituted by a string of extremely short events.

As early as 1920 Whitehead proposed the term 'event-particle' for the ultimate elements of matter. Schrödinger's remark many years later that "Properly speaking, one never observes the *same* particle a second time – very much as Heraclitus says of the river –" suggests also the momentary, event-like (though not instantaneous!) character of the microphysical entities.[12] But we cannot meaningfully say that we can observe a particle *even once* as long as we understand the term itself in the classical sense. For what else is the classical corpuscle but a *definite correlation of certain position with certain momentum – mv*? But precisely such correlation, according to Heisenberg's principle of indeterminacy, is impossible to be found in nature; and – as Eddington and after him Philipp Frank stressed – it cannot be found because it *does not exist*.[13] Are we not witnessing, in this case, how the very concept of corpuscle, that is, the conjunction of the definite values of momentum and position, *breaks apart* under the pressure of recalcitrant microphysical experience? For this reason even Whitehead's bold term 'event-particle' has a subtly misleading connotation. The term 'event' or 'spatio-temporal pulsation' is far more appropriate. As we have seen, Bergson characterized matter as consisting of elementary "vibrations of vanishing duration." Even the term 'vibration' is misleading as long as it suggests something which vibrates – a sort of material or semi-material medium. As we shall see, Bergson's 'vibrations' are in this sense imageless since, according to him, change does not need any substantial vehicle.

This will be the object of the next chapter. But first, let us state briefly the epistemological significance of our previous discussion. On the level of sensory perception the opposition between 'full' and 'empty' expresses only the opposition between our biological interest and its absence. Space can be empty to our sense and with respect to our action, but not absolutely. Similarly, bodies are regarded as 'full' or 'compact' when their inner porousness is either not perceptible or when it does not interfere with our capacity to treat them as manipulable units; but they are never full in an absolute sense. Yet, atomism, overlooking the practical origin of the opposition between 'the full' and 'the empty', and seeing correctly that such absolute opposition cannot be found in our macroscopic perception, hoped to find it on the subsensory, microphysical level. But it has not found it there either.

### NOTES

[1] M. Čapek, *The Philosophical Impact of Contemporary Physics*, Part I.
[2] E. Meyerson, *Identité et réalité*, Paris, 1951, esp. Ch. II–VII; *De l'explication dans les sciences*, Paris, 1931, I, Ch. V.
[3] Cf. my article 'The Myth of Frozen Passage: the Status of Becoming in the Physical World', in *Boston Studies in the Philosophy of Science*, II, Humanities Press, 1965, pp. 458–460. Cf. also my book, pp. 158–165.
[4] Cf. Ch. 8 of Part I; also Jean Piaget, 'A propos de la psychologie de l'atomisme', *Thales* V (1949), pp. 3–7.
[5] *C.E.*, pp. 206–207.
[6] Jean Perrin, *Les atomes*, 2nd Ed., 1927, p. 225.
[7] *E.C.*, p. 222.
[8] Cf. Ch. Huygens, *Traité de lumière*, Gauthiers-Villars, Paris, 1920, p. 16; *Leibnizens Mathematische Schriften* (ed. by G. J. Gerhardt), Halle, 1850–1863, VI, p. 228.
[9] Cf. David H. Frisch and Alan M. Thorndike, *Elementary Particles*, Van Nostrand, Princeton, 1964, p. 140.
[10] N. Bohr, *Quantum d'action et noyaux atomiques*, Actualités scientifiques et industrielles, No. 807, Hermann, Paris, 1939, p. 12.

[11] G. Bachelard, *Le nouvel esprit scientifique*, 4th Ed., Paris, 1946, p. 85.
[12] A. N. Whitehead, *The Concept of Nature*, p. 86; E. Schrödinger, *What is Life? and Other Scientific Essays*, Doubleday, Garden City, 1958, p. 175.
[13] A. E. Eddington, *The Nature of the Physical World*, p. 225; P. Frank, *Philosophy of Science: The Link Betweens Science and Philosophy*, Prentice-Hall, 1957, p. 215.

CHAPTER 10

CHANGE WITHOUT VEHICLE AND CONTAINER.
FALLACY OF SIMPLE LOCATION

The revision of the distinction between "the full" and "the empty" also requires a rejection of the classical, Newtonian space, separable from its changing physical content and containing all changes without changing itself. Although this step is merely another aspect of the rejection of the classical, corpuscular models, it is considerably more difficult to take it, and to realize all its implications.

The habit of imagining physical reality – whether it is conceived of in an atomistic or a continuous fashion or, more abstractly, as energy – located in a passive, all-embracing, homogeneous container is so inveterate and so intimately bound up with our whole mental constitution, that it is extremely difficult to get rid of it. A continual, fresh effort is required to overcome its tyrannical power and to discover that a great number of intellectual inhibitions are due to its indirect influence. Even contemporary physicists do not entirely escape it when they still continue to talk about "particles" without realizing that the corpuscular language borrowed from our macroscopic experience subtly and insidiously affects their thought. Contrary to their intentions, and verbal declarations notwithstanding, the idea of corpuscle and with it the classical distinction between 'full' and 'empty' creeps into their mind and remains a part of their intellectual subconscious. Even the substitution of non-Euclidian space for that of Euclid does not destroy this habit. We often read, and not only in popular expositions of the general theory of relativity, that "the curvature of space is due to the presence of matter". It is obvious that such linguistic usage reflects the traditional dualism of matter and space, of the content and of the container, precisely what the general theory of relativity *did* eliminate.

Bergson was opposed to the idea of an inert, unchanging container of physical becoming for one reason already mentioned: if physical reality is truly characterized by the mutual penetration of its various components, then it must necessarily be foreign to the relations of mutual exteriority which are characteristic of classical space. His second reason is only another form of the first one: spatial relations are *ex definitione* timeless. The timelessness of spatial relations is for Bergson only an unreal limit which physical reality approaches without ever attaining it. Physical duration, even in its most diluted form, is constituted by the mutual immanence of its component parts, and its diversity is never assimilable to the externality of the juxtaposed units. He could not claim in support of his view the impressing array of facts and theories which appeared after the publication of *Matter and Memory* and *Creative Evolution*. We dealt with these new facts and theories in the three previous chapters: the negation of the simultaneity of distant events implies the negation of the relations of juxtaposition, which are the essence of classical space; the changing and elastic structure of the non-Euclidian continuum is in striking contrast to the alleged rigid and timeless structure of the Newtonian space. In other words, the concept of distance was so thoroughly transformed that its original meaning is now irretrievably lost. There are no longer bare, spatial distances in the world, but only *spatio-temporal distances*. The term 'space-like connections' which the physics of relativity continues to use is a mere misnomer for *the absence of causal relations*. In this respect again Bergson's conclusions coincide with those of contemporary physics.

In light of previous considerations it is natural that Bergson also had to challenge the classical concept of *motion*. Both classical science and classical philosophy defined it as a *displacement*, that is, as a *change of position*. Defined in this way, the concept of motion implied both that of space and that of material body distinguished from space. In other words, every motion is a motion *of something* – which was called matter – and *in something* – which was called space. While matter was, so to speak, *an immutable vehicle of motion*, space was its *equally immutable* container. While the belief in the immutability of the moving particles, which was the cornerstone of every kind of atomism, was strengthened by the verification of the law of conservation of matter, a few scientists or philosophers – Ostwald was apparently one exception – spoke of "conservation of space",[1] probably because the immutability of space, though stressed explicitly by Newton, was always tacitly assumed. It is clear that after the elimination of the distinction between the physical content and its container, nothing, literally nothing, is left of the classical concept of displacement which is based on this distinction. If the word still survives, it is due exclusively to the inertia of language, which is a depository of the petrified habits of our thought.

This negative conclusion will appear less paradoxical in the light of the biological theory of knowledge. According to this theory, the solidity and immutability of macroscopic bodies is merely a surface phenomenon resulting from the selective and utilitarian character of our sensory perception. For our perception of the external world not only ignores all practically useless discontinuities which constitute the prodigiously complex and porous structure of the apparently compact bodies, but also *all minute changes* which in the light of the kinetic theory of matter exist under their apparently solid surface. Long before the discovery of the Brownian motion, Lucretius anticipated the existence of invisible molecular motion in one beautiful passage of his poem in which he illustrated the illusion of the solidity of macroscopic bodies in the following way. Herds of sheep on the slopes of distant hills appear to us as motionless specks, although we know that there is a perpetual and disorderly motion of individual sheep within the herd itself.[2] In the same way a piece of macroscopic matter appears to us as a bit of solid material without any intrinsic motion. The metaphor is remarkably precise: when the apparent amplitude of physical motions does not surpass a certain angular magnitude – whether it is because of their distance or because of their minuteness – the motions themselves cannot be registered by our sight. As long as molecular oscillations do not endanger the solidity of the shape of the macroscopic body, the inaccuracy of our perception is a useful economy of nature. For it is the cohesion and the shape of macroscopic bodies which makes them potential tools with which we can operate effectively.

Today we know that perpetual minute changes take place even in the most stable materials; 'the fatigue of metals' is no more a metaphor, but a microscopic reality detectible by the optical means. These minute modifications are no longer negligible for long range technical planning; this only shows how the extension of our perception goes hand in hand with the extension of the reality which is practically significant for us. Microscopic vision differs only in degree from vision by a naked eye. In either case changes that are too small remain unnoticed. In either case perception means a convenient, useful immobilization. *"Percevoir signifie immobiliser."* [3]

What is the "mobile" to which our eyes attach movement as to a vehicle? Simply a colored spot which we know perfectly well amounts, in itself, to a series of extremely rapid vibrations. This alleged movement of a thing is in reality only a movement of movements. [4]

On this point we must expect an objection analogous to that raised previously: what Bergson asserts is true of the *macroscopic aggregates* whose changeability is due precisely to their composition; but it is not true of the elementary particles which remain permanent and immutable and are thus the true vehicles of motion. It is true that beneath the sensation of solidity which I experience in touching the surface of a hard body there is an ultramicroscopic whirling; but are there not the molecules which are whirling? Was not the purpose of Lucretius' metaphor precisely to show that there are individual sheep, that is, *identifiable individuals,* within the disorderly mingling of the flock? It is true that behind the qualities of color which our eyes attribute to bodies there are millions of electromagnetic vibrations; but these vibrations must be – as Arthur Compton observed long after the classical period of physics was over [5] – vibrations of *something,* that is, of the electromagnetic aether. This shows that the spontaneous intellectual attitude of even the twentieth century physicist is not basically different from that of John Tyndall of the last century when he insisted on the impossibility of motion without something which is moving or from that of David Hume two centuries before who claimed that "the idea of motion necessarily presupposes that of a body moving". [6] (The example of Hume shows how little even the most radical epistemological scepticism affects established habits of common sense.) If this were true, then the immutable vehicle of motion would be found in sub-microscopic, elementary particles on the electronic or aetheral level.

There is hardly any need to deal with this objection after our previous discussion of the concept of particle and its inadequacy in the light of modern discoveries. Objections of this kind were, indeed, raised against Bergson by René Berthelot in 1913 and by A. O. Lovejoy one year later. [7] But that was in the period when the triumphs of the electron theory made it possible to speak of the "renaissance of the atomistic theories". [8] This makes understandable Lovejoy's claim that in the physical universe there is no genuine novelty, since "according to the prevailing physical hypo-

theses" the universe is made of old materials. But this was prior to the
general theory of relativity, prior to wave mechanics, prior to the discovery
of the materialization and dematerialization of the particles. It is needless
to repeat what was said about it before, how "particles" of modern physics
are neither indestructible nor uncreatable, neither immutable through
time nor sharply localizable in space. In one word, *they are not particles
at all*. Similarly, the space-time of the general theory of relativity with
its spatially and temporally variable curvature is *toto caelo* different
from the classical, immutable, container-like Euclidian void. Furthermore, the reabsorption of matter into the local curvature of time-space
eliminates the classical distinction between the vehicle of motion and its
container altogether. Thus Bergson was on a far more correct track than
both Berthelot and Lovejoy when he wrote nearly at the same time:

> The more it [i.e. physical science] progresses, the more it resolves matter in actions
> moving through space, into movements dashing back and forth in a constant vibration
> so that mobility becomes reality itself. No doubt science begins by assigning a support
> to this mobility. But as it advances, the support recedes; masses are pulverized into
> molecules, molecules into atoms, atoms into electrons or corpuscles; finally the support
> assigned to movement appears merely as a convenient schema – *a simple concession
> of the scientist to the habits of our visual imagination*. (Italics added.) [9]

Such a view was a mere consequence of the general principle which
Bergson, after establishing it first in psychology, generalized by extending
it to physics as well:

> There are changes, but there are underneath the change no things which change;
> change has no need of a support. There are movements, but there is no inert or in-
> variable object which moves: *movement does not imply a mobile*. (Italics added.) [10]

In other words, the concept of immutable, material substance is as
illegitimate as that of immutable, mental subject. The immutable particles
in physics are as fictitious as the mental atoms of Hume, Herbart and
Mach in psychology. Finally, the ideas of a homogeneous, inert container
separable from its content – homogeneous space in physics, homogeneous
spatialized time in psychology – are equally artificial.

As the distinction between 'the full' and 'the empty' was abolished,
motion cannot be merely a displacement of a visible and isolated particle
in space, but also a displacement of the whole complex of subtle gravitational and electromagnetic links which join 'the particle' to the rest of
the universe. In the language of the general theory of relativity, matter

became a local deformation of the spatio-temporal medium. More accurately, what was called a material corpuscle is nothing but a *center of this deformation*. The deformation spreads itself with decreasing intensity in all directions, producing thus the phenomenon of the gravitational or the electromagnetic field. The movement of *the center* of this deformation implies the concomitant motion of *the whole deformation*.

This has one important consequence. It becomes meaningless to say that the spatio-temporal deformation is *moving in something* – unless it means that motion of the spatio-temporal deformation, i.e. of time-space itself, is in another container-like space with an additional dimension *in respect to which* the motion would take place. This would set us on the path of Zeno: our space is located in a hyperspace which may be located in still another one and so forth. But from an epistemological standpoint such reification of mathematical abstractions is highly suspect, especially since it is motivated by our unconscious desire to save the classical concept of motion as a *displacement of something with respect to something*. There is not the slightest shadow of evidence for the existence of such an imaginary hyper-space.[11] What is given empirically is the *relative* displacement of various regions of time-space with respect to each other – nothing more. What is *suggested* by experience is the abolition of the distinction between 'full' and 'empty'. As soon as this distinction is given up, the revision of the classical concept of motion is inevitable and it is so radical that the old word itself should be dropped entirely.

For the word 'motion' is imbued with misleading visual associations. It fails to suggest that what appears to us as 'displacement' is a *mere surface phenomenon* for which a more general term like 'change,' 'transformation,' or 'modification' would be less misleading. In my previous book I tried to show that in light of new physical discoveries the traditional concept of "change of place" yields place to the more comprehensive concept of qualitative transformation in the Aristotelian or Heraclitean sense. In truth, neither Russell nor Reichenbach nor Schrödinger hesitated to refer specifically to Heraclitus when characterizing the physical processes with which contemporary physics deals.[12]

As early as 1896, Bergson claimed that "real movement is rather the transference of a state than of a thing".[13] The first term is synonymous with qualitative change and as such it retains its absolute character. Only the latter is relative, depending on the frame of reference. When

Bergson speaks of "real motion" or "real mobility", he understands by it the first meaning. The misunderstanding of this misled d'Abro to believe that Bergson was trying to defend absolute motion in the Newtonian sense. The very opposite was true. Bergson welcomed the principle of constant velocity of light as evidence that light, as Bridgman stressed only a few years later, is not a "travelling thing".[14] The distinction made in *Durée et simultanéité* between "*le transport*" and "*la propagation*" is a mere terminological modification of the distinction made in *Matter and Memory* mentioned above.

Thus we must conclude that change is far more pervasive than classical physics believed and that it affects both the allegedly changeless space and the allegedly changeless corpuscles. What else should we expect if matter was reabsorbed into the non-Euclidian modifications of time-space and the "motion of matter" itself is, as W. K. Clifford clearly anticipated[15], nothing but a variation of this local modification? Since this local, spatio-temporal modification is not marked off by sharp boundaries from the wider spatio-temporal context, what was viewed as a displacement of a quantity of matter in passive, immutable space is in truth a transformation of a considerable portion of the universe. After Clifford this was anticipated by Bergson in 1896 when the first signs of the twentieth century revolution in physics were barely discernible:

The change is everywhere, but inward; we localize it here and there, but outwardly; and thus we constitute bodies which are both stable as to their qualities and mobile as to their positions, a mere change of place summing up in itself, to our eyes, *the universal transformation.* (Italics mine.)[16]

The same approach led Whitehead to the same conclusion when he rejected *the fallacy of simple location*:

In a certain sense, everything is everywhere at all times. For every location involves an aspect of itself in every other location. Thus every spatio-temporal standpoint mirrors the world.[17]

As already mentioned, this view of motion is historically related to the Aristotelian, or rather Heraclitean, opposition to the atomistic-kinetic tradition, since it interprets the phenomenon of spatial displacement as an outward, sensory manifestation of a far more pervasive qualitative transformation. In the modern post-Renaissance period, Leibniz was one of a few who insisted that motion is a mere phenomenal aspect of inner qualitative change and that "the action of substance on another" – to use his own language – cannot be interpreted as a sort of "emission or transplantation of an entity as the vulgar opinion believes".[18] That is, it cannot be reduced to a change of location only. But this was only Leibniz-as-metaphysician; as a physicist, Leibniz was as mechanistic as Descartes and his vigorous opposition to Newton in his *Antibarbarus physicus*[19] stemmed from his belief that even the action of gravity must be explainable by some mechanical model.

## NOTES

[1] 'Das Gesetz der Erhaltung des Raumes'. Cf. W. Ostwald, *Vorlesungen über die Naturphilosophie*, 2nd Ed., Leipzig, 1902, p. 280.

[2] *De rerum natura*, Book II, vv. 310–333.

[3] *Matière et Mémoire*, 28th Ed., Paris, 1934, p. 232 (p. 204 of the English translation).

[4] *C.M.*, p. 175.

[5] *Electrons et photons, Rapports et discussions*, Institut International de Physique, Paris, 1928, pp. 86–87.

[6] D. Hume, *Treatise on Human Nature*, Part IV, § 4; J. Tyndall, *Light and Electricity*, Appleton, New York, 1873, pp. 123–124.

[7] R. Berthelot, *Un romantisme utilitaire*, II, pp. 235–237; A. O. Lovejoy, *Bergson and Romantic Evolution*, The University of California Press, Berkeley, 1914, p. 18.

[8] Cf. Part I, Chapter IV, Note 4.

[9] *C.M.*, p. 175.

[10] *Ibid.*, p. 173.

[11] This is what is correctly stressed by d'Abro: " L'Espace-Temps courbe à quatre dimensions n'est pas suspendu dans l'espace euclidien à cinq dimensions." (*Bergson ou Einstein*, p. 265.)

[12] B. Russell, *The Analysis of Matter*, Dover, New York, 1954, pp. 61–62; H. Reichenbach, *The Philosophy of Space and Time* (Dover, New York, 1957, p. 263; E. Schrödinger, 'Our Conception of Matter', in *What is Life? and Other Scientific Essays*, Doubleday-Anchor Book, 1956, p. 175.

[13] *M.M.*, p. 197f; *D.S.*, p. 34: "Reciprocity of displacement is the visual manifestation of an absolute internal change occurring somewhere in space." Cf. also pp. 39–40.

[14] P. W. Bridgman, Macmillan, New York, 1927, p. 164; *D.S.*, p. 40.

[15] W. K. Clifford, 'On the Space Theory of Matter', *Proceedings of the Cambridge Philosophical Society* II (1876) 157–158.

[16] *M.M.*, p. 205.

[17] *S.M.W.*, p. 133.

[18] G. W. Leibniz, 'Système nouveau de la nature et de la communication des substances': "Aussi l'action d'une substance sur l'autre n'est pas une emission ou une transplantation d'une entité, comme le vulgaire le conçoit...".

[19] On these two contrary trends in Leibniz's thought cf. M. Čapek, *The Philosophical Impact of Contemporary Physics*, pp. 274–275.

CHAPTER 11

# LIMITS OF THE CRITICISM OF SIMPLE LOCATION:
## CONTEMPORARY INDEPENDENCE

Two important qualifications concerning the criticism of the fallacy of simple location are necessary. Although it is true that what we used to call the 'displacement of a particle' is in truth a change far transcending the local region in which a particle seems to be located, the Bergsonian term '*la transformation universelle*' should be understood only in a metaphorical sense. In using this term Bergson apparently overlooked the finite velocity of gravitational and electromagnetic disturbances. The spatio-temporal deformation of which the particle is a center is precisely spatio-*temporal*, that is, it is a *process* and not a *thing*. It spreads itself in all directions, but this spreading is not instantaneous and thus it *cannot be ubiquitous* in the strict sense of the word. Both Bergson and Whitehead were probably misled by Leibniz's famous saying about each monad mirroring the whole universe. While they were both correct when they regarded Leibniz as rightly anticipating the fact of universal cosmic interaction, they apparently forgot, as Leibniz himself did, that this 'mirroring' *takes time*. It is true that faint gravitational and electro-magnetic actions penetrate into even the darkest corners of the universe, but this does not mean that the actions of *each* spatio-temporal region pervade *each* other region. The finite velocity of all causal actions – we can as well say the finite velocity of every field – simply forbids it.[1]

Besides, is the expression 'the totality of the universe' physically meaningful? Certainly not in the light of the theory of relativity, which forbids speaking of 'the universe at a given instant'. Nor is it meaningful from Bergson's own point of view. We have only to remember how vigorously he criticized "the cinematographic mechanism of thought" which produces the illusion that the cosmic becoming may be conceived of as a succession of instantaneous states of the world. The very concept of a "state of the world at an instant" is for Bergson merely an artificial cut in the ever-growing becoming, a kind of instantaneous picture in which the flight of time is artificially suspended.

Leibniz's statement should then be corrected in the following sense: not the whole universe, *but a considerable and an ever-growing region* of it is reflected by each 'monad' or, as we should say today, by each spatio-temporal region. Bergson evidently exaggerated his emphasis on the continuity of the universe. His critical reaction against what his disciple E. Le Roy called *"l'illusion du morcelage"* [2] led him to forget his own implicit rejection of instantaneous relations within the world. As René Berthelot correctly observed, there is a significant difference of view, or at least of emphasis, in this respect between Bergson and James.[3] James, in his rejection of the "bloc-universe," did more justice to the fact of *contemporary independence* which is one of the consequences of the theory of relativity. It is wrong, then, to say that the motion of each particle is at *the same time* a transformation of the *whole* surrounding field. This would imply a sort of rigid connection between the center of the spatio-temporal deformation and the deformation itself. As Paul Langevin pointed out as early as 1911[4], to admit such rigid connections in the universe is equivalent to the admission of the absolute simultaneity of distant events, which is eliminated by the special principle of relativity.

The second qualification is closely related to the first one, if it is not merely another aspect of it. The statement that the individuality of the particle is artificially carved out of the dynamic continuity of the universal interaction may easily be misinterpreted in the sense that the distinction between the field and its center is completely abolished and physical individualities are simply dissolved in the continuity of the cosmic medium. There is more than one reason to believe that this would be a serious oversimplification. First, even the classical field theories, which were then called 'aether theories', faced the problem of accounting for the empirically given distinction between matter and its surrounding space; and if this distinction is only apparent, they must explain the *appearance* of this distinction. In the classical aether theories, which had their roots in the plenist theories of Hobbes and Descartes, 'corpuscles' were regarded not as ultimate units, but either as structural or as kinetic complications of the aethereal medium. Thus Denis Papin explained the cohesion of the so-called atoms by *"motus conspirans"*,[5] i.e. by the pressure of the aether, and even William Thomson's nineteenth century theory of "vortex-atoms" was clearly Cartesian in its spirit. Although physics gradually moved away from such crude, intuitive, hydrodynamical models, the tendency to regard the difference between 'particle' and 'field' as a mere difference of structural complication still persisted and even persists today on a more abstract level. A more abstract term, 'field,' replaced the old *materia subtilis* of Descartes, and physicists, instead of speaking of condensations of the aether or aetherial vortices, began to speak of "condensations of energy" or "non-Euclidian humps" in the four-dimensional continuum. But this distinction apart, there is basically the same ideal common to the classical fluid theories of matter and the modern "unitary field theories", as Einstein himself conceded in his lecture *Relativity and Aether*. This ideal is to reduce matter to a local complication of the field, the task in which Descartes failed, since from his homogeneous and Euclidian space he was unable to obtain logically the phenomenal difference between ponderable matter and its surrounding. With the concept of non-Euclidian space, with locally and temporally varying curvature, the task ceased to be impossible.

Since according to our present-day notions the primary particles of matter are also, at bottom, nothing but condensations of the electromagnetic field, our modern schema of the cosmos recognizes two realities which are conceptually quite independent of each other even though they may be causally connected, namely the gravitational aether and the electromagnetic field or – as one might call them – space and matter. It would, of course, be a great step forward if we succeeded in combining the gravitational field and the electromagnetic field into a single structure. Only so could the era in theoretical physics inaugurated by Faraday and Maxwell be brought to a satisfactory close. The antithesis of aether and matter would then fade away, and the whole of physics would become a completely enclosed intellectual system, like geometry, kinematics and the theory of gravitation, through the general theory of relativity....[6]

This ideal of a "unitary field theory" fascinated Einstein for the rest of his life, although he anticipated the difficulties of its realization when he added prophetically to the passage quoted above: "in thinking about the immediate future of theoretical physics we cannot unconditionally dismiss the possibility that the facts summarized in the quantum theory may set impassable limits to the field theory."

But what is important in this context is the fact that even within the field theory of matter the distinction between 'matter' and 'field', though far less sharp, is far from obliterated. It is less sharp since, unlike the particles of classical physics, the microphysical 'particles' known today can be converted into radiation and *vice versa*. While radiation itself, unlike the electromagnetic energy of the Maxwell era, has a granular structure, the "particles" on the other hand have a wave-like character. These are important corrections. Yet, differences still remain: the 'rest mass' of the photons is zero while that of all other particles except neutrinos and hypothetical gravitons is always different from zero. Thus precisely those phenomena which suggest continuity between 'the particles' and their field – the materialization of the radiation and the converse process – indicate clearly that a *genuine structural change* takes place when one is converted into another.

In what sense is this second qualification related to the first one? As already stated, if the distinction between the ultimate physical elements and their dynamic surrounding were *purely* fictitious, then there would be only *one single undifferentiated field* and any change within any point of this field would immediately and instantaneously affect the totality of this field. Within such a universe no "delayed potentials" would be possible. Both light and gravitation would move with infinite velocity. In truth, such was the world of Descartes in which light was the pressure in the aether, and communicated instantaneously; and the world of Newton in which gravitation was spreading with an infinite speed. But such is not the world of modern physics, for in it no physical action can move faster than the quanta of radiation. The world of classical physics cohered rigidly along its *transversal*, i.e. spatial, dimensions. 'The movement of time', as long as it was reluctantly admitted and inadequately described by kinematic metaphors, had, so to speak, *one single wave-front* on which all simultaneous events were carried ahead to the future.

The universe of modern physics is of a different type. Not only are there definite indications that the cohesion of its successive phases is different from the rigid necessitarian bond which in its ultimate consequences makes genuine succession spurious and impossible; but the theory of relativity also loosened the connexion of the universe along its *transversal*, i.e. spatial, cross-section when it replaced the classical concept of simultaneity with that of *causal independence*. The network of cosmic interaction is not spread instantaneously over the whole universe nor is it infinitely dense. There are gaps in it which represent the *absence* of causal relations. Thus when I for instance, perceive, Polaris in the sky, my retina, my nervous system, my mind are affected by events fifty light-years away. Using Leibniz's, Bergson's, and Whitehead's language, I can say that my consciousness 'mirrors' a very extensive spatio-temporal region – what the relativity theory calls my 'causal past' affecting my present moment – but *not* the whole universe. Also, my very same present moment *will remain forever independent* from the photons which will reach my eye tomorrow, after a week, a month or a year. The light from Polaris tomorrow will not reach my present "Now," but only its causal successors. All future effects of Polaris will be *forever absent* from this particular present moment.

This is what Jean Wahl called "the principle of absence", even though he coined this term independently of relativistic considerations.[7] William James apparently saw it more clearly than Bergson and Whitehead. If there is a Jamesian universe whose parts "have a certain amount of loose play on one another, so that laying down of one of them does not necessarily determine what the others shall be", then it is the relativistic universe. Bergson's distrust for the artificial fragmentation of reality caused him to overlook the fact that cosmic interaction and continuity can never be of the static instantaneous type. Similarly, his emphasis on the absence of any rigid cohesion along the temporal dimension, made him overlook that there is no such thing as a rigid *transversal* connection. In this point lay his misunderstanding of the relativity theory.

But the above reservations apart, Bergson was right when he did not regard the distinction between "particles" and their fields as *altogether* artificial. This follows naturally from his biological theory of knowledge. We have seen that according to this theory, the human perception *simplifies*, but *does not falsify* the reality. Its practical and selective nature consists in ignoring certain tenuous connections as well as certain discontinuities in nature which do not interest our action in a biologically significant way. Hence the illusion of the *isolated* and *compact* body, in other words, the classical distinction between the 'full' and the 'empty' which exerted such a lasting fascination on physicists – even those who were nominally opposed to atomism. But this distinction was only an *exaggeration* of the objectively existing situation. The isolated, compact bodies are phenomena, but to use Leibniz's words, "well-founded phenomena". On such phenomena we act; and, according to Bergson's profound words, "our action cannot move in the unreal".[8]

Another reason why the Bergsonian philosophy of the physical world was incompatible with the field theory of matter was that the latter, both in its classical as well as its modern form, was essentially *deterministic*. Bergson's view of matter implied the contingency of the elementary physical events. Matter, being durational in its character, cannot be exhaustively described by the Laplacean type of determinism which the unitary field theories share with the classical, corpuscular-kinetic view of reality. The corpuscular as well as the field model of matter assumed that it was meaningful to speak of a definite state of the world at a given instant. While this state was describable in the atomistic model as a huge instantaneous configuration of particles with definite velocities, in the plenist or the field models it was characterized by an instantaneous set of all energy values.

In either case, all posterior as well as anterior states were unambiguously determined by this state. The very concept of 'direction of time' becomes vague in such a scheme in which causation is reduced to a logical co-implication. One of the main reasons why Einstein and eventually also Louis de Broglie adopted a negative attitude toward microphysical indeterminism was their confidence that some modern version of the field theory would successfully account for the atomicity of both matter and action, and thus also *explain away* microphysical indeterminacy as a *mere appearance*. We shall return to this point later.

## NOTES

[1] Cf. my article 'Simple Location and Fragmentation of Reality', *The Monist* **48** (1964) 195–218.
[2] E. Le Roy, 'Continu et discontinu dans la matière: le problème du morcelage', in *Continu et discontinu. Cahiers de la nouvelle journée*, No. 15, Paris, 1929.
[3] R. Berthelot, *op. cit.*, p. 232.
[4] P. Langevin, 'Le temps, l'espace et la causalité dans la physique moderne', *Bulletin de la Société française de philosophie* (séance du 19 octobre 1911), pp. 23–24.
[5] Papin's letter to Huygens, June 18, 1690 (*Oeuvres complètes*, IX, pp. 429–430).
[6] A. Einstein, *Essays in Science*, Philosophical Library, New York, 1934, pp. 110–111.
[7] J. Wahl, *Les philosophes pluralistes d'Angleterre et d'Amérique*, Paris, 1920, p. 126.
[8] *C.E.*, p. XXI.

# CHAPTER 12

# THE INDETERMINACY OF MICROPHYSICAL EVENTS.
# BERGSON AND BOUTROUX

The impossibility of instantaneous cuts across the spatio-temporal becoming, an impossibility asserted both by Bergson and modern physics, sheds light on the last feature of the ultimate physical events postulated by Bergson: *their elementary indeterminacy*. The relativity theory rejected the possibility of instantaneous cuts on the macrophysical level, but remained deterministic precisely because of its macroscopic character. Quantum physics, however, since 1927 questions the possibility of instantaneous cuts even on the microphysical level as it shows itself in the second form of Heisenberg's indeterminacy principle:

$$\Delta E \cdot \Delta t \geqslant h \quad \begin{aligned} \Delta E &= \text{the imprecision of energy} \\ \Delta t &= \text{imprecision in time} \\ h &= \text{Planck's constant of action} \end{aligned}$$

For Bergson basically the same conclusion followed from his claim that the *emergence of novelty* characterizes every duration – including the duration of physical reality. If we accept his previously expressed view that the terms 'present' and 'novelty' are synonymous, then the postulate of the elementary indeterminacy of microphysical events follows naturally. Like every temporal process, the physical duration grows not by infinitely thin instants, but *by concrete drops of novelty*. This is obscured by the fact that the duration of the physical present is negligible. In our myopic, macrochronic perspective the temporal thickness of the basic, microphysical events shrinks practically to the "zero-length" of mathematical instants. But, according to Bergson, such shrinkage is *never complete*. The correlated assertions of the instantaneity and of the absence of novelty in matter are mere exaggerations of our intellect which, encouraged by repeatedly successful applications of both concepts on the macrophysical level, imposes them on the microphysical scale as well. In doing this, it merely exaggerates the trend which physical reality really exhibits.

In the words of the passage previously quoted, the intellect transports itself "to the terminal point of the movement (i.e. of the tendency) of which matter simply indicates the direction".[1] The narrower the span of the present, the greater is the restriction of novelty. As long as the present does not shrink to a static instant, that is as long as 'the flight of time' is not suspended entirely, indeterminacy cannot be altogether absent. In the suggestive words of the Polish historian of science Zygmunt Zawirski, this is what the second form of the Heisenberg principle shows:

If the instantaneous cut of the temporal flow according to Heisenberg's formula leaves energy completely undetermined, does not this prove *that the universe needs a certain time to take on precise forms*?[2]

In Bergson's words, to which a literal translation hardly does justice, "the delay of duration on instantaneity" ("*le retard de la durée sur l'instantanéité*") represents a "certain hesitation or indetermination." In other words, "time is this very hesitation or it is nothing".[3] From this point of view matter consists of elementary events which in spite of their brevity *still endure* and thus never coincide with mathematical instants. The novelty inherent in each elementary event is so insignificant that *for practical purposes* physical time is homogeneous, i.e. devoid of qualitative differences and infinitely divisible. Thus *the tendency toward determinism* and *the tendency toward homogeneity and repetition* are two aspects of one and the same fact. In 1903 Bergson pointed out that in reducing the mnemic span of our psychological duration:

we advance toward a duration more and more scattered, whose palpitations, more rapid than ours, dividing our simple sensation, dilute its quality into quantity; at the limit would be the pure homogeneous, the pure *repetition*, by means of which we define materiality.[4]

Needless to stress again that such a limit is never attained and *cannot* ever be attained, since a *complete* repetition resulting in a *complete* identity of the subsequent moments would destroy their very succession.[5] Three decades later Bergson expressed the same idea with greater clarity when he characterized the inorganic world as "a series of extremely rapid repetitions or quasi-repetitions which in their sum constitute visible and predictable changes."[6] This was clearly written in the light of the knowledge absent at the beginning of the century, that the microphysical indeterminism passes over asymptotically to a macrophysical, *practically rigorous* causality when very large aggregates of events are considered. But, as mentioned many times before, the affirmation of novelty on the physical realm is an *essential* part of Bergson's theory of duration, explicitly stated as early as 1896 in *Matter and Memory*.

This does not mean that Bergson was the first to suggest that physical determinism is only *approximately* valid even in the physical world. Émile Boutroux explicitly proposed it in the fourth chapter of his courageous book *De la contingence des lois de la nature* in 1874. He prophetically submitted that the indeterminacy of the microphysical processes may be too small to be detected by means of nineteenth century physics.[7] Like C. S. Peirce in his article 'The Doctrine of Necessity Examined' nearly two decades later[8], Boutroux pointed out that experimental measurement can never determine an absolutely sharp value since the measured values are always distributed within certain range, no matter how narrow and practically negligible it may be. But he also pointed out that if an effect were *completely* identical with its cause, it would be meaningless to speak of any difference between them.[9] Causation itself would be resolved in a bare identity. Anticipating Bergson, he insisted that "bare and simple repetition nowhere in nature exists" and "homogeneous quantity is merely an ideal surface of reality".[10] In other words, like Bergson later, he claimed that there must be an element of *qualitative heterogeneity* in matter itself which resists any attempt at reducing it to bare quantitative relations:

Everything that is possesses qualities, and consequently participates in that radical indetermination and variability which belong to the essence of quality.
    Thus the principle of the absolute permanence of quantity does not apply exactly to real things: these latter have a substratum of life and change, which never becomes exhausted. The singular certainty presented by mathematics as an abstract science does not authorize us to look upon mathematical abstractions themselves, in their rigid monotonous forms, as the exact image of reality.[11]

This was also the reason why Boutroux, again like Bergson much later[12], denied the absolute validity of the law of conservation of energy in which the last century's physicists, and in particular philosophers, saw the most fundamental law of physics and the most accurate formulation of the deterministic principle.

Only occasionally is Boutroux praised for his remarkable anticipation, for instance, by the French mathematician Jacques Hadamard in his introduction to the French translation of Birtwistle's book on quantum mechanics. Yet, in his affirmation of the contingency of microphysical events he was as definite as Peirce, and far more definite and less hesitant than Bergson. (So far as I know, only E. T. Whittaker praised Peirce for his anticipation.)[13]

We have pointed out several times that in his first book Bergson tended toward a sharp dualism of the physical and the mental. To the psychological realm of duration, indetermination and quality he opposed the spatial realm of determinism and quantity. It is true that even then he realized that the purely private freedom of our psychological 'stream of consciousness' would have very little pragmatic value without some degree of indetermination in the physical world, indetermination without which all our 'overt actions' would remain as rigorously determined as other parts of the physical world.[14] Hence his doubts, even in 1889, about the strict validity of the law of conservation of energy. This was another indication that the original dualism of his first book was far from definitive.

In *Matter and Memory* the tendency to overcome the original dualism became quite explicit. In the concluding part of the book he stressed that its main idea is to overcome the triple opposition – that of time and space, that of quality and quantity, and finally, that of freedom and determinism. Yet, the denial of the last opposition was still phrased very cautiously and rather shyly, undoubtedly because it implied the affirmation of the microphysical contingency which in 1896, because of the tremendous prestige of the Newtonian determinism, was regarded as a despicable heresy:

Can each moment be mathematically deduced from the preceding moment? We thought in this work, and *for the convenience of study, supposed that it was really so*; and such is, in fact, the distance between the rhythm of our duration and that of the flow of things (*l'écoulement des choses*), that the contingency of the course of nature, so profoundly studied in recent philosophy, must, for us, be practically equivalent to necessity. So let us keep to our hypothesis, though it might have to be attenuated.[15]

This is the only reference to Boutroux which can be found in Bergson's writings. Although Boutroux is not named explicitly, there is hardly any doubt that it is his philosophy of contingency to which Bergson refers here.

René Berthelot, who in spite of his negative attitude to Bergson's philosophy understood it well – we may add that he was opposed to it precisely *because* he understood it so well, and thus fully realized its incompatibility with classical physics to which he remained loyal until the end of his life – was correct when in 1913 he regarded the postulate of the elementary indeterminacy of physical processes as an *essential* part of Bergson's theory of matter.[16]

This becomes obvious if we analyze the further development of Bergson's thought on this point. In truth, there was no further development in this respect. What happened was that Bergson's language became less timid and more explicit after the historical year of 1927. It is true that his essay *De la position des problèmes* was written in 1922. But according to the author himself, the passages dealing with contemporary physics were added later, certainly *after* the formulation of the indeterminacy principle. What he wrote then does not differ in substance from what he wrote earlier:

One can, then, and even should continue to speak of physical determinism, even when postulating, with the most recent physics, the indeterminism of elementary events which make up the physical fact. For this physical fact is perceived by us as submitted to inflexible determinism, and by that fact is radically distinguished from acts we accomplish when we feel ourselves free.[17]

This is nothing but a reaffirmation of the *practical* validity of determinism in physics, a validity which is due to the macroscopic and *macrochronic* character of our perception. It also shows that, unlike Pascual Jordan, Bergson did *not* confuse contingency and freedom.

The continuation of the passage indicates how this view was related to Bergson's biological theory of knowledge. It was this passage which Louis de Broglie quoted with justified admiration:

One might ask himself if it is not precisely in order to pour matter into this determinism (*pour couler la matière dans ce determinisme*), to obtain in the phenomena which surround us a regularity of succession, thus permitting us to act upon them, that our perception stops at a certain degree of condensation of elementary events.[18]

It is now clear how this passage is related to the one quoted from *Matter and Memory*. The rhythm of physical duration is so remote from our own that *for practical purposes* its element of novelty (synonymous with the span of the physical present) is negligible and can be safely disregarded on the macrochronic scale. As Charles Eugene Guye said, it is *the scale of observation which creates the phenomenon*.[19] In Bergson's words, it is *a degree of condensation of duration* which creates the useful illusion of the discontinuity of sensory qualities and of the isolated bodies as well as the illusion of strict determinism:

What would become of the table upon which I am at this moment writing if my perception, and consequently my action, was made for the order of greatness to which the elements, or rather the events, which go to make up its materiality, correspond? My action would be dissolved; my perception would embrace, at the place where I see my table and in the short moment I have to look at it, an immense universe and a no less interminable history. It would be impossible for me to understand how this moving immensity can become, so that I may act upon it, a simple rectangle, motionless and solid. It would be the same for all things and events: the world in which we live, with the actions and reactions of its parts upon each other, is what it is by virtue of a certain choice in the scale of greatness, a choice which is itself determined by our power of acting.[20]

This is a restatement of one of the basic ideas of *Creative Evolution*, that the place in the hierarchy of living beings is determined by their *mnemic span* of their consciousness, which in its turn determines the width of their choice and the extent of both their sensory field and their possible action on their surroundings. As was explained in Part I, both vary from one zoological species to another. "How different must be the world in the consciousness of ant, cuttle-fish and crab!" – wrote William James in one section of his *Principles of Psychology*.[21] This also accounts for the difference separating the perception and action of man from that of those microorganisms which, as Bergson says with some exaggeration, "vibrate almost in unison with the oscillations of the aether" and thus are almost "caught up in the mechanism of things". The passage is worth being quoted *in extenso*:

The primal function of perception is precisely to grasp a series of elementary changes under the form of a quality or of a simple state, by a work of condensation. The greater the power of acting bestowed upon an animal species, the more numerous, probably, are the elementary changes that its faculty of perceiving concentrates into one of its instants. And the progress must be continuous in nature, from the beings that vibrate almost in unison with the oscillations of aether, up to those that embrace trillions of these oscillations in the shortest of their simple perceptions. The first feel hardly anything but movements; the others perceive quality. The first are almost caught up in the running-gear of things (*l'engrenage des choses*); the others, react, and the tension of their faculty of acting is probably proportional to the concentration of their faculty of perceiving. The progress goes on even in humanity itself. A man is so much the more "man of action" as he can embrace in a glance a greater number of events: he who perceives successive events one by one will allow himself to be led by them; he who grasps them as a whole will dominate them.[22]

This passage is merely an application of the view expounded in *Matter* and *Memory* according to which *"la perception dispose de l'espace dans l'exacte proportion où l'action dispose de temps"*.[23] For a sufficiently reduced mnemic span the sensory qualities which on the human level result from 'the condensation of the past' inevitably disappear. Thus the simple quality of sound would be diluted in an enormous succession of sub-sensations of individual impulses as they are vaguely discerned in the sensation of the lowest audible tone. This is still far off from the microphysical level. But it is understandable how with a further narrowing of the temporal span the hypothetical organism would not be capable of any effective action, since macroscopic determinism, which is a function of a large number of elementary events, disappears on the level at which Planck's constant $h$ is no longer negligible. In other words, no effective action is possible for the beings with a vanishing temporal span, since such action requires choice and, consequently, anticipation. Anticipation requires a certain degree of memory span as well as a certain degree of predictable order, which *ex definitione* is absent on the level of individual physical events.

## NOTES

[1] Cf. Note 14 of Chapter 5 of this part.
[2] Z. Zawirski, 'L'Evolution de la notion du temps', *Scientia* **28** (1934) 260.
[3] *D.S.*, p. 63; *C.M.*, pp. 109–110.
[4] *C.M.*, p. 221.
[5] Cf. a penetrating analysis of Louis Weber 'La répétition et le temps', *Revue Philosophique* **36** (1893) 263–286, esp. pp. 270–273.
[6] *C.M.*, p. 109.
[7] E. Boutroux, *The Contingency of the Laws of Nature* (transl. by F. Rothwell), The Open Court, Chicago, 1920, p. 28: "Supposing that phenomena were indeterminate, though only in a certain measure insuperably transcending the range of our rough methods of reckoning (*'evaluation'* in the original), appearances would none the less be exactly as we see them."
[8] *The Monist* **II** (April 1892), pp. 321–337.
[9] Boutroux, *op. cit.*, p. 29: "Would this also be a consequent, an effect, a change, if it differed from its antecedent neither in quantity nor in quality?"
[10] *Ibid.*, p. 29.
[11] *Ibid.*, pp. 68–69.
[12] *Ibid.*, pp. 59–69; *T.F.W.*, pp. 143–152: *C.E.*, p. 127, 264.
[13] E. T. Whittaker, 'Change, Free Will and Necessity in the Scientific Conception of the Universe', *Proc. Phys. Soc. London* **55** (1943) 459–471; the reference to Peirce's 'tychism' on p. 465.
[14] *T.F.W.*, pp. 149–150.
[15] *M.M.*, pp. 244–245.

[16] R. Berthelot, *Un romantisme utilitaire*, II, Ch. X. Cf. also p. 133 on the affinity of Boutroux's contingentism with the thought of Bergson.

[17] *C.M.*, p. 303.

[18] *Ibid.*, p. 303; Louis de Broglie, *Physics and Microphysics* (transl. by Martin Davidson; with a foreword of A. Einstein), Pantheon Books, 1955, p. 193. In the English translation the Chapter IX 'Conceptions of Contemporary Physics and the Ideas of Bergson on Time and Motion' is merely summarized according to the wish of the author who meanwhile returned to classical determinism.

[19] Ch. E. Guye, *L'Evolution physico-chimique*, 3rd Ed., Hermann, Paris, 1947, pp. 147–149.

[20] *C.M.*, p. 69.

[21] *The Principles of Psychology*, I, p. 289.

[22] *C.E.*, pp. 327–328.

[23] *Matière et mémoire*, p. 19.

# CHAPTER 13

# BERGSON AND LOUIS DE BROGLIE

It is true that all previous considerations presuppose that microphysical indeterminacy has an objective status and therefore is not a mere result of temporary technological limitations of our present measurements. I discussed this particular problem at length in a chapter entitled 'The End of the Laplacean Illusion'[1] in my previous book in which I listed all the facts supporting the objective status of indeterminacy: not only the general bankruptcy of *all* the ideas constituting the classical deterministic model of the physical reality, but also the peculiar character of radioactive explosions, whose statistical character and indeterminacy cannot by their very nature depend on the intervention of the observer. I also pointed out that this character is not confined to radioactive processes only. The emissions of photons have essentially the same 'radioactive' character, and this is true of spontaneous disintegrations of all recently discovered "particles" as well. I also pointed out that resistance to the concept of the objective contingency of microphysical events is due mainly to the tenacity of certain classical beliefs, including the belief in the Laplacean-Spinozist concept of causality, which is still wrongly regarded as the *only* type of rational order.

In this context I shall confine myself to a brief discussion of the philosophical thought of Louis de Broglie, whose discovery of the un-dulatory nature of matter had as revolutionary a significance for the development of physics as did the early discovery by Planck and Einstein of the corpuscular character of light. I am choosing de Broglie not only because he shows for a physicist a rather uncommon knowledge of Bergson's work, but also because his successive hesitations between a subjectivist and objectivist interpretation of quantum indeterminacy ex-emplify in an abbreviated form the present controversy about it.

The philosophy or, more particularly, the philosophy of science of Louis de Broglie can be gathered from a number of essays contained in five of his books: *La physique nouvelle et les quanta* (1937), *Matière et lumière* (1937), *Continu et discontinu en physique moderne* (1941), *Physique et microphysique* (1947) and *Nouvelles perspectives en microphysique* (1956). With the exception of the last book, he consistently maintained the indeterministic interpretation of microphysics, not hesitating also to use philosophical and epistemological arguments against the plausibility of any return to classical deterministic models. In the first four books mentioned above he consistently maintained that the principle of indeterminacy of Heisenberg is an inevitable consequence of the indivisibility of the atom of action $h$, and that the very existence of this constant makes it impossible to speak of the precise localization of physical entities in space and time.[2] This is what makes both the concepts of the classical particle as well as of the classical wave inapplicable on the microphysical level. Thus the extension of the classical principle of determinism to microphysical phenomena is nothing but an illicit extrapolation of macroscopic experience beyond its legitimate limits.

He also pointed out that his own interpretation, though incompatible with rigorous Laplacean causality, nevertheless remains compatible with a *widened* or generalized notion of causality which would still preserve some continuity or connection between successive events. In this type of connection the antecedent "cause" would be a *necessary, though not a sufficient, condition of the "effect"*. Such causal continuity is describable, *not* by the classical dictum *"causa aequat effectum"*, but by its emended version, *sublata causa tollitur effectus*.[3]

His view, then, was fairly close to Reichenbach's view of "ambiguous causal nexuses" and to the prevailing tendency among physicists to regard each event as merely *probably* implied by its antecedents instead of being deducible from them in strict Laplacean fashion. This makes it understandable that in this period he regarded his original commitment to determinism as a result of mere habit rather than of conviction. He firmly rejected the hypothesis of "hidden variables" by means of which determinism could be restored on the sub-quantum level. He concluded that "the laws of mechanics, with their apparent rigour, are nothing but a macroscopic illusion due to the complexity of the objects on which our direct experience bears and not to a lack of precision of our measurements".[4]

These views of Louis de Broglie showed themselves in his attitude toward the three twentieth century French thinkers who were concerned about the problem of physical determinism, even though they approached it from different angles and arrived at different conclusions: Émile Meyerson, Léon Brunschvicg and Henri Bergson. Louis de Broglie dealt with each of them in separate essays and expressed, with his habitual tact and gentleness, his critical reservations with respect to each of them. Yet, when we re-read these essays now, we see clearly that his maximum disagreement was with Brunschvicg, who in various places stated his firm belief that microphysical indeterminacy is due exclusively to the disturbing intervention of the observer in the phenomenon observed.[5] As far as Meyerson was concerned, he merely observed that, had he had more time to study the revolutionary implications of modern physics, he might have modified some of his views. (He probably meant Meyerson's main thesis about the immutability of human reason in the sense that the search for explanation in science will always be synonymous with the search for identification in space and time.)[6] As far as Bergson's view of physical indeterminacy was concerned, we already mentioned de Broglie's approval and admiration of the passage quoted above, according to which macroscopic determinism is due to our *macrochronic* perspective, that is, to the fact that our mnemic span is incomparably wider than the temporal span of the elementary physical events. After quoting it, Louis de Broglie concludes his essay with the following remark:

A curious suggestion according to which living beings would necessarily have a 'macroscopic' perception, for in the macroscopic only there prevails the apparent determinism which renders possible their action on things. In reading this isolated text, how much one regrets that the great philosopher was unable to survey with his piercing glance the unforeseen horizons of the new physics![7]

Several pages before this conclusion de Broglie observed that:

If Bergson had been able to study the quantum theories in detail, he would undoubtedly have joyfully stated that in the picture they offer us of the evolution of the physical world, they show at each instant nature hesitating between several possibilities, and without doubt he would have repeated, as in his last work, that "time is this hesitation itself or it is nothing."[8]

In the concluding footnote de Broglie calls attention to the notion of "weakened causality" (we would prefer to call it 'generalized causality'), formulated by Bergson in his very first book. According to this notion, "the effect will no longer be given in the cause. It will reside there only in a state of possibility...." After quoting this Louis de Broglie concludes:

"The analogy with the probabilistic concept of causal relation of the present quantum physics is obvious."[9] The agreement between Bergson and Louis de Broglie on this point could hardly have been more complete.

Another interesting rapprochement which Louis de Broglie makes in this essay is that between Bergson's analysis of Zeno's paradoxes and the Heisenberg principle. Already in his book *Matière et lumière* in the essay "Les idées nouvelles introduites par la mecanique quantique" de Broglie pointed out that the impossibility of correlating a precise spatio-temporal location together with a definite dynamic state of a moving body has a certain affinity with Zeno's famous paradoxes of motion. The essence of Zeno's argument is that if a 'flying' arrow really *is* at each particular instant at a certain place, that is, if it really *coincides* with its position, then it cannot be moving, it cannot fly at all. De Broglie pointed out that the traditional, and still current, answer to this paradox is that based on the concept of spatio-temporal continuity. Consideration of the successive positions infinitely close to each other makes it possible to define the instantaneous velocity at each point in the same way as we determine the derivation of the usual continuous functions at each of its points. But the basic assumption of this rebuttal of Zeno, that of the spatio-temporal continuity, is now threatened by the discovery of the constant *h* – the atom of action – which introduces an element of discontinuity absent in the classical theories:

Without wishing, then, to make Zeno a precursor of Heisenberg, and without forgetting the part played in this problem at the present day by the finite value of Planck's constant, we may still say that the impossibility, which recent theories reveal, of assigning simultaneously to a moving body an exact spatio-temporal localization and also a completely defined dynamic state, appears to have some kinship with a philosophical difficulty which has long been familiar. To use Bohr's terminology, the exact location in Space and Time is one "idealization," and the concept of a completely defined state of motion is another, so that the two "complementary idealizations," while almost quite compatible on the macroscopic scale, are not strictly so on the microscopic scale.[10]

It was nearly a decade later when Louis de Broglie brought forth this reflection on Zeno and Heisenberg's principle in relation to Bergson's criticism of Zeno. Both Bergson and Zeno believed that motion cannot be built of immobilities. While Zeno concluded from this that motion does not exist, Bergson concluded that motion – and, more generally, any change – is real and therefore does *not* consist of immobilities.

A moving body *is* as little at any particular point of its trajectory as, according to James, the stream of thought *is* at any particular idea or as, according to Koffka, a melody *is* at any particular tone. Louis de Broglie then underlines the fact that forty years before the formulation of the indeterminacy principle Bergson insisted on the impossibility of motion *being* at any static point of its trajectory. In truth, borrowing the language of the quantum theories, Bergson could have said: "If one seeks to localize the moving object in a point of space, one will obtain nothing but a position, and the state of motion will completely escape." De Broglie then added that, if Bergson erred, it was by excess of caution.[11] He still spoke of a continuous trajectory, although such a concept loses its definiteness on the microphysical level. (In fairness to Bergson, however, it must be pointed out that not only were the instances in which he referred to trajectories macroscopic in nature, but also that he regarded spatial displacement as a mere outward manifestation of qualitative becoming. See Chapter 9 of this part.)

We know today that Louis de Broglie does not uphold the objectivity of microphysical indeterminacy any longer. In the session of the French Philosophical Society on April 25, 1953, he announced his return to *basically* the same philosophical position which he upheld prior to 1928.[12] For twenty-five years he defended the indeterministic position of Heisenberg, Born, Bohr and the majority of physicists. In truth, in his article, 'Souvenirs personnels sur les débuts de la mécanique ondulatoire', the concluding paragraph describes his "final conversion" to the probabilistic interpretation of microphysical phenomena.[13] It was not as 'final' as he believed when he wrote it in 1947; for his reconversion to determinism came only five years later.

Our main concern in the present context is the role of a *philosophical* motive in Louis de Broglie's reconversion to determinism. That philosophical motives played the main and almost exclusive role in it, was conceded by de Broglie himself. Certainly no new discovery in physics induced him to abandon the view which he held for a quarter of a century. His two main reasons were both of a philosophic nature: first, distrust of the phenomenalistic and even idealistic tendencies present in the orthodox, probabilistic interpretation of quantum physics; second, nostalgia for "*clarté cartésienne*" of the classical, deterministic models.[14]

It is true that de Broglie's reconsideration of his own position was triggered by two external factors: the articles of David Bohm in *The Physical Review* in 1952 and the theoretical work of Jean-Pierre Vigier. But it would hardly be appropriate to call these factors 'external'. David Bohm resumed de Broglie's original efforts in 1927 to construct a quasi-hydrodynamical model of wave-mechanical phenomena while J. P. Vigier called de Broglie's attention to the affinity between his original hypothesis of 1927 and the general theory of relativity.[15] It is precisely this affinity which sheds light on both Einstein's and de Broglie's negative attitude toward physical indeterminism as well as on the Cartesian character of their thought.[16]

Needless to say that this character is 'Cartesian' only in a very broad and sophisticated sense which can be conspicuous only to a historian of ideas acquainted with both the history of physics and philosophy. What is common to both Einstein's unitary field theory and de Broglie's "hypothesis of hidden parameters" is, besides the firm belief in rigorous determinism, also the idea that the existence of corpuscles and, more generally, the grain-like structure of the fields, should be explained by the structure of the corresponding field itself.[17]

In this respect, this view is akin to the plenist or fluid theories of matter which from Descartes and Hobbes to William Thomson regarded the particles as mere structural complications of the all-pervading aether. It is true that the concrete idea of the aetherial medium was gradually replaced by a more abstract idea of the field which in the general theory of relativity was identified with a still less intuitive concept of the non-Euclidian continuum – with locally variable curvature. Correspondingly, the concrete aetherial "vortex-atoms" of William Thomson were finally replaced by the singularities of the field or the local "humps" of the non-Euclidian continuum. The motion of the "particles" in de Broglie's theory was guided by their dynamic surroundings (i.e. the so-called "pilot waves") as fully as the motion of the Cliffordian "humps" is by their non-Euclidian medium or the motion of the planets by the rotation of the aether in the original, naive Cartesian model.

It is interesting that de Broglie, who prior to 1953 stressed the *macroscopic* character of the general theory of relativity (and of the field theories in general) which cannot do justice to the quantum phenomena, reversed his position: instead of regarding the continuity of fields as a macroscopic appearance resulting from basic discontinuities, he now regards the spatio-temporal continuity of fields as *basic* and the discontinuity of particles, including photons, as mere local irregularities in the underlying continuum. In this respect both Einstein's unitary field theory and de Broglie's hypothesis of "hidden parameters" are Cartesian in a broad sense.

In fairness to Louis de Broglie it must be stressed that the term 'reconversion to determinism' is perhaps too strong. While not hiding his preference for a deterministic re-interpretation of the quantum phenomena, he is aware of the great difficulties of its concrete and convincing elaboration. All that he asks is an open-minded attitude toward exploring this possibility.[18] This cautious attitude is even more evident in the thought of David Bohm. Although he considers the possibility of re-establishing physical determinism on the sub-quantum level, he explicitly rejects any kind of Laplacean determinism which would regard the postulated, causal, sub-microscopic levels as "the ultimate bottom of nature". This aspect of Bohm's thought, which he himself calls "the principle of qualitative infinity of nature", is not always stressed in cursory references to his criticism of the "orthodox" indeterministic interpretation.[19]

It is interesting and even amusing to contrast the cautious attitude of de Broglie and Bohm with the enthusiastic welcome which they received from classically oriented rationalistic philosophers. Nothing shows their *a priori* commitment to determinism more than the jubilation with which they hailed de Broglie's "reconversion" to determinism at the meeting of *Société française de philosophie* referred to above. Thus M. Salzi praised Louis de Broglie for accepting the Cartesian view, according to which nature, instead of containing a surd irrational element, as Plato erroneously believed, is completely transparent to rationality. That he means by 'rationality' the narrow, Spinozist-Laplacean type of rationality is clear from his illustrations borrowed from classical physics: the kinetic theory of gases shows that statistical laws are always reducible to hidden causal processes. His concluding words have a ring both naive and bombastic:

Such, Mr. Louis de Broglie, are some of the reasons why philosophers, no less than scientists, rejoice to learn that you are closing your ears to the song of the Sirens of the probabilistic contingentism. Your researches reacquired the fruitfulness by which you made the country of Descartes glorious.[20]

It is the same deterministic attitude as that of Léon Brunschvicg in 1940, mentioned above. Understandably it was the attitude of one of Bergson's severe critics, René Berthelot in 1934.[21] More surprisingly, it was also the attitude of Edouard Le Roy in 1935 in spite of his Bergsonism.[22] It was also the attitude of Bertrand Russell, at least in his book *Scientific Outlook*, when he wrote: "The principle of Indeterminacy has to do with measurement, not with causation... There is nothing whatever in the Principle of Indeterminacy to show that any physical event is uncaused." It is then only natural that Brand Blanshard, whose commitment to rigorous determinism approaches a religious fervor, approvingly quotes this passage from Russell.[23] There would hardly be any quarrel with this quotation, provided that the term 'causation' were not understood in a narrow, Laplacean sense. This is what the traditional rationalists are constitutionally unable to do. Neither Boutroux nor Peirce nor Whitehead nor Bergson ever claimed that any event is "uncaused" as Russell insinuates. All they claim is that it is *not* equivalent with, and reducible to, its antecedents.

It is also unfortunately true that the current, indeterministic interpretation is most frequently associated with, and justified by phenomenalistic and even idealistic tendencies. On this point Louis de Broglie's reminder that a physicist, as Émile Meyerson stressed, is always a realist is fully justified.[24]

As I tried to show in my previous book, it is the usage of phenomenalistic language which, in the official interpretation, makes the denial of determinism so confusingly ambiguous. If indeterminacy is an *objective* feature of nature, it must be understood and stressed as such, that is, without exaggerated emphasis on the mind of the observer and the 'disturbing' effect of the act of observation. Very few philosophies offer an appropriate framework for such an objectivist interpretation: the contingentism of Boutroux, tychism of Peirce, Whitehead's ideas of the "self-creativity of actual occasions" are among them. Needless to say that Bergson's emphasis on the elementary indetermination of physical processes belongs with this group.

Finally, one interesting concluding footnote. We said that Einstein also – like Lorentz, Planck, Schrödinger and de Broglie – retained the hope that microphysical indeterminism is not the last word of physics. It is interesting, however, that in his preface to the English translation of *Physique et Microphysique* he wrote:

What impressed me most, however, is the sincere presentation of the struggle for a logical concept of the basis of physics which finally led de Broglie to the firm conviction that all elementary processes are of statistical nature. I found the consideration of Bergson's and Zeno's philosophy from the point of view of the newly acquired concepts highly fascinating.[25]

This was written a few months before Einstein's death. It is significant that in spite of the brevity of the preface, he did not fail to include the passage quoted above. This passage constitutes nearly a half of the whole preface! It hardly seems to be a mere conventional politeness. Could it be that the great physicist in encountering Bergson's criticism of Zeno in the light of Louis de Broglie's comment began to have some second thoughts about his life-long commitment to classical determinism?

## NOTES

[1] *The Philosophical Impact of Contemporary Physics*, Ch. XVI.

[2] L. de Broglie, *Continu et discontinu en physique moderne* (Paris, 1940), p. 61.

[3] *Ibid.*, p. 64.

[4] L. de Broglie, *Physics and Microphysics*, pp. 199–200.

[5] L. de Broglie, 'Léon Brunschwicg et l'évolution des sciences', *Revue de métaphysique et de morale* **50** (1945) 72–76. Here the author gently, but firmly rejected the deterministic interpretation of the Heisenberg's uncertainty relations in Brunschwicg's book *La physique nouvelle et la philosophie*, Hermann, Paris, 1936.

[6] Cf. 'To the Memory of Émile Meyerson' in de Broglie's book *Matter and Light. New Physics* (transl. by W. H. Johnston), Dover, New York, 1939, p. 286.

[7] *Physics and Microphysics*, p. 193.

[8] *Ibid.*, pp. 191–192.

[9] The quotation from Bergson is from *T.F.W.*, p. 212. The footnote of de Broglie is not translated in Davidson's translation: its full text is in the French original version *Physique et microphysique*, Paris, 1947, p. 211.

[10] *Matter and Light*, p. 255.

[11] *Physics and Microphysics*, pp. 189–200.

[12] *Bulletin de la Société française de philosophie* (séance du 25 avril 1953). Cf. also *Nouvelles perspective en microphysique*, Paris, 1956, pp. 199–201.

[13] *Physique et microphysique*, pp. 181–190. (In the English translation the term 'conversion finale' is translated rather obscurely as 'final transformation', p. 156).

[14] *Nouvelles perspectives en microphysique*, p. 140.

[15] *Ibid.*, p. 116.

[16] Cf. *La physique quantique restera-t-elle indeterministe?*, Gauthiers-Villars, Paris, 1953, esp. pp. 65–111 with several contributions of J.-P. Vigier. On Einstein's satisfaction with which he reacted to the attempts ro revive the deterministic interpretation of quantum physics cf. *Nouvelles perspectives en microphysique*, pp. 200–201.

[17] Cf. Note 6 of Chapter 11 of this part; also *The Philosophical Impact of Contemporary Physics*, pp. 316–322.

[18] *Nouvelles perspectives en microphysique*, pp. 142–143.

[19] D. Bohm, *Causality and Chance in Modern Physics*, Harper Torchbook, 1961, pp. 158–160.

[20] Cf. Salzi's intervention in the discussion following de Broglie's lecture in *Bulletin de la Société française de philosophie*. See Note 12 of this chapter.

[21] *Bulletin de la Société française de philosophie* **34** No. 5 (octobre-december 1934), 172–183.

[22] Edouard le Roy in his article 'Ce que la microphysique apporte et suggère a la philosophie', *Revue de métaphysique et de morale* **42** (1935), 345–347 rejects – like Brunschwicg – any "reification" of quantum indetermination.

[23] Brand Blanshard, *The Nature of Thought*, George Allen & Unwin, London, 1948, II, p. 493. Russell's attitude to this problem was far from consistent; while in the passage quoted by Blanshard he claimed the principle of indeterminacy has to do with measurement not with causation (*Scientific Outlook*, 109–110), he was far more cautious in *Analysis of Matter* where he conceded that "there may be the limits to physical determinism" and that this possibility 'interposes a veto on materialistic dogmatism" (p. 393). In *Human Knowledge*, Simon and Schuster, New York, 1962, he explicitly concedes the objective character of microphysical uncertainty, but belittles its significance since it, according to him, does not affect the macrophysical determinism. (Cf. pp. 23–24.)

[24] L. de Broglie, *op. cit.* in Note 18, p. 141, J. Ullmo, *La crise de la physique quantique*, Paris, 1955, accuses Heisenberg of "pure subjectivism" (p. 36) and the language of certain passages of Heisenberg's *Physics and Philosophy*, New York, Harper & Brothers 1958, easily yields to this interpretation (see in particular pp. 133, 144.)

[25] L. de Broglie, *Physics and Microphysics*, p. 7.

CHAPTER 14


# PHYSICAL EVENTS AS PROTO-MENTAL ENTITIES. BERGSON, WHITEHEAD AND BOHM


Although the element of novelty differentiating two successive events of physical duration is negligible in our macroscopic perspective, it cannot be completely absent. In other words, there is an *element of heterogeneity* even in the physical world. For we must remember the result of our previous analysis: if the differentiating element of novelty is due precisely to the survival of the antecedent moment within the present, then there must be an element of memory, that is, a certain degree of interpenetration of successive phases even in physical duration. Without such an element of memory there would be no duration at all. Here is the basis of Bergson's panpsychism:

What we wish to establish is that we cannot speak of a reality that endures without inserting consciousness into it. The metaphysician will have a universal consciousness intervene directly. Common sense will vaguely ponder it. The mathematician, it is true, will not have to occupy himself with it, since he is concerned with the measurement of things, not with their nature. But... if he were to fix attention upon time itself, he would necessarily picture succession, and therefore before and after, and consequently a bridge between the two (otherwise, there would be only one of the two, a mere snapshot); but, once again, *it is impossible to imagine or conceive a connecting link between the before and after without an element of memory and, consequently, of consciousness.* (Italics added.)[1]

Bergson was well aware that such a statement would expose him to the charge of anthropomorphism or hylozoism and thus he hastened to clarify it:

We may perhaps feel averse to the use of the word "consciousness" if an anthropomorphic sense is attached to it. But to imagine a thing that endures, there is no need to take one's own memory and to transport it, even attenuated, into the interior of the thing .. It is the opposite course we must follow. We shall have to consider a moment in the unfolding of the universe, that is, a snapshot that exists independently of any consciousness, then we shall try conjointly to summon another moment brought as close as possible to the first, and thus have a minimum of time enter the world without allowing the faintest glimmer of memory to go with it. *We shall see that this is impossible. Without an elementary memory that connects the two moments, there will be only one or the other, consequently a single instant, no before and after, no time.* We can bestow upon this memory just what is needed to make the connection; it will be, if we like, this very connection; a mere continuing of the 'before' into the immediate 'after' with a perpetually renewed forgetfulness of what is not the immediately prior moment. We shall nonetheless have introduced memory. To tell the truth, it is impossible to distinguish between the duration, however short it may be, that separates two instants and memory that connects them, because duration is essentially a continuation of what no longer exists into what does exist. This is a real time, perceived and lived. *This is also any conceived time,* because we cannot conceive a time without imagining it as perceived and lived. *Duration therefore implies consciousness;* and we place consciousness at the heart of things for the very reason that we credit them with a time that endures. (Italics added.)[2]

I have already pointed out a similarity of Bergson's view of matter, as *"l'oubli perpetuellement renouvellé,"* with Leibniz's characterization of matter as "instantaneous mind" (*mens momentanea*), the main difference being that, according to Bergson, even physical events are not *strictly* instantaneous. (See Chapter 5 of this part.) At the same time I mentioned that Whitehead's view of the physical world was not only similar to, but the same as that of Bergson. The following passage shows how strikingly identical Whitehead's and Bergson's views were:

When memory and anticipation are completely absent, there is complete conformity to the average influence of the immediate past. There is no conscious confrontation of memory with possibility. Such a situation produces the activity of mere matter. When there is memory, however feeble and short-lived, the average influence of the immediate past, or future, ceases to dominate exclusively. There is then reaction against mere average material domination. Thus the universe is material in proportion to the restriction of memory and anticipation.

And further:

According to this account of the World of Activity, there is no need to postulate two essentially different types of Active Entities, namely, the purely material entities and the entities alive with various modes of experiencing. The latter type is sufficient to account for the characteristics of that World, when we allow for variety of recessivenes and dominance among the basic factors of experience, namely consciousness, memory and anticipation.[3]

In the light of the previous, documented exposition of Bergson's theory of matter as being constituted by events of vanishing, but non-zero duration, hardly any additional comment is necessary. Both thinkers define the difference between the "physical" and the "mental" in terms of differences of the mnemic, temporal span. They both insist that the physical present *approaches* the mathematical, durationless present, *without ever coinciding with it.* There are numerous passages in Whitehead's writings in which the reality of durationless instants is resolutely denied.[4] For any genuine process-oriented thinker this is the only course which can consistently be adopted.

The passage quoted above was among the last Whitehead wrote. But what may be called a *vibratory theory of matter*, which is so strikingly Bergsonian in its character, may be traced at least as far back as 1925, the date of the publication of *Science and the Modern World*. Commenting at the end of Chapter II on some paradoxical features of the quantum theory, he concluded that "we have to revise all our notions of the ultimate character of material existence".[5] More specifically, he thinks that certain paradoxes are due to our classical habit of regarding the existence of matter as occupying time in a continuous way so that the material "is fully itself in any sub-period however short". No matter how far we subdivide any interval of time, matter remains the same; it remains so even if we consider it at any durationless instant. "In other words, dividing the time does not divide the material." Whitehead thus shows how the whole notion of uninterrupted existence of matter through time – which was one of the basic assumptions of atomism – is based on the classical concept of homogeneous, infinitely divisible time together with the idea that "the lapse of time is an accident, rather than of the essence of the material".[6] (Whitehead purposely uses a more general term, "material," to indicate that the same assumption underlay the concept of aether as that of ponderable matter.) But these two correlated assumptions must be drastically revised. We must:

consent to apply to the apparently steady undifferentiated endurance of matter the same principles as those now accepted for sound and light. A steadily sounding note is explained as the outcome of vibrations in the air; a steady color is explained as the outcome of vibrations in aether. If we explain the steady endurance of matter on the same principle, we shall conceive each primordial principle as a vibratory ebb and flow of an underlying energy, or activity.[7]

In other words, 'the ultimate elements of matter are in their essence vibratory' not in the sense that they are vibrations of some more minute particles which would endure in an undifferentiated way through time, but in a more radical, imageless sense. Time – or more accurately time-space itself, is *pulsational in its nature*. The same must be true of matter itself, since it is *inseparable* from time-space and not merely "located in it". From this point of view a material system "is nothing at any instant.

It requires its whole period in which to manifest itself"[8] – a conclusion which sheds light on the second form of Heisenberg's indeterminacy principle which excludes the possibility of an absolutely precise dating, i.e. of locating any physical entity at a durationless instant.

Is there any need to press the striking similarities between Whitehead's and Bergson's views of matter? They both stress the pulsational nature of physical duration. In other words, they both claim that matter is constituted by events of very short temporal span. They both reject the reification of mathematical instants which they regard as mere artificial cuts across concrete processes. They both deny a homogeneous, infinitely divisible time, separable from its physical content. They both, though in different words, in incorporating matter into the *pulsational becoming*, restore the causal efficacy of time. They both insist that the postulated "vibrations" are not the vibrations of some permanent particles; like Bergson, Whitehead regards events as basic and change as requiring neither substrate nor container:

Thus the endurance of the pattern now means the reiteration of its succession of contrasts. This is obviously the most general notion of endurance on the organic theory, and "reiteration" is perhaps the word which expresses it with most directness. But when we translate this notion into the abstractions of physics, it at once becomes the technical notion of "vibration". *This vibration is not the vibratory locomotion: it is vibration of organic deformation.*[9]

It is therefore naive to interpret the undulatory nature of matter in a sense similar to the undulatory character of sound or the classical undulatory character of light. In both these cases there was a periodic spatial displacement of some particles – whether of air or of aether – persisting through space and time. In the light of wave mechanics, the material "particles" themselves are constituted by the imageless, rhythmical patterns. When this rhythmic pattern – 'reiteration' to use Whitehead's term, 'repétition' using Bergson's term – is translated into the abstractions of physics, it becomes a technical term of periodical disturbance with all the misleading associations, including that of an oscillating particle and of the elastic medium in the Cartesian-Kelvinian sense.

The passages quoted above were written *after* the discovery of the undulatory character of matter. Although this discovery was then not generally known, I doubt that Whitehead was not aware of it.[10] Furthermore, the whole trend in quantum physics pointed in this direction and

Whitehead's early doubts about the mathematical continuity of time and about the reification of durationless instants certainly enabled him to discern this trend. In truth, the whole concluding chapter 'Rhythms' of *The Principles of Natural Knowledge* clearly foreshadows his later vibratory theory of matter. How much more remarkable is the fact that Bergson's thought clearly was moving in this direction *thirty years before*, in 1896, when barely the first cracks in the imposing construction of mechanistic physics began to appear.

This raises the question of the possible influence of Bergson on Whitehead. I doubt that any influence took place with respect to this particular point. Even as late as 1929 Whitehead was not acquainted with Bergson's theory of matter; or at least he professed not to understand it. He accuses Bergson of not giving "any explanatory insight" into the origin and the nature of matter.[11] Misled by the metaphors of *Creative Evolution* he speaks of "the relapse of *élan vital* into matter," apparently not realizing how complex a notion is hidden behind this deceptively picturesque metaphor. He apparently was not an avid reader of *Matter and Memory* and thus did not realize that the metaphor mentioned above is intelligible only in the context of the concept we discussed in Chapters 2 and 5 of this part – that of the variability of the temporal span. Only then do such apparently meaningless terms like '*tension de la durée*', '*condensation de la durée*', '*relâchement de la tension de la durée*' become intelligible. If Whitehead still arrived at a doctrine of matter which in its essential features is, as we have just seen, indistinguishable from that of Bergson, then the agreement of both thinkers is even more remarkable.

But such agreement could not have been *entirely* accidental and becomes intelligible if we do not disregard Bergson's influence on Whitehead altogether. This influence is beyond doubt and was very rarely denied.[12] We know on the authority of Bertrand Russell himself that his intellectual divorce from Whitehead began when the latter began to be influenced by Bergson.[13] But we know it also from Whitehead himself. In a number of places he acknowledged his agreement with Bergson, even though he always tried to define its limits. The influence of Bergson is acknowledged as early as 1919 in the preface to *An Enquiry Concerning the Principles of Natural Knowledge*. Whitehead is more specific on this point in *The Concept of Nature* where (p. 54) he states his full agreement with Bergson about the all-pervasive reality of "the passage of nature" of which metrical time is merely one aspect. In *Science and the Modern World* he shares Bergson's protest against the spatializing tendency of the human intellect, but he adds that he disagrees with his view that "such a distortion is a vice necessary to the intellectual apprehension of nature".[14] (This disagreement, by the way, was merely semantic, since by 'intellect' Bergson meant merely its prevailing Newtonian-Euclidian form. It is unnecessary to repeat how Bergson insisted on the flexibility of intellectual modes and the possibility of their widening.) In another reference from the same book, Bergson is credited with introducing the organic philosophy of nature into philosophy and praised as having moved "most completely away from the static materialism of the seventeenth century".[15]

But the greatest number of references occur in *Process and Reality*. Besides the acknowledgement of Whitehead's general debt to Bergson in the preface there are no less than seven references to Bergson in the text. In two of them he repeats the same reservation about the alleged "incorrigibility" of intellect[16]; the view which he erroneously imputed to Bergson already in his *Science and the Modern World*. It would go beyond the scope of this chapter to discuss to what extent Whitehead's assimilation of his concept of "physical purpose" to Bergson's intuition is legitimate.[17] (In truth a more detailed discussion of the relation of both thinkers would require another book.) But similarities in general as well as in their views of matter are rather striking and definitely prevail over the differences which often loom much bigger because of the differences in style.[18]

We may conclude that although the Whiteheadian "vibratory theory of matter" was not a result of direct Bergsonian influence, it was still a result of a broad philosophical agreement, that is, of Whitehead's sharing major metaphysical theses with Bergson, among which the insistence on the primacy of process and its irreducibility to static factors was of primary importance. It was this over-all process orientation which Whitehead undoubtedly owed to Bergson; the community of their views was even such that it led them both to erroneous conclusions, when they overlooked the limits imposed by the relativity theory on their criticism of simple location. It must also be mentioned that Bergson in two places referred with satisfaction to the similarity of his views with those of Whitehead: in *Durée et simultanéité* where he referred to Whitehead's concept of the "creative advance of nature"[19]; and in his last book, in his comment on the undulatory nature of matter which he regarded as a confirmation of his (and Whitehead's) view that change is the very substance of matter.[20]

Thus Whitehead's concession that Bergson introduced the organic view of nature seems to be more than a mere polite phrase, but rather the admission of the at least initial stimulating influence of Bergson on him. Undoubtedly, there were other factors which influenced him, among them the crisis of classical mechanistic physics under the impact of the relativity theory and the theory of quanta; also an intensive reading of classical philosophers whom Whitehead apparently began to read rather late in his life. This explains why he was able to discern even the faintest anticipations of his (and Bergson's) organic view of nature even in the works of the thinkers where we would expect them the least – such as Bacon and Locke.[21]

Whether we apply the term 'organic view of nature' or 'panpsychism' to Bergson's and Whitehead's philosophy, there is no question that they both regarded physical events as "proto-mental entities" to use a term coined recently by Professor Abner Shimony.[22] This 'proto-mental' character of microphysical entities was also mentioned by other contemporary thinkers, though more cursorily; such as Bertrand Russell when he observed in 1950 that matter in modern physics becomes "as elusive as fleeting thoughts"[23] or by David Bohm when only one year later he pointed out some striking resemblances between the mental processes and quantum phenomena. What Bohm stressed in particular was the *indivisibility* of thought and quantum processes, that is, the impossibility of their unlimited subdivision. "In any thought process, the component ideas are not separate but flow steadily and indivisibly. An attempt to analyze them into separate parts destroys or changes their meaning."[24] This is as close as possible to the central idea of Bergson: to use psychological duration, especially its indivisible, pulsational continuity as a conceptual model for understanding the nature of the basic microphysical processes. Similarly, I can hardly find a more accurate description of "the logic of solid bodies" than the following words of David Bohm:

Logically definable concepts play the same fundamental role in abstract and precise thinking as do separable objects and phenomena in our customary description of the world.[25]

Bergson also repeatedly stressed that the discontinuity of concepts has its roots in the apparent discontinuity of perceptual objects; in the same way as a refined introspection restores the continuity among what James called "sensory nuclei" or "places of rest" in the stream of thought, a refined physical observation restores the continuity among apparently discontinuous sensory objects. Like Bergson and Whitehead, Bohm regards the concept of isolated entity to be nothing but an artificial abstraction which prevents us from recognizing that each concrete event is inseparable from its context. Numerous are the passages in Bohm which are strikingly similar to Whitehead's criticism of "simple location". Similarly, Bohm's constant emphasis on ceaseless becoming, on its basic irreversibility, on the uniqueness of every event and on "the qualitative infinity" (i.e. the logical inexhaustibility) of nature[26] shows clearly his affinity with process philosophy of the type discussed in this book. Needless to say that this affinity should be differentiated from 'influence'; I do not see any evidence of the influence of either Bergson or Whitehead on Bohm's thought, which apparently developed independently of them. But it is this independence which makes the affinity even more significant.

It is an apparent paradox that the very individuality and uniqueness of each event is based on its connection with its wider cosmical context. In atomism, whether of the physical or logical kind, the very opposite is true. The individuality of the atom is based precisely on its ontological separation from other simply located entities. Hence the importance of the idea of *void* in atomism and of the *sharp logical separation* which underlies the traditional definition of "well discernible object" in set theory.[27] This is not so in the organic theory of nature which has its roots in Leibniz's monadology. The monad owes its own individuality to the fact that it 'mirrors the whole universe'; and since each monad reflects the universe from its own 'point of view', which is unique and unexchangeable with any other monad, there are no two monads which are completely identical. I pointed out how deeply both Bergson and Whitehead were Leibnizian in this respect. In truth, they yielded to this influence too much – so much that they overlooked two closely correlated ideas of the relativistic physics – the impossibility of speaking of the *totality* of the universe at any particular instant and the impossibility of instantaneous actions. We dealt with this limitation of the criticism of simple location in Chapter 8 of this part. Yet, the basic idea of Leibniz that the essence of individuality lies in the irreducible difference with which the wider cosmic context is 'reflected' in each particular event, remains unaffected by this reservation.

Yet, can this claim be seriously maintained in the face of what modern physics tells us? True, the atoms are not identical; but are not their differences due to the *differences of configuration* of more basic units – electrons, all of which have the same charge and the same mass? And is not the same thing true of other nucleons as well? In other words, is not nature in its ultimate elements *homogeneous* and *quantifiable* in a way not basically different from what classical atomists claimed? Bohm does not think so. Consistently applying his view that a "definite and unvarying mode of being" is an artificial abstraction, valid only approximately, he questions the view that the electrons are *completely* and in *all* respects identical.[28] F. A. Lindemann, drawing all the consequences from the Heisenberg uncertainty relations, raised a similar doubt several decades ago. I dealt with his doubts extensively in my previous book and pointed out that *the very essence of the quantitative view of nature* now is at stake.[29] The idea of a sharp-edged quantity loses its meaning on the microphysical level. It is my belief that the deeper meaning of the indeterminacy principle consists in the fact that our instantaneous snapshots of nature result always in fuzzy pictures.

This may sound like a mere metaphor, but it applies literally to the second form of Heisenberg's principle. No process can be pinned down at an instant for the simple reason that *there are no instants at all.* More generally, if there is no sharp-edge, if no absolutely exact quantity can be found in nature, it is because *there is no such thing.* For this reason even the term 'fluctuation,' still retained by Bohm, is hardly adequate, since it still suggests two definite *sharply defined* quantities between which "fluctuation" supposedly takes place. Peirce, as we noticed, correctly saw that determinism, which was regularly associated with the quantitative view of nature, artificially imposes the schema of mathematical continuity on nature.[30] Boutroux stated it in different words when he wrote that "homogeneous quantity is merely an ideal surface of the reality; in depth everything lives, moves and evolves". Bergson does not say anything different when he claims that matter *tends* toward homogeneity without ever being completely homogeneous; the residual element of qualitative difference which, as we saw, is synonymous with "the elementary memory", persists on the deepest level.

We are thus returning to the same conclusion which we reached in Chapter 7 of Part I ('Limitations of Panmathematism'). Not all mathematicians are unhappy about the inapplicability of the mathematical continuum. Hermann Weyl, Norbert Wiener and Karl Menger are outstanding examples. After all, there is a fairly general agreement that the mathematical continuum is inapplicable to a qualitative continuum of the sensory and introspective type. Why should it then apply *exactly* and *literally* to other strata of experience? Even David Hilbert recognized that to believe so is nothing but an "enormous extrapolation" of our limited macroscopic experience.[31] To believe that this concept *exactly* applies to nature while being inapplicable to mind is another instance of the artificial "bifurcation of reality" which both Bergson and Whitehead – and before them Peirce and Boutroux – tried to overcome.

## NOTES

[1] *D.S.*, pp. 47–48.
[2] *Ibid.*, pp. 48–49.
[3] Cf. the essay, 'Immortality', in A. N. Whitehead, *The Interpretations of Science* (ed. by A. H. Johnson), Bobbs-Merrill, New York, 1961, p. 262.
[4] For instance: *An Enquiry Concerning the Principles of Natural Knowledge*, pp. 2–3, 6–8; *The Concept of Nature*, pp. 72–73; *S.M.W.*, pp. 54, 172, 185–186 and the whole chapter VIII; *Process and Reality*, pp. 103–107; *Modes of Thought*, Capricorn Books, New York, 1958, pp. 198–201.
[5] *S.M.W.*, p. 53.
[6] *Op. cit.*, p. 73.
[7] *Op. cit.*, p. 53.
[8] *Op. cit.*, p. 54.
[9] *Op. cit.*, p. 193.
[10] Contrary to the view of Abner Shimony who claims that it was only the 'old' quantum theory which provided the conceptual frame work. (Cf. his article 'Quantum Physics and the Philosophy of Whitehead', *Boston Studies in the Philosophy of Science*, vol. II, p. 307.) Cf. the following explicit reference to the undulatory character of the corpuscles in *S.M.W.*, p. 193: "There are certain indications in modern physics that for the role of corpuscular organisms at the base of the physical field, we require vibratory entities."
[11] Cf. Note 5 to Chapter 5 of this part.
[12] Cf. Victor Lowe, 'Whitehead's Philosophical Development', in *The Philosophy of A. N. Whitehead* (ed. by Paul Schilpp), Evanston, 1941, pp. 89–90.
[13] B. Russell, *Portraits from Memory and Other Essays*, Simon & Schuster, New York, 1965, p. 101.
[14] *S.M.W.*, p. 74.
[15] *S.M.W.*, p. 212.
[16] *Process and Reality*, p. 319, 489.
[17] *Op. cit.*, pp. 49–50. On p. 428 Whitehead identifies Bergson's intuition not with

"pure physical purpose", but "pure and instinctive intuition". This second characterization is grossly misleading. Compare it with Bergson's rejection of the identification of his intuition with "instinct" or "feeling" in *C.M.*, p. 103: "Not one line of what I have written could lend itself to such an interpretation. And in everything I have written there is assurance to the contrary: my intuition is reflection." Cf. also Chapter 7 of Part I of this book.

18 Cf. Whitehead's acknowledgement of his debt to Bergson in Preface to *Process and Reality* (p. VII).

19 *D.S.*, p. 62 note. Here Bergson calls *The Concept of Nature* "one of the most profound books ever written on the philosophy of nature."

20 *C.M.*, p. 85.

21 *S.M.W.*, pp. 60–63; *Process and Reality*, p. 86.

22 A. Shimony, *art. cit.*, p. 309.

23 B. Russell, 'Materialism, Past and Present'. Introduction by F. A. Lange to a new edition of *The History of Materialism*, Humanities Press, New York, 1950, p. XII.

24 David Bohm, *Quantum Theory*, Prentice Hall, 1951, p. 170.

25 *Ibid.*, p. 170.

26 D. Bohm, *Causality and Chance in Modern Physics*, p. 138; 143–153, 160–164. Cf. also the Heraclitean metaphor of the universe as "a flame which exists by feeding on the old structures, and by thus creating new orders and structures" in his *Problems in the Basic Concepts of Physics*, An Inaugural Lecture, Birkbeck College, London, 1963, pp. 41–42.

27 Cf. Georg Cantor, *Contributions to Founding of the Theory of Transfinite Numbers* (Dover, New York, n.d.), p. 85: "By an 'aggregate' (*Menge*) we are to understand any collection into a whole (*Zusammenfassung zu einem Ganze*) of definite and separate objects *m* of our intuition or our thought " (p. 85).

28 D. Bohm, *Causality and Chance in Modern Physics*, pp. 153–158.

29 *The Philosophical Impact of Contemporary Physics*, pp. 322–329; F. A. Lindemann, *The Physical Significance of the Quantum Theory*, Clarendon Press, Oxford, 1932, p. 110.

30 Cf. Note 8 of Chapter 12 of this part; or p. 322 of my book. The relevant passage is as follows: "For the essence of the necessitarian position is that certain continuous quantities have certain exact values. Now, how can observation determine the value of such a quantity with a probable zero absolutely *nil*?"

31 H. Weyl, *Das Kontinuum*, pp. 69–71 (cf. Note 14 Chapter 9, Part II of this book): Karl Menger, "Topology without Points", *Rice Inst. Pamphlets*, **XXVII**, No. 1 (Jan. 1940) 260; D. Hilbert and P. Bernays, *Grundlagen der Mathematik*, Jena, 1931, pp. 15–17; Norbert Wiener, *I am a Mathematician* (M.I.T. Press, Cambridge, Mass., 1964), p. 107: "the infinite divisibility of the universe cannot any longer be applied without serious qualifications."

# CHAPTER 15

# THE SIGNIFICANCE AND THE LIMITATIONS OF AUDITORY MODELS.
## BERGSON AND STRAWSON

The claim that microphysical events are 'proto-mental entities' may well cause alarm. Nothing sounds more suspicious and irrational than the phrase 'introspective models in physics'. The fear of being charged with relapsing into primitive animism or pre-Socratic hylozoism prevents many serious thinkers from considering such an approach at all. The previous trend in physics, which for the last three and one-half centuries tended progressively to eliminate *all* subjective elements from our scheme of the physical world, seemingly excludes any usefulness of the introspective models in physics or in the philosophy of physics.

Let us state quite explicitly that such an attitude is fully justified. From what has been said previously, it is quite clear that Bergson fully accepted the exclusion of immediately perceived qualities from the physical world. It is true that his phrasing in *Matter and Memory*, especially in its first chapter, is often ambiguous, especially when he speaks of "pure perception" or when he says that the sensory data are "the very qualities of things, perceived first in the things rather than in us".[1] But it would be a serious error to take such sentences out of context and to assimilate Bergson's view to that of the American neorealists or of John Dewey, who claimed that "things are what they are experienced as".[2] Only a few lines before the quoted phrase Bergson says explicitly that our perception and the physical world do *not* coincide, since our sensory perception consists in the selective registration of biologically significant physical processes; to perceive everything would be "to descend to the condition of a material object"[3] which reflects indiscriminately all physical influences passing through it.

The difficulty of understanding Bergson's view consists in the traditional philosophical tendency to think in one of two radically opposite ways: *either* we assume with all the materialists from Democritus to J. J. C. Smart that perception is located *in* the brain, an assumption which would provide for the extensive character of sensations, but leads to enormous difficulties which the critics of materialism pointed out long ago; *or* we assume that perception, being a part of the inextensive mind, is itself inextensive, residing in a mathematical point as Descartes and the young Leibniz did believe. Such a dilemma is not exhaustive for Bergson whose main innovation is his acceptance of the whole spectrum of *different degrees of extension*, of which the altogether inextensive mind of Descartes and the geometrical space with its completely exteriorised parts are two extreme, but equally fictitious, limits. These different degrees of extension or of externality are correlated with different degrees of the mnemic or temporal span. We have seen that the greater temporal thickness of sensory qualities accounts for their subjectivity, that is, for their difference from the underlying physical events. In other words, 'pure perception' (*la perception pure*) is a mere methodological fiction. Concrete human perception is always 'tinged with memory', i.e. it has a greater degree of 'condensation of the past in the present' than the physical events themselves. This accounts for the fact that, for instance, the perceived quality of any musical tone contrasts by its own indivisible oneness with the enormous number of corresponding air vibrations. Thus any fear that this kind of panpsychism is a relapse into primitive hylozoism is altogether unfounded. In truth, the very opposite is true. Bergson excludes not only the secondary, *but even the primary qualities* from the extra-mental realm; hence his characterization of the classical atom as a "reified sensation of touch"[4] and his early rejection of the tactile and visual models.

This warning against confusing Bergsonism with hylozoism is essential for understanding the true significance of the *auditory metaphors* which occur in Bergson's philosophy, including his philosophy of the physical world. But for an attentive reader of the previous chapter – in truth of all previous chapters – this warning was hardly necessary. Bergson's "panpsychism", as well as Whitehead's "organic philosophy of nature", claims nothing more, and nothing less, than that the physical events share with psychological processes *some basic general features of temporality*. In other words, although physical events cannot be perceived in their immediacy and qualitative specificity, they have *certain structural features* in common with our stream of consciousness.

Now, although temporal character pervades the whole of our experience, sensory as well as introspective, it is present *with different degrees of obviousness* in its different strata. In the visual and tactile components of our sensory perception it is least conspicuous. In truth, without the kinesthetic sensations connected with the movements of our muscles – including the eye muscles – there would not be any perception of motion at all. This explains why in the classical model of matter, which was built of the sensory elements of touch and sight, the reality of time was obscured. The only change compatible with the classical, corpuscular-kinetic model was *change of place*. Since such change affects neither matter nor space – Piaget showed how early the idea of constancy of matter is formed in a child's mind – it appears as something less substantial and less important. When the development of mechanics later showed that motion itself can be quantified either in the concept of momentum or that of kinetic energy and that both these quantities are preserved, this very quantification and *substantialization* of motion pushed its reality still farther into the background and thus further strengthened the "category preference" – to use Strawson's words[5] – for the static view of reality.

The situation is different when the auditory components of our sensory perception are considered. Because of their dynamic, transient character they were excluded from 'the nature of things' by the atomists as early as the fifth century B.C. From the early claims of Lucretius that the atoms are deprived of sound – '*sonitu sterila*'[6] – up to Helmholtz's investigations in the last century about the complex physiological conditions of the sensations of sound, the central idea remained the same, even though it was expressed with increased sophistication and supported by a still greater number of observations: *the auditory qualities have no place in the objective physical world.*

This trend toward the elimination of the secondary qualities from the objective world – the claims of the neo-realists notwithstanding – cannot be reversed; there are no auditory qualities without the auditory organs and the conscious percipient. But neither are the primary qualities real if the conscious percipient is subtracted. 'Impenetrability' is merely an objectified sensation of touch which remains meaningful only within the tactile, conscious experience. Now since both the primary and secondary qualities have basically the same status, the time comes to inquire to what extent analysis of the auditory qualities can help in our understanding of that region of physical reality which our visual-tactile imagination failed to grasp. More specifically, the analysis of the dynamic structure of the auditory data, besides freeing our imagination from the obtrusive influence of mechanistic models, may perhaps provide us with a helpful analogy for understanding the paradoxical nature of the physical time-space and of the physical processes which resist traditional interpretations.

This is obviously something quite different from any neo-realistic objectification of the auditory qualities; and only in this sense should Bergson's occasional use of auditory metaphors be understood. We have seen the significance which the auditory experience has in his analysis of psychological time. In the perception of melody a number of the essential features of *durée réelle*, that is, of the temporal awareness uncontaminated by the spatial and visual imagery, show themselves especially clearly:

Let us listen to a melody, allowing ourselves to be lulled by it: do we not have the clear perception of a movement which is not attached to a mobile, of a change without anything changing? This change is self-sufficient, it is the reality itself. And even if it takes time, it is still indivisible; if the melody stopped sooner it would no longer be the same mass of sounds (*la même masse sonore*), it would be another, equally indivisible. We have, no doubt, a tendency to divide it and to picture, instead of the uninterrupted continuity of a melody, a juxtaposition of distinct notes. But why? Because we are thinking of the discontinuous series of efforts we should be making to recompose approximately the sound heard if we were doing the singing, and also because our auditory perception had acquired the habit of absorbing visual images... We picture notes placed next to one another upon an imaginary piece of paper. We think of a keyboard upon which someone is playing, of the bow going up and down... If we do not dwell upon these spatial images, pure change remains sufficient to itself, in no way divided, in no way attached to a "thing" that changes.[7]

The *open* or *incomplete* character of every temporal process, its indivisible, heterogeneous continuity by which successive phases cohere without ceasing to be qualitatively diverse, continuity which is altogether different from spurious, 'mathematical continuity', and diversity which is completely different from arithmetical multiplicity, as well as the merging of the past with the present – all are thus found in the perception of melody. Bergson returns to the image of melody again a few pages later when he insists that the continuity of becoming "has nothing common with immutability, nor this indivisibility with instantaneity",[8] as Lovejoy and Ushenko erroneously assumed when they criticized a similar passage.

The above quotation is from Bergson's Oxford lecture of 1911. What is interesting to our present context is that he then used the image of melody also to illustrate the fact that change *does not need to be a change of something*, since it does not need any static support or vehicle, *not even in the physical world*:

*There are changes, but there are underneath the change no things which change: change has no need of support. There are movements, but there is no inert or invariable object which moves: movement does not imply a mobile.* (Italics in the text.)[9]

If such a view appears difficult, it is so only because of the preponderance of the optical elements in our thought, "because the eye developed the habit of separating in the visual field the relatively invariable figures which are then supposed to change place without changing form, movement is taken as superadded to the mobile as an accident".[10] (Hence the traditional common sense idea incorporated in the Aristotelian metaphysics, that every motion requires an explanation and justification by external 'moving force'.) However, the idea of *mobility without a mobile*, or better, of *change without any changeless substratum* loses its paradoxical character when we turn from the *visual* to *auditory* or introspective experience; "nowhere is substantiality of change so visible, so palpable".[11] This is especially true of the perception of melody described above. Within it "change without vehicle and container" is *directly and concretely intuited* and loses altogether the paradoxical and absurd character which it retains in our visually oriented common sense. In this sense the musical *Zeitgestalt* – to use Ehrenfels's word – may prove a helpful aid for understanding the new concept of physical change discussed in Chapter 10.

It is thus only natural that in his last book Bergson pointed out with satisfaction that the development of physics bore out his anticipation that sooner or later "the idea of an immutable vehicle of motion would have to be given up". In the wave-mechanical fusion of corpuscle and wave he saw the ultimate confirmation of the substantiality of change which he had advocated since 1896. He then concluded by stressing the kinship of his view with that of Whitehead, who came to regard a piece of iron as "a melodic continuity".[12]

Bergson could have referred to another passage in which Whitehead used the image of melody to illustrate the physical impossibility of mathematical instants and of "the instantaneous states" of the world. Commenting on the wave character of matter in *Science and the Modern*

*World*, according to which "the primordial elements" of matter are vibratory patterns, Whitehead made the following comment:

> This system, forming the primordial element, is *nothing at any instant*. It requires its whole period in which to manifest itself. In an analogous way, a note of music *is nothing at an instant*, but also requires its whole period in which to manifest itself... If we divide time into smaller elements, the vibratory system as one electronic entity *has no existence*. (Italics added.)[13]

This passage, in particular the last sentence, is a precise and rarely encountered formulation of the philosophical significance of the second form of the Heisenberg principle, discussed in the previous chapters. It constitutes an effective answer to those who, guided by subconscious visual habits, insist on the necessity of the unlimited divisibility of temporal intervals. Such "necessity" not only disappears when we turn our attention to the auditory data, but it becomes a straight impossibility. Considered at a durationless instant, a melody is literally *nothing at all*, not even a single present tone; for even such a tone, in possessing a certain temporal thickness, is *not* instantaneous in a rigorous mathematical sense. Whitehead's comment is even more remarkable since it appeared prior to Heisenberg's famous article, although *after* the first theoretical investigations of Louis de Broglie. In truth, Whitehead's views about the vibratory character of matter and the correlated denial of the infinite divisibility of time go back to his first writings on the philosophy of nature. Not only did he speak about the possible existence of "quanta of time" as early as 1920, but even the year before when, after insisting that durationless instants do not exist anywhere in concrete experience, he concluded: "It is equally true of a molecule of iron or of a musical phrase." In truth, it is this sentence to which Bergson referred above.[14]

But neither Bergson nor Whitehead fully exploited the fruitfulness of the auditory models. There is one formidable objection to such models with which neither of them dealt – not because they were unable to but because they simply ignored it: in which way can any auditory pattern – which is intrinsicially *temporal* and *nothing but temporal* – serve as a meaningful model of the physical reality which is *extended* in space?

This objection is similar to that we had already faced in Chapter 3, and the answer given to this objection in Chapter 4 virtually contains that which is going to be given here. The objection is based on one plausible, but very questionable premise: that auditory experience is 'nothing but temporal' in its structure; consequently, that it is *completely* foreign to any kind of space, including what we call 'psychological' or 'representational' space. Such a belief has its roots in the Cartesian notion of a completely extensionless mind – *res inextensa* – as sharply opposed to *res extensa* of the physical world. It was taken over by the British associationists of the last century who regarded all sensations as intrinsically unextended and explained the origin of the idea of space by a sort of mental chemistry, as 'emerging' from the reversible associative series of the originally inextensive sensations. In a less extreme form this view was retained by those who affirmed the extensive character of visual and tactile sensations, but insisted on the intrinsic spacelessness of other sensations, including the auditory ones. We have seen that Bergson resolutely rejected this view: "the idea of unextended sensations, artificially located in space, is a mere view of the mind, suggested by an unconscious metaphysics more than by psychological observation."[15] We know that in this respect Bergson was not alone. He is in the same company with William James, James Ward, Ernst Mach and others. He notes that on this point he is also in agreement with Kant "since *The Transcendental Aesthetic* allows no difference between the data of the different senses as far as their extension in space is concerned".[16]

The problem of a purely auditory world was recently discussed by P. F. Strawson. His conclusion is that such a world is impossible, since it is both *spaceless* and *solipsistic*.[17] It is, of course, impossible to do justice to Strawson's complex and subtle argumentation within a few short paragraphs. But one thing is clear after the very first reading: that Strawson, whether he knows it or not, is deeply and irrevocably committed to the deliberately hazy scheme which nevertheless possesses the essential features of the classical corpuscular-kinetic scheme of nature. I say 'deliberately hazy', since Strawson deliberately confines his thought within the limits of the supposedly ageless common sense ("a massive central core of human thinking which has no history").[18] But his explicit statements do not leave us in any doubt that this common sense view of nature contains the essential ingredients of the classical Newtonian scheme, even though it does not contain its articulateness and clarity. He tells us explicitly that "ontological priority" belongs to "the basic particulars," conceived of as "three-dimensional bodies enduring through time";[19] that there is a "single unified spatio-temporal system" within which it is meaningful to speak of the same re-identifiable place [20]; finally, that motion consists in a change of position – in his terminology, "things pass through places".[21]

It is true that, as was pointed out by one critic [22], Strawson hesitates between the absolutist and the relational theory of space and in this respect his thought seems to move in a circle; but this apart, there is no question that his 'conceptual scheme' is basically Newtonian. All the basic features of the classical scheme are present here: absolute space as a single, immutable container of matter; a single, container-like time in which all events take place; absolute place as a portion of absolute space, immutable and indifferent to the displacements of the bodies; finally, substantial matter divided into individual bodies which persist through time while they change their places; their displacements affect neither their permanence nor the homogeneity and immutability of space.

Nor can it be doubted that Strawson accurately described the common sense scheme of reality and that out of such a scheme the classical Newtonian model developed. But he does not raise the question of how this scheme originated; in truth, he excludes this question by claiming that "it has no history". Such a claim is certainly incompatible with the accepted evolutionary approach toward the present psychophysiological organization of man. The biological theory of knowledge is, as we have seen, a mere consistent application of this approach. Furthermore, Piaget's studies have shown how this allegedly ageless conceptual scheme is formed at a very early age in the individual mind under the pressure of macroscopic experience. "Ontogeny recapitulates phylogeny" in a condensed way not only on the biological, but also in the psychological sense. As we pointed out in the First Part the present 'crisis' in physics is due to the fact that the enormous widening of our experience made obsolete the supposedly 'ageless scheme' of Strawson which was so decisively fashioned by our original, narrow, macroscopic experience. And the most important problem now is how to *widen* this narrow scheme.

This, for Strawson, would be a "revisionary metaphysics" and he does not want to do anything of this kind.[23] But then he faces two possibilities: either his so-called "descriptive metaphysics" is nothing but a faithful phenomenological account of the way we think and talk without any metaphysical pretensions – and then it would be meaningless to speak of the "ontological priority" of the three-dimensional solid bodies moving in the three-dimensional Euclidian space. *Or* he is trying to give something more than a mere description of our thinking and talking – and then he is committing himself to a certain metaphysics which, because of its claim to be ageless, can be properly characterized as *conservative* or *anti-revisionary*. Since the development of physics made such anti-revisionary metaphysics obsolete, we must consider the first possibility and ask whether and to what extent his *phenomenological account* of our experience is complete.

There is no question that Strawson describes correctly the conceptual scheme which people use and of which our common language is a more or less faithful expression. But again he is handicapped by the limits which he imposed upon himself. His scheme has neither the immediacy of spontaneous sensory perception stripped of its inferential and conceptual elements nor the articulateness and precision of the classical, corpuscular-kinetic model. It is somewhere between, and this intermediate position accounts for its haziness and, I would venture to say, for its limited usefulness for both psychology and physics. Thus his 'unifying spatio-temporal scheme' is clearly similar to the space of Euclid and Newton; but Strawson does not stress this similarity since both Euclid and Newton went *beyond* the hazy modes of ordinary language. On the other hand, this 'unifying spatio-temporal scheme' is *toto coelo* different from the representational space which psychologists analyzed and which, prior to its conceptualization and associative correlation with tactual experience, does *not* have, as Ernst Mach pointed out, the characteristics of Euclidian space.[24] But although Strawson's space is something between the space of Euclid and Newton and the heterogeneous, anisotropic, psychological extension, it is clearly closer to the former.

His physical 'individuals' show a similar haziness. They are clearly akin to the classical atoms, although they do not have the same permanence and definiteness because, being macroscopic in nature, they are composed. Yet, Strawson hardly ever stresses their composite and consequently impermanent character. In endowing them with "ontological priority", he clearly favors the model which differs from the classical corpuscular-kinetic scheme only by its haziness. His description of the common conceptual scheme certainly renders accurately its haziness as well as its kinship with the classical atomistic model.

The lack of distinction between geometrical-physical space and psychological extension also vitiates Strawson's otherwise subtle analysis of auditory experience. When he regards "purely auditory world" as "No-space world", he is unquestionably correct, but only as long as we identify 'space' with the conceptualized, geometrical space of Euclid. Strawson is apparently not fully aware of the distinction between geometrical space and qualitative extension; nor is he aware of James's, Ward's and Mach's view that "crude extensity" belongs to *all* sensations, including the auditory ones. Nor does he realize that this was the view of Kant, as Bergson correctly recognized. He claims that if we differentiate direction and distance in our auditory experience, it is so only because of "the existence of correlations between the variations of which sound is intrinsically capable and other non-auditory features of experience".[25] In other words, the auditory sensations in their immediacy are allegedly devoid of spatiality and their spatial features are due exclusively to their associations with visual and kinesthetic sensations. For this reason there is no purely auditory analogue of spatial distance.

But Strawson overlooks the fact that we do not have a *purely* visual experience of distance either. Without the kinesthetic sensations resulting from the movements of our body by which we approach the object seen; also without the kinesthetic sensations resulting from the movements of the eye-muscles in binocular vision, we would hardly have any conception of depth or the third dimension. The observations of blind persons who, after acquiring their sight by operation, at first claim that 'the things touch their own eyes'[26] show how in an *uninterpreted* visual experience the feeling of depth is extremely vague. It arises only after a long process in which visual data are gradually associated with kinesthetic and tactile sensations. But do we not have an immediate visual perception of distance at least in the *monocular* vision of a *motionless* object? Even such vision is not 'purely' visual, since the characteristic muscular sensations resulting from the focussing of the lens are *not* absent. All that is immediately given within an uninterpreted visual perception is a *certain juxtaposition of various* colored patches. (It would not be accurate to characterize it as the perception of a colored 'surface', since the idea of a two-dimensional surface is a highly abstract concept which is formed only *after* the concept of a three-dimensional body is formed; only then by abstracting one dimension can the concept of 'surface without depth' be formed. This is in agreement with James's view, according to which in the original feeling of extensity no dimensions are at first differentiated.)

In fairness to Strawson it must be pointed out that by 'distance' he most often means not that in the third dimension, but that in the plane *parallel* to our forehead – in the plane in which the distinction between 'right' and 'left' and also between 'above' and 'below' are immediately differentiated.[27] But could even such perception be completely devoid of kinesthetic elements? Can we be aware of the distinction between 'right' and 'left', 'above' and 'below' without the slight motions of the eyeball which are necessary for making the two spots to be compared fall successively on the spot of sharpest vision on the retina? While this may not be true of two *immediately adjacent* visual spots, it must be true of any other spots within a single visual field; for only such spots can appear more or less *distant* from each other. Thus there is no such thing as *purely visual* distance; the immediate visual datum, which exists for hardly more than a few months of our conscious life after birth, is a *pre-dimensional* and *pre-metrical* feeling of crude extensity consisting of the juxtaposition of colored patches centered around the clearest spot, that of the sharpest vision.

In the light of previous analysis, Strawson's claim that there is no auditory analogue to the spatial relations will appear far less convincing. For as there are no purely visual data, there are also no purely auditory data; and as there is *no purely visual distance* there is also no purely auditory distance. The difference between two realms of data is only that of degree. What is even more significant is that there is certainly a precise auditory counterpart of the visual *neighborhood*. For Strawson's most plausible argument for the non-spatiality of the auditory elements is that they always appear in succession. Their relations are always relations in time and whenever the sounds acquire spatial characteristics, it is always by means of visual associations. Thus only the visual elements can be presented *simultaneously*, that is, in *spatial relations*. But this is true only if we consider a melody or, more generally, a mere succession of sounds. *It ceases to be true* if we take into account the simultaneous blending of tones as in a chord, or of individual simultaneous melodies in a polyphonic phrase. In such cases we have the auditory elements which, besides their temporal character, are with respect to each other also in *simultaneous* relations. These relations differ from the simultaneous relations of the visual elements only by their *lesser* stability; they change from moment to moment. (But it would not be difficult to find a visual analogue to it.

Let us consider instead of quiet scenery or a group of stable objects some dramatic scene, such as an explosion, a violent storm or some surrealistic movie and we would have instances of visual elements whose simultaneous relations are changing from one moment to another as quickly.) Thus while awareness of *auditory distance* can be gained only indirectly, by association with the non-auditory data (the same is true of 'visual distance' as well), 'auditory neighborhood' is a primary datum as much as visual neighborhood.

Before considering this important and neglected aspect of auditory experience, it will be useful to mention one more significant fact or rather a *group* of facts. Although Strawson's attention is exclusively focussed on normal human experience, he – unlike the behaviorists – explicitly concedes that the term 'experiences of animals' is not meaningless.[28] Now recent observations of the behavior of certain animals definitely suggest that in certain conditions their perception of space is *auditory*. This is true in particular of the ultrasonic 'seeing' or echolocation by bats. The lesser horseshoe bats 'see' – or rather *hear* – wires only 0.2 millimeter thick in the darkness and will avoid them no matter how many and how tangled they are. They are able to locate and pass through a small hole in a network partition dividing a huge university lecture hall without the slightest hesitation. In truth, they move straight toward a hole as soon as they are released from a wooden box. This indicates that their auditory picture of a darkened room is superior to human vision of the same objects in daylight. The wires of the thickness mentioned above are seen by man only under favorable optical conditions, that is, when they glisten in reflected light; the hole in the network partition is found by man only in a good light and after much searching. What is even more amazing is that the 'auditory memory of space' in bats is as accurate as their auditory perception. On this point the experiments of Professor D. R. Griffin are both spectacular and convincing.[29]

All this indicates that the bats possess a very accurate "auditory" picture of surrounding space and of the location of objects in it. While man and the majority of other mammals construct their visual picture of space by binocular vision when the rays of light reflected from the object are focussed on the retina, bats achieve their 'hearing of space and of the objects in it' by binaural hearing of the ultrasonic waves reflected by the objects. It is obvious that bats do not have a 'purely auditory picture' of space either; such a picture would be biologically useless without being correlated with the complex kinesthetic sensations resulting from the movements of their bodies in flight. As Vitus B. Dröscher concludes, "the animal, through its locomotion during flight and through the double impression in the two ears, obtains an accurate notion of space".[30]

Evidently the above reflexions would be dismissed by dogmatic behaviorists who, ironically enough, are committed to the conceited Cartesian myth of '*l'automatisme des brutes*' and thus dismiss the very existence of animal psychology altogether. As mentioned above, Strawson does not share this anthropocentric myth.

It is time now to return to the main problem of this chapter: the significance and the limitations of auditory models in physics. The two features of the auditory model which can make the following two features of contemporary physics less paradoxical have already been mentioned: the impossibility of mathematical instants and the superfluousness of any immutable vehicle of motion. These two features which remain unintelligible and even absurd in any visual geometric scheme are directly and concretely intuited within the perception of melody. Yet, the structure of melody seems to be completely foreign to the *extensive* character of the physical world. Nothing in a bare succession of tones corresponds even remotely to relations in space. It is on this point that the more complex structure of polyphony is far more illuminating. In a contrapuntal composition two or several melodically independent movements, whether harmonious or dissonant, are going on. The component melodic patterns, besides being each unfolded successively, are also *contemporary* or *alongside* each other; and this relation of 'alongside' or 'beside' is clearly analogous to the relation of 'beside' in space, as the French psychologist Theodule Ribot observed a long time ago.[31] This is only a different way of expressing the similarity of 'auditory neighborhood' and spatial neighborhood as I tried to show it in the previous discussion of Strawson's views.

This similarity has its limitations; but, as we shall see, these limitations make it even more significant and more useful for philosophy of nature in the light of present physics. The spatial relation of juxtaposition implies a complete mutual *externality* of the static elements, whereas in polyphonic movements the component melodies not only proceed, so to speak, parallel to each other – 'in the direction of the future', but also overlap 'transversally' without, however, losing their melodic individuality. But while the dynamic togetherness of the component melodies is different from the static relations of spatial *externality*, it is on the other hand akin to what we called the coexistence or rather *co-becoming* or *co-fluidity* of the world lines constituting the relativistic time-space. The transversal overlapping of the component melodies illustrates the causal interaction of the world-lines – the interaction which inspired Whitehead to his criticism of the fallacy of simple location. The plurality of co-becoming causal lines – or, rather, of causal 'tubes' – cannot be derived from the bare relation of succession. In my previous book I tried to point out that Carnap's attempt to do this was based on a hidden circle.[32] Relativistic physics does not exclude the *transversal* width or *extension* of the four-dimensional world-process, even though this width cannot be conceived of as an instantaneous three-dimensional cross section as in the physics of Newton. Thus the polyphonic pattern is a concrete exemplification of what Whitehead called by the term, significantly borrowed from the language of music also, "unison of becoming" and Bergson "the simultaneity of fluxes".[33]

This simultaneity of fluxes, as Bergson correctly observed, can never pass over into the simultaneity of juxtaposed instants; such instantaneous cross sections in the four-dimensional process are as impossible as instantaneous cuts across a polyphonic or any temporal pattern. Like the relativistic world-lines, the component melodic movements are *contemporary*, but never *co-instantaneous*; the time-space of Einstein, as much as Bergson's extensive becoming, in Whitehead's words, is literally "nothing at an instant". Thus the non-existence of the simultaneity of distant events which appears so absurd in any spatial scheme becomes intuitively clear and evident in this auditory model. From this point of view Strawson's objection that there is no auditory counterpart of spatial distance, instead of being directed against auditory models, becomes their recommendation. For we know today that there are no purely spatial distances; every concrete physical distance is *spatio-temporal*, that is, it stretches through time.

Similarly, the antinomy 'waves versus particles' appears intellectually intolerable only as long as we interpret both terms in their classical, that is, visual-geometric sense. It is obvious that a physical entity cannot be both a corpuscle and an undulatory process at the same time. Hence the attempts to regard one of them as physically primary and the other as derivative. None of these attempts was successful and it is safe to predict that any such effort will fail, since it is based on the illegitimate extrapolation of macroscopic imagery to microphysical processes. Physical entities are neither waves nor corpuscles, although their behavior in a *certain respect* reminds us of the behavior of macroscopic bodies and in another respect of the periodic processes which we observed in the macroscopic media.

But the wave-particle antagonism loses its sharpness when we transpose it to the imageless level. We pointed out that in the concrete experience of duration individuality and continuity do not appear antithetic as long as we do not try to geometrize them. Psychological duration is both continuous and pulsational as described in Part II; not only is the individuality of events not incompatible with their continuity, but it makes sense *only within this dynamic continuity*. On this point the auditory model suggested above shows its fruitfulness again; the temporal pattern of melody or polyphony is not a mere arithmetic addition or mere aggregate of the individual tones, and thus it beautifully suggests the organic, moving totality of nature. For the same reason, an individual tone, while possessing its specific quality, does not have the individuality of a bit of classical matter persisting through time. While each tone perishes in a (non-instantaneous) moment, it tinges by its "mnemic persistence" the whole subsequent melodical pattern. In other words, the individual tones and the total melodic *Gestalt* are as inseparable as the quantum and the field; more accurately, we should say as inseparable *and as diverse*, provided we do not confuse inseparability with logical identity, and diversity with complete externality. Even a single and "isolated" tone is not isolated, since it is perceived in the context of antecedent silence as William James observed. This pulsational continuity of physical events, when translated into abstractions of physics, as Whitehead wrote, "at once becomes the technical notion of vibration"[34] with all the misleading visual associations of distinct oscillating particles.

On this point, again, what Strawson regarded as a shortcoming of the auditory models becomes their recommendation. Sounds – 'sonorous quanta' we might say – are impermanent by their own nature and thus do not qualify as 'basic particulars' which, according to him, can only be physical bodies in the classical atomistic sense. Such impermanence is the general feature of microphysical individualities; as David Bohm observed, "every element, however fundamental it may seem to be, has always been found under suitable conditions to change even in its basic qualities".[35] Needless to repeat what has been said about the creation and annihilation of 'particles' – more accurately, about their mutual transformation. This only shows how much we have to be on guard against our instinctive 'logic of solid bodies' of which Strawson's anti-revisionary metaphysics is one of the last editions.

Yet it must be stressed again that the image of melody or polyphony has its limitations and thus it is rather a *metaphor* than a model. It cannot properly convey the homogeneity or quasi-homogeneity of physical duration. Nor does it express its deterministic or almost deterministic character. Each auditory pattern is a mere *exemplification* of the concrete universal of duration and this is why Bergson insisted that no image can ever replace the imageless awareness of *durée réelle*. He pointed out that the role of various metaphors consists in the fact that their inadequacies *cancel each other* and thus make transparent the dynamic imageless pattern underlying them all.[36] We have seen that the criticisms of Lovejoy and Ushenko were based on the misunderstanding of this point when they failed to grasp the distinction made by Bergson between *imageless qualitative diversity* and the sensory diversity of successive tones.

Thus we must conclude – as I did in my previous book – that the significance of auditory 'models' lies mainly in the fact that they free our imagination from the tyrannical sway of visualizing tendencies which tinge even some of the most abstract concepts of classical physics. They thus open the way to new imageless patterns which will supersede concrete mechanistic models of classical physics and complement the mathematical formalism of the present physics. In this way a future philosophy of nature would escape the dilemma – or rather the trilemma – which we face today: either to accept with the phenomenalists and positivists that physical reality in itself, i.e. independent of human observation and experimentation, has *no definite structure at all*; or, if it has, it is of an abstract, mathematical kind, as objective idealists haunted by the old Pythagorean dream believe; or, finally, as the steadily diminishing posterity of Descartes and Lord Kelvin hopes, that it is basically similar to the structure of macroscopic phenomena. The fourth way was outlined by Louis de Broglie prior to his reconversion to determinism:

Besides the questions of mathematical formalism which are already partially settled, there are still difficult problems of interpretation the solution of which will require long efforts for those whose main concern is to understand as much as it is possible the nature of the physical world.[37]

In this sense the present task of philosophy of science is to create dynamic, imageless patterns, more adequate than old, mechanistic models, but also sufficiently flexible and sufficiently general to be filled by the empirical content of future discoveries. In this way it can avoid both hazardous and presumptuous predictions *à la* Hegel as well as a dogmatic agnosticism bordering on intellectual apathy.

## NOTES

[1] *M.M.*, p. 36.

[2] John Dewey, *The Influence of Darwin on Philosophy and Other Essays in Contemporary Thought*, Holt, New York, 1910, p. 226f.

[3] *M.M.*, p. 35.

[4] *M.M.*, p. 213: "Amorphous space, atoms jostling against each other, are only our tactile perceptions made objective, set apart from all our other perceptions on account of the special importance which we attribute to them, and made into independent realities".

[5] P. F. Strawson, *Individuals, An Essay in Descriptive Metaphysics* (Anchor Books, 1959), p. 50.

[6] Lucretius, *De rerum natura*, II, v. 844.

[7] *C.M.*, p. 174.

[8] *Ibid.*, p. 180.

[9] *Ibid.*, p. 173.

[10] *Ibid.*, p. 173.

[11] *Ibid.*, p. 175. In a footnote to *Introduction to Metaphysics* Bergson insists that he does not reject the idea of substance provided it is understood in a dynamic, temporalistic sense. He rejects characterized as a Heraclitean. "On the contrary, I affirm the persistence of existences." (*C.M.*, p. 305.)

[12] *C.M.*, p. 85.

[13] *S.M.W.*, p. 54. In several places of *Process and Reality* (pp. 189–195; 487, 517, 523–524, 531) Whitehead uses a semi-auditory term "unison of becoming" for what Bergson calls "simultaneity of flows" and modern physics "contemporary independence".

[14] *An Enquiry Concerning the Principles of Natural Knowledge*, p. 196.

[15] *C.E.*, p. 221.

[16] Cf. Chapter 4 of this Part; also *M.M.*, pp. 213–214 n.

[17] P. F. Strawson, *op. cit.*, p. 57.

[18] *Ibid.*, p. XIV.

[19] *Ibid.*, pp. 28–29; 50.

[20] *Ibid.*, p. 25.

[21] *Ibid.*, p. 45.

[22] B. A. O. Williams, 'Mr. Strawson on Individuals', *Philosophy* 36 (1961) 319. Cf. also the extensive review-article of Ivor Leclerc 'Individuals', *Philosophy* 38 (1963).

[23] *Op. cit.*, p. XIII.

[24] E. Mach, *Space and Geometry* (Open Court, Lasalle, Ill., 1960), Chapter I 'On Physiological, as Distinguished from Geometrical, Space'.

[25] *Op. cit.*, p. 57.

[26] H. Taine, *De l'intelligence*, 16th Ed., II, pp. 155–160; John Stuart Mill, *Examination of Sir Willam Hamilton's Philosophy*, London, 1867, Chapter 3.

[27] *Op. cit.*, p. 72.

[28] *Ibid.*, p. 31.

[29] Griffin, D. R., *Listening in the Dark*, Yale Univ. Press, New Haven, 1958; 'More about Bat "Radar"', *Scientific American* **199** (July 1958) 40–44.

[30] Vitus B. Dröscher, *The Mysterious Senses of Animals*, E. P. Dutton & Co., New York, 1965, pp. 14–15.

[31] Théodule Ribot, *La logique des sentiments*, 5th Ed., Paris, 1905, p. 151: "la musique ne peut se developper que dans le temps: mais par l'harmonie et la polyphonie, elle a réalisé une simultanéité des successions qui lui donne un plus large champ et qui est comme un succedané de l'espace."

[32] *The Philosophical Impact of Contemporary Physics*, p. 219.

[33] *Process and Reality*, pp. 189–195; 487, 517, 523–524, 531.

[34] A. N. Whitehead, *S.M.W.*, p. 193. Cf. also R. Blanché, 'Psychologie de la durée et physique du champ'. *Journal de psychologie* **44** (1951) 411–424.

[35] D. Bohm, *op. cit.*, p. 153; cf. also Bohm's inaugural lecture at Birkbeck College 'Problems in the Basic Concepts of Physics' (1963), in particular pp. 29–33.

[36] *C.M.*, pp. 195–196.

[37] L. de Broglie, *Continu et discontinu en physique moderne*, Paris, 1949, p. 129.

# CHAPTER 16

# CONCLUDING REMARKS: THE WORLD OF LAPLACE AND THE WORLD OF BERGSON

We have now to indicate briefly how the paradoxical structure of the microphysical – and microchronical – world is related to the matter of our daily experience as well as to the matter of classical physics which, according to the expressive abbreviative formula of Edouard le Roy, is "a more refined common sense, a more penetrating sensory perception (*un sens commun plus affiné, une perception plus aiguë*)".[1] Implicitly, and in part even explicitly, this has been discussed before. All that I am going to do here is systematically restate and summarize it.

It has been shown that the elementary pulsations of physical reality ("*ébranlements*" of Bergson, "*vibrations*" of Whitehead) have a duration so insignificant in comparison to the intervals of time to which we are used and which are of the order of our specious present, that for all *practical* purposes they may be considered instantaneous. Thus in our usual macroscopic and macrochronic perspective the physical time-space appears to be continuous. It remains practically divisible *ad infinitum*.

If we adopt another *practically* justified perspective, that is, considering velocities small with respect to the velocity of light and the spatio-temporal distances occurring in our daily environment and even on the planetary scale, it is permissible at each moment to split the dynamic chronotopic process into two distinct components which may be considered independent: the static space, independent of duration, and continuous time, altogether foreign to spatiality. For the physical interactions in our daily surrounding and even on the terrestrial scale are *practically* instantaneous and thus suggest the idea of *geometrical*, i.e. *instantaneous* relations between the juxtaposed entities; the network of such distances constitutes the instantaneous geometrical space of Newton. This idea is then extrapolated to the whole universe and thus the concept of "Everywhere Now" or "World-wide Instant" is created. The world process is then regarded as a succession of instantaneous spaces.

It is thus clear that the idea of durationless space and that of mathematical instant are correlated. With the discovery of the finite velocity of light by Olaf Römer in 1675 and especially after the coming of relativity it was found that there are no purely geometrical distances, but only *time-consuming, chrono-geometrical* relations. Their temporal aspect can be disregarded for small distances; even for the distance between the Earth and the moon the corresponding dynamical links – luminous as well as gravitational – do not last more than one second. It is more difficult to disregard them for Neptune where their duration is four hours, and still more difficult for Alpha Centauri, Polaris, and the Andromeda nebula where these links last four, fifty, and more than a million years respectively. When we still continue to stretch the network of purely geometrical distances underneath these concrete dynamical links, we merely persist in the habits created by our limited macroscopic experience. Thus the biologically useful illusion leads to a distorting view of reality in general; but within the original limits of our experience the illusion fully retains its useful economic character.

For the same reason, the elementary indeterminations, too minute from the macrochronic point of view, are negligible on that scale. Macroscopic matter fits very approximately into the frame of classical determinism and the resulting inaccuracy does not interfere with the practical accuracy of our technical calculations. For once the concept of spatio-temporal continuity is at our disposal, nothing prevents us from considering all physical changes as infinitely divisible, though in reality they consist of fine pulsations. In other words, their mathematical continuity is only apparent and hides their individuality, just as the individuality of water-drops disappears in the spurious and superficial continuity of water jets. Their spatio-temporal continuity renders possible the application of differential equations by which the successive states of the material system are bound together.

The concept of isolated system is formed by a similar approximation: in disregarding all tenuous physical links which join an individual body to the rest of the universe, we obtain the concept of *isolated body* and, eventually, by a further idealization, the concept of *material point*. A material system, then, is nothing but an assemblage of material points. Thus all the components of the Laplacean model of reality are present as well as its basic significance: rigid determinism with its implicit tendency to eliminate the reality of succession. While this is clearly not the world of contemporary physics, it still represents an *approximately valid* picture of the world of middle dimensions; otherwise, in building a bridge, we would be unable to compute its maximum allowed load; nor could we compute the ballistic trajectories of artillery shells or of space rockets. In other words, on the human and terrestrial scale the world of Laplace still remains an *economic simplification* and *practical approximation* of the world of Bergson.

In this way we are returning – by a long detour – to the biological theory of knowledge discussed in Part I, without which a complete understanding of Bergson's thought, and of his philosophy of matter in particular, is clearly impossible. But if the macroscopic determinism remains intact, does this not mean that all recent modifications of our views concerning the structure of matter are without any practical significance? We know that this is not true. As pointed out in Part I, the opening up of the world beyond the limits of our spontaneous sensory perception at the same time widens the zone of the practically significant for man. It is unnecessary to repeat what has been said about this before. Today we can hardly anticipate the extent to which human life will be changed by the interaction between man and the microcosmos – the interaction which the physics and the technology of this century established and which even now is both magnificent and frightening. It is magnificent because never in the history of the human species has man penetrated so far beyond the limits of his spontaneous sensory perception; nor has he manipulated forces so minute and remote from his original manual grasp. It is frightening since in creating the artificial amplifying mechanisms by which the forces of microcosmos can produce enormous macroscopic effects, he exposed his own destiny to risks never faced before. To use Bergson's words in *Matter and Memory*, "the promises and the threats" to human life increased on a gigantic scale with the discovery of the microcosmos and its technical manipulation. Bergson seemed to anticipate it when he wrote in 1932: "mankind lies groaning, half crushed beneath the weight of its own progress." But he added characteristically: "Men do not sufficiently realize that their future is in their own hands. Theirs is the task of determining first of all whether they want to go on living or not."[2] His whole philosophy is one great effort to justify the activistic and anti-fatalistic attitude which is reflected in the sentences just quoted.

## NOTES

[1] E. Le Roy, 'Continu et discontinu dans la matière: le problème du morcelage', in *Continu et Discontinu. Cahiers de la nouvelle journée*, No. 15, Paris, 1929, p. 136.
[2] Cf. the concluding paragraph of *The Two Sources of Morality and Religion* (transl. by R. Ashley Audra and Cloudesley Brereton), H. Holt, New York, 1935.

APPENDIX I

# RUSSELL'S HIDDEN BERGSONISM

The title above certainly sounds strange and even facetious; for Russell's attitude to Bergson was not only that of philosophical disagreement, but of positive, almost personal dislike. This dislike accounts for Russell's frequent misunderstandings and misrepresentations of Bergson's thought – the misrepresentations which often border on caricature. It is true that this caricaturing was due more to Russell's inattentive reading than to a conscious desire to ridicule. Russell's own reading of Bergson was accurately characterized by Russell himself when he wrote that "to read an author in order to refute him is not the way to understand him." (OKEW, 47)*). Sometimes, however, the desire to ridicule is clearly discernible; for instance when, ignoring all the distinctions which the author of *Creative Evolution* draws between instinct and intuition, he confuses them, adding with humor that intuition is strongest "in ants, bees and Bergson." (PB, 3.) In any case, inattentive reading is as much a sign of intellectual indifference or hostility as a distorting caricature. Whether the touch of personal animosity in Russell's attitude was due, as it was submitted, to his suspicion that Bergson "lured" Whitehead away from him, is not certain[1], but it would not be too surprising; philosophers are human beings too, Russell more than any other.

Although we pointed out a number of times the deep differences separating Bergson's thought from that of Russell, let us briefly recall those which are the most basic. In this way we shall have a contrasting backdrop against which the unintentional agreements between them will appear even more striking. One of Russell's sentences in *Our Knowledge of the External World* (1914) summarizes the contrast between his and Bergson's philosophy in the most concise way: "Both in thought and in feeling, to realize the unimportance of time is the gate of wisdom." (OKEW, 167). In *Mysticism and Logic* he repeated the same sentence, but somehow more cautiously: "Both in thought and in feeling, *even though time be real*, to realize the unimportance of time is the gate of wisdom." (ML 21–22; italics mine). But this note of caution disappears altogether in the sentences which immediately follow:

That this is the case may be seen at once by asking ourselves why our feelings toward the past are so different from our feelings toward the future. The reason for this difference is wholly practical: our wishes can affect the future, not the past, the future is to some extent subject to our power, while the past is unalterably fixed. But every future will some day be past: if we see the past truly now, it must, when it was still future, have been just what we now see it to be, and what is now future must be just what we shall see it to be when it has become past. The felt difference of quality between past and future, therefore, is not an intrinsic difference, but only a difference in relation to us; to impartial contemplation, it ceases to exist. And impartiality of contemplation is, in the intellectual sphere, that very same virtue of disinterestedness which, in the sphere of action, appears as justice and unselfishness. Whoever wishes to see the world truly, to rise in thought above the tyranny of practical desires, must learn to overcome the difference of attitude towards past and future and to survey the whole stream of time in one comprehensive vision.

It certainly would be difficult to find in the philosophical literature a passage which would be more anti-Bergsonian in spirit as well as in letter. It is a perfect illustration of the view that "all is given" (*tout est donné*) – the view which eliminates becoming, transforms the future into a concealed present and wipes out the qualitative differences between the successive phases of time. It is the view of all strict determinists from Democritus to Laplace, and Russell is merely consistent when he says that "it is a mere accident that we have no memory of the future". (OKEW, 234). It eliminates the concept of causation in its original and dynamical sense by substituting for it the relation of logical co-implication in which the future is deducible from the past and *vice versa*; thus there is not such a thing as "direction of time" or "asymmetry of becoming". In Russell's words: "We shall do better to allow the effect to be before the cause or simultaneous with it, *because nothing of any scientific importance depends upon its being after the cause*." (OKEW, 226.)

This fundamental difference between Russell's and Bergson's views shows itself clearly in their attitude toward Plato and Zeno. While for Bergson, "the intelligible world" of ideas resembles the world of solids in its essential character except that its constitutive elements are "lighter, more diaphanous, easier for the intellect to deal with than the image of concrete things", for Russell in 1912 Plato's doctrine of ideas is one of the most successful attempts to solve the problem of the universals which he accepted with some terminological modifications. He was aware that this view leads to a very sharp kind of dualism:

Thus thoughts and feelings, minds and physical objects *exist*. But universals do not exist in this sense; we shall say that they *subsist* or *have being*, where 'being' is opposed to 'existence' as being timeless. The world of universals, therefore, may be also described as the world of being.

It is true that he somehow softens his commitment to Platonism by the following rather sober and remarkably impartial passage:

The world of being is unchangeable, rigid, exact, delightful to the mathematician, the logician, the builder of metaphysical systems, and all who love perfection more than life. The world of existence is fleeting, vague, without sharp boundaries, without clear plan or arrangement, but it contains all thoughts and feelings, all the data of sense, and all physical objects, everything that can do either good or harm, everything that makes any difference to the value of life and the world. *According to our temperaments*, we shall prefer the contemplation of the one or of the other. The one we do not prefer will probably seem to us a pale shadow of the one we prefer, and hardly worthy to be regarded as in any sense real. But the truth is that both have the same claim on our impartial attention, *both are real*, and both are important to the metaphysician. (P. 100; italics added.)

The passages just quoted show clearly the complexity of Russell's mind as well as the resulting instability of his views. While explicitly admitting his preference for the realm of being, he still conceded then the reality of becoming only to deny – only after two years – the reality of time altogether while at the same time ridiculing Kant for degrading time to a mere appearance. (OKEW, 116–117). But in the passage just quoted another note creeps in – an uneasy awareness that the metaphysical preferences for either being or becoming are perhaps mere personal idiosyncracies, due to individual differences in temperament. A tendency to prefer the metaphysics of Being, together with an underlying note of radical scepticism are two characteristic features of Russell's thought.

Comparison of Bergson's and Russell's attitude toward Zeno's paradoxes will show again the basic contrast in their philosophical views, but at the same time will bring out in the most unexpected way certain hidden affinities. Bergson's view of Zeno's paradoxes was consistently held through all his books: the paradoxes arise from the fallacious assumption that motion and time are divisible *in infinitum*, that is, that the only parts of them which are indivisible are geometrical points and durationless instants. This assumption is based on the confusion of the movement itself with its motionless trace in space; it is this motionless trace, not the act of moving (*la mobilité, le mouvant*), which is infinitely divisible. "At bottom, the illusion arises from this, that the movement, *once effected*, has laid along its course a motionless trajectory on which we can count as many immobilities as we will. From this we conclude that the movement, *while being effected*, lays at each instant beneath it a position with which it coincides. We do not see that the trajectory is created in one stroke, although a certain time is required for it; and that though we can divide at will the trajectory once created, we cannot divide its creation, which is an act in progress and not a thing." [2]) Russell's comment on Zeno in his *Principles of Mathematics* (1903) was significantly different:

After two thousand years of continual refutation, these sophisms were reinstated, and made the foundation of a mathematical renaissance, by a German professor, who probably never dreamed of any connection between himself and Zeno. Weierstrass, by strictly banishing infinitesimals, has at last shown that *we live in an unchanging world*, and that the arrow, at every moment of its flight, is truly at rest. The only point where Zeno *probably* erred was in inferring (if he did infer) that, because there is no change, therefore the world must be in the same state at one time as at another. This consequence by no means follows. (PM, 347; italics mine.)

In other words, Russell agrees with Zeno that we are living in "an unchanging world"; but against Zeno he claims that the world is not in the same state at every moment. How an unchanging world can be different at different successive moments, he does not explain. The only plausible explanation of what appears to be a glaring contradiction is that by 'change' Russell meant the dynamic passage, the transition, the overflow of one moment into the subsequent one; he rejected 'change' understood in this sense, since it is incompatible with the mutual externality of instants in the mathematically continuous time, and since in mathematical continuum there is no "next" element with respect to the "preceding" one. In any case, it is certain that in 1903 Russell regarded the mathematically continuous space and time as the *truly real*, as "the world in which we are living", as he says. What we call 'change' was for him nothing but "diversity in time", time being unconsciously regarded by him in a mathematical fashion as the axis of independent variables on which "successive" instants with corresponding different states of the world exist or rather *coexist*. This reconstruction of Russell's thought is the only possible way in which his strange view that "the world is unchanging without being identical in its successive moments" could be made at least *psychologically* understandable – without becoming any more convincing. Needless to stress that the adjective 'successive' loses its meaning in Russell's scheme which is a perfect illustration of what Bergson called the "fallacy of spatialization". This came up again eleven years later in a strangely ambiguous passage of *Our Knowledge of the External World* where Russell wrote:

The contention that time is unreal and that the world of sense is illusory must, I think, be regarded as based on fallacious reasoning.

Was Bergson then right? Not quite; here is the immediately following sentence:

Nevertheless, there is some sense – *easier to feel than to state* – in which time is an unimportant and superficial characteristic of reality. *Past and future must be acknowledged to be as real as the present,* and a certain emancipation from the slavery to time is essential to philosophic thought. The importance of time is rather practical than theoretical, rather in relation to our desires than in relation to truth. ... But unimportance is not unreality... (OKEW, 166–167; italics added.)

One must agree with Russell: it is "easier to feel than to state" how time can be real – though unimportant – when past and future are as real as the present. It is ironical to see the thinker who so severely accused Bergson of vagueness, make an appeal to such a diffused feeling and to get entangled in transparent contradictions.

But in 1914, when the book just mentioned above was published, Russell's views on Zeno were to some extent modified. While in 1903 he agreed with Zeno that the arrow at every instant of its flight is "truly at rest", in 1914 he disclaims it: "we cannot say it is at rest at the instant, since the instant does not last for a finite time... Rest consists in being in the same position at all the instants through a certain period, however short..." (OKEW, 136). Furthermore, there is another modification of Russell's view: he does not insist any longer that we live in the unchanging world of Zeno. He admits explicitly that "the theory of mathematical continuity is an abstract logical theory, not dependent for its validity upon any properties of actual space and time". But he is clearly aware that the applicability of such continuous series to the world of experience is another matter. He concedes that "interpenetration", that is, the "transition which is not a matter of discrete units" is a datum of immediate experience, but he attempts to escape this uncomfortable fact by differentiating fictitiously "appearance" from "reality". Such distinction is clearly meaningless on the level of immediate experience where – as

Russell himself conceded at another place of the same book (72, 85–86) appearance and reality coincide. But while insisting that the world of senses *may* be continuous, he concedes that there is no sufficient reason for it either. In other words, while the theory of mathematically continuous series is *compatible* with experience, it is not *necessitated* by it.

From what has just been said it follows that the nature of sense data cannot be validly used to prove that they are not composed of mutually external units. It may be admitted, on the other hand, that nothing in their empirical character specially necessitates the view that they are composed of mutually external units. This view, if it is held, must be held on logical, not on empirical grounds. I believe that the logical grounds are adequate to the conclusion. They rest, at bottom, upon the impossibility of explaining complexity without assuming constituents. (OKEW, 145).

In other words: Russell's *logical atomism* still makes him lean toward accepting the infinite divisibility of space and time and the actual existence of points and instants, even though he is aware that they are unverifiable empirically. While he was then closer to Bergson in admitting the conceptual constructive nature of "instants", his insistence that every multiplicity must be of the *atomistic* type, built of the mutually external units, ignores completely the qualitative multiplicity which constitutes immediate experience.

It was one year later, in two articles which appeared in *The Monist*, that Russell – without realizing it – came closest to Bergson's thought. In the article 'On the Experience of Time' he explicitly gave up the reality of the durationless mathematical present in psychology:

Suppose, to fix our ideas, that I look steadily at a motionless object while I hear a succession of sounds. The sounds A and B, though successive, may be experienced together, and therefore my seeing of the object while I hear these sounds need not be supposed to constitute two direct experiences. But the same applies to what I see while I hear the sounds B and C. Thus the experience of seeing the given object will be the same at the time of the sound A and at the time of the sound C, although these two times may well not be parts of one specious present. Thus our definition will show that the hearing of A and the hearing of C form parts of one experience, which is plainly contrary to what we mean by one experience. Suppose, to escape this conclusion, we say that my seeing the object is a different experience while I am hearing A from what it is while I am hearing B. Then we shall be forced to deny that the hearing of A and the hearing of B form parts of one experience. In that case, the perception of change will become inexplicable, and we shall be driven to greater and greater subdivision, owing to the fact that changes are constantly occurring. We shall thus be forced to conclude that one experience cannot last for more than one mathematical instant, *which is absurd.* (ET, 216–217); italics mine.)

This is the very opposite of the view held by Russell only a year before when he claimed that "it is perfectly reasonable to suppose that the sense data of a given type ... really form a compact series". The passage above could well have been written by Bergson himself; in truth, it is not difficult to find similar passages in Bergson's works. It is not only the concept of durationless instant which Russell here rejects; he concedes also, however reluctantly, the inapplicability of the concept of arithmetic multiplicity to sensory and introspective experience. It is true that he tries valiantly to define "one momentary total experience", as well as the "specious present"; but as it is clear from the text above, he arrives at the conclusion that the relation "belonging to the present" is not transitive and "that two presents may overlap without coinciding". But what else is this than Bergson's *"pénétration mutuelle"*, i.e. "qualitative multiplicity"? (ET, 223, 214).

No less Bergsonian, but equally consistent was Russell's acceptance of the direct perception of succession and change, in other words, of "immediate memory". Thus he says that "succession may be directly experienced between parts of one sense datum, for example, in the case of a swift movement" which is the object of one sensation. (227) "It is indubitable that we have knowledge of the past, and it would seem, though this is not logically demonstrable, that such knowledge arises from acquaintance with past objects in a way enabling us to know that they are past." (222) He admits that we know the past by acquaintance, that is, directly; like Bergson, he claims that "immediate memory is intrinsically distinguishable from sensation" (226) and therefore should not be confused with the present trace of the past sensation:

There is first what may be called "physiological" memory, which is simply the persistance of a sensation for a short time after the stimulus is removed... This fact is irrelevant to us, since it has nothing to do with anything discoverable by introspection alone. Throughout the period of "physiological memory", the sense datum is actually *present* ... We will give the name of "immediate memory" to the relation which we have to an object which has recently been a sense datum, but is now felt as past, though *still given in acquaintance*. (224–225; last italics added).

He also fully realizes that the direct knowledge of the past presupposes the Bergsonian *immanence of the past within the present*:

At first sight, we should naturally say that what is past cannot also be present; but this would be to assume that no particular can exist at two different times, or endure throughout a finite period of time. It would be a mistake to make such an assumption, and therefore we shall not say that what is past cannot also be present... *The present has no sharp boundaries*, and no constituent of it can be picked out as certainly the earliest. (222–223; italics mine).

All these passages have their nearly exact counterparts in Bergson's writings and they have been quoted in this book. It is true that Russell's language is still static, atomistic and spatializing in its characteristics; the terms like 'object', 'constituent', 'part of an object' show it clearly. Thus instead of speaking of "duration", he speaks of "one particular existing at two times". But the substance of his view is the same. The fact that Bergson is hardly quoted in this article should not deceive us; Russell refers to William James's analysis of the perception of time (225) and we do not need to repeat how close James's "stream of thought" was to the Bergsonian *durée réelle*.

In the second essay, 'The Ultimate Constituents of Matter' (reprinted in *M.L.*), Russell gave up the applicability of the concept of durationless instant even to physics. He attacks the assumption that the ultimate constituents of matter must be permanent and indestructible; the allegedly permanent "thing" is a logical construct based on the perceptual illusion of fusing together the successive series of momentary states:

Each of these [i.e. of tables, chairs, the sun, moon, stars] is to be regarded, not as one single persistent entity, but as a series of entities succeeding each other in time, each lasting for a very brief period, though probably not for a mere mathematical instant.... A true theory of matter requires a division of things into time-corpuscles as well as into space-corpuscles. (M. L., 129).

Russell thus constructs the world out of momentary particulars which, he insists, "are to be conceived, not on the analogy of bricks in a building, but rather on the analogy of notes in a symphony. The ultimate constituents of a symphony (apart from relations) are the notes, each of which lasts only for a very short time. We may collect together all the notes played by one instrument: these may be regarded as the analogues of the successive particulars which common sense regard as successive states of one 'thing'" (*M.L.* 129–130). It is the same auditory model as that used by Bergson and Whitehead. This attack on the basic assumptions of classical atomism and the tendency to reduce physical existence to the succession of events is obviously similar to Bergson's and Whitehead's "vibratory theory of matter" whose consonance with the philosophical implications of the quantum theory we stressed before. Russell took it over again in his *Analysis of Matter* where he spoke about the atomicity of space-time and "the quantized geodesic routes between two events". (A.Ma., 304, 341).

A close analysis of Russell's view will show the limits of its affinity with the views of both Bergson and Whitehead. This can be stated in one single sentence: Russell gave up *physical*, not *logical* atomism. Russell's idea of corpuscular time may appropriately be characterized as an atomistic translation of the pulsational time of James and Bergson; Russell's "corpuscles of time" are apparently externally related. It is true that this view was not consistent with Russell's view upheld nearly at the same time that the present does not have sharp boundaries. But he apparently overlooked it. Otherwise he would not have characterized his view of the universe as "cinematographic". He recalls that his first visit to a cinema was motivated by his desire to verify Bergson's statement that the mathematician conceived the world after the analogy of a cinematograph, and he found it "completely true" (M.L. 128). There was, however, a significant difference between both philosophers. For Bergson the successive projection of the static picture on the screens symbolizes "the cinematographic mechanism of thought" which tries to reconstruct change out of changeless entities; for Russell, it is a correct and adequate analysis of the physical processes. For Bergson the experienced continuity of change, succession and motion is real, the successive static "moments" are illusory; for Russell the very opposite is true, "the cinema is a better metaphysician than common sense, physics or philosophy". (M.L. 129) Bergson would agree with Russell that the persistence of the physical objects is an illusion since Bergson's matter is also constituted by the momentary (though not instantaneous) events.

Another basic difference is that Russell's "atoms of time" apparently have the same, no matter how minute, duration; the idea of variable temporal span, which is the cornerstone of Bergson's metaphysics, is altogether absent from his thought. This explains his view that he – again unlike Bergson – does not recognize the genuine continuity of psychological duration:

The real man too, I believe, however the police may swear to his identity, is really a series of momentary men, each different one from the other, and bound together, not by a numerical identity, but by continuity and certain intrinsic causal laws. (M.L. 129)

This is nearly altogether a Humean view; I say 'nearly', since in the Humean framework, there is no place for 'intrinsic causal laws' at all.

In the subsequent development of Russell's thought, his 'logical atomism' came to the fore far more clearly. In *The Analysis of Mind* he upheld the associationism of Hume with hardly any improvement. Memory is reduced by him to the occurrence of a *present* image accompanied by the belief in past existence: "this existed". He insists on the absolute externality of the present remembering and the event remembered; for this reason he condemns "Bergson's interpenetration of the present by the past, Hegelian continuity and identity-in-diversity, and a host of other notions which are thought to be profound because they are obscure and confused" (A.Mi. 180), apparently forgetting all the subtle analyses of his essay 'On Our Experience of Time'. His amnesia on this point continues in *The Analysis of Matter* (1927): "For my part, I do not think logical interpenetration can be defined without obvious self-contradiction; Bergson, who advocates it, does not define it." (A.Ma. 387) In his *Human Knowledge* he does not stop short of the most extreme form of "the fallacy of simple location in time", as Whitehead would call it, when he claimed that not only is a man private from other people, but he is also private from his own past. It is not "here" alone that is private, but also "now". (HK. 90) Obviously, if there is such complete externality of the past with respect to the present, the supposition that the world might have come into existence a few moments ago with all things as they are now, including my fallacious recollections, is not absurd and is, strictly speaking, irrefutable. (HK. 212) Russell, however, concedes that nobody takes such supposition seriously. But if it is so, if nobody takes seriously what cannot 'logically' be refuted, then there must be something radically wrong with a logic of this kind.

It is precisely, such atomistic logic, – an inadequate conceptual tool for dealing with the dynamic and elusive structure of time – which prevented Russell from agreeing with Bergson. For this reason his "incipient Bergsonism" of the year 1915 remained so well hidden – to him as well as to his commentators.

## NOTES

* The references and the abbreviations refer to the following of Russell's books:
  *The Principles of Mathematics* (PM), W. W. Norton & Co., New York, 1964.
  *The Problems of Philosophy* (PP), Galaxy Book, Oxford Univ. Press, New York, 1959.

*The Philosophy of Bergson* (PB), Macmillan, London, 1914; originally in *The Monist*
**22** (1912).
*Our Knowledge of the External World* (OKEW), Allen & Unwin, London, 1914.
'On the Experience of Time' (ET), *The Monist*, 1915.
*Mysticism and Logic* (ML), W. W. Norton, New York, 1929.
*The Analysis of Mind* (AMi), Allen & Unwin, London, 1921.
*The Analysis of Matter* (AMa), Dover, New York, 1954.
*Human Knowledge* (HK), Simon & Schuster, New York, 1962.
[1] Cf. H. C. McElroy, *Modern Philosophers: Western Thought since Kant*, Russell F.
Moore Co., New York, 1950, p. 141. Bergson referred to Russell's attack in his conver-
sation with Jacques Chevalier on May 30, 1933 when he expressed a different theory
about the origin of Russell's animosity toward him: "Bertrand Russell has never for-
given me the refutation which I made once of his too material interpretation of the
Platonic Ideas. He revenged himself by saying that the evolution culminated on one
side in the intellect which found its complete development in mathematicians, and on
the other side, in instinct which is at best in ants, bees and Bergson." (J. Chevalier,
*Entretiens avec Bergson*, Paris, 1959, p. 197.) Bergson's mildly ironic quotation of
Russell clearly referred to the discussion which took place in the Aristotelian Society,
1911, and to which Russell refers in *My Philosophical Development*, Simon & Schuster,
New York, 1959, p. 161. The unfairness of Russell's attack was recognized by Alan
Wood in his book *Bertrand Russell, the Passionate Sceptic*, Simon & Schuster, New
York, 1958, pp. 197–198, where he pointed out that while Russell criticized Bergson
for the confusion of subject and object, in the next chapter he praised William James
for denying the same distinction!
[2] *C.E.*, p. 336.

# MICROPHYSICAL INDETERMINACY AND FREEDOM,
# BERGSON AND PEIRCE

Many of man's technical discoveries have been anticipated – sometimes with truly astonishing precision – by various organisms, animal as well as vegetable. L. Cuénot's book, *Invention et finalité en biologie* surveys facts of this kind in a systematic way.[1] These facts are not unknown to biologists, – and even to some philosophers who, however, under the impact of the magic formula of 'natural selection' have lost the capacity for wonder which Aristotle correctly identified as the main source of philosophical meditation. The twentieth century man invented extremely complex and ingenious mechanisms by which the effect of microphysical events can be amplified to produce a sizeable macroscopic effect. We thus face the following fascinating question: Has not organic nature also anticipated this human achievement? More specifically: Is not *the very structure of the organism* an embodiment of such an amplifying device?

This, indeed, is the view of Niels Bohr in his 1957 lecture 'Physical Science and the Problem of Life'. When he then remarked that "amplification effects similar to those permitting observation of individual atomic particles play a decisive role in many functions of the organism", he stated more explicitly what he had said nearly three decades before in the 1929 lecture which was later incorporated into his book *Atomtheorie und Naturbeschreibung*. He then very cautiously concluded that while there is no basic difference between organic and inorganic matter, a deeper understanding of biological problems, in particular of the freedom and capacity of adjustment ("die Freiheit und Anpassungsvermögen") of organisms in their reactions to external stimuli, requires taking into account the limits of causal description in atomic processes.[2] Bohr's ideas stimulated another outstanding physicist, Pascual Jordan, to outline more explicitly in a series of articles in *Naturwissenschaften* and *Erkenntnis* the theory that organisms are the amplifiers of microphysical indetermination.[3] More recently, another physicist, Walter Elsasser, formulated a similar theory in the light of new neurological data and related it explicitly to Bergson's views.[4]

There is unquestionably a certain similarity between Jordan's, Elsasser's and Bergson's views, even though – as we shall see later – this similarity has definite limits. One of the central ideas of *Creative Evolution* is precisely that the organism, in particular the nervous system, is an amplifying device by which "the slightest quantity of indetermination, by continually adding to itself, will make as much freedom as possible". This was how Bergson summarized its leading ideas four years after the publication of *Creative Evolution* in his Huxley lecture at the University of Birmingham on May 29, 1911.[5]

When we investigate the way in which a living body goes to work to execute movements, we find that the method it employs is always the same. This consists in utilizing certain unstable substances which, like gunpowder, need only a spark to explode them. I refer to foodstuffs, especially to ternary substances, carbo-hydrates and fats. A considerable sum of potential energy, accumulated in them, is ready to be converted into movement. That energy has been slowly and gradually borrowed from the sun by plants; and the animal which feeds on a plant, or on an animal which has been fed on a plant, or on an animal which has fed on an animal which has been fed on a plant, and so on, simply receives into its body an explosive which life has fabricated by storing solar energy. To execute a movement, the imprisoned energy is liberated. All that is required is, as it were, to press a button, touch a hair trigger, apply a spark: the explosion occurs, and the movement in the chosen direction is accomplished.[6]

Hence the significance of very complex and unstable organic molecules containing a great amount of potential energy which can be released by a quantitatively minute stimulus. As the complexity of the organic body is increased, the effect of the triggering action would be greater as the quantity of the bound energy will be correspondingly larger. In this way – in the words of *Creative Evolution* – there will be introduced into the physical world "the largest possible amount of indetermination" and liberty[7], or, in Elsasser's words, "the main function of the organism is the creation of a new physical reality".[8] Life is thus characterized as "an effort to engraft on to the necessity of physical forces the largest amount of indetermination":

This effort cannot result in the creation of energy, or, if it does, the quantity created does not belong to the order of magnitude apprehended by our senses and instruments of measurement, our experience and science. All that the effort can do, then, is to make the best of a pre-existing energy which it finds at its disposal. Now, it finds only one way of succeeding in this, namely to secure such an accumulation of potential energy from matter, that it can get, at any moment, the amount of work needed for its action, simply by pulling a trigger. The effort itself possesses only that power of releasing. But the work of releasing, although always the same and always smaller than any given quantity, will be the more effective the heavier the weight it makes fall and the greater the height – or, in other words, the greater the sum of potential energy accumulated and disposable. As a matter of fact, the principal source of energy usable on the surface of our planet is the sun. So the problem was this: to obtain from the sun that it should provisionally and partially suspend, here and there, on the surface of the earth, its continual outpour of usable energy, and store a certain quantity of it in the form of unused energy, in appropriate reservoirs, whence it could be drawn at *the desired moment, at the desired spot, in the desired direction.* The substances forming the food of animals are just such reservoirs. Made of very complex molecules holding a considerable amount of chemical energy in the potential state, they are like explosives which only need a spark to set free the energy stored within them. (Italics mine.) [9]

Two central ideas of *Creative Evolution* are present in this passage: first, that life seems to operate in a direction opposed to the second law of thermodynamics by creating structures more and more complex and more and more *improbable* and thus suspending, at least temporarily and locally, a continuous dissipation of energy; second, that these complex organic structures are the means by which the microphysical indeterminacy is transmitted and amplified to become macrophysically effective. It is beyond the scope of this book to discuss the first point, since it would involve us inevitably in a very extensive critical comment on *Creative Evolution* itself. The discussion of the second point will show that the similarity of Bergson's view to that of Jordan and Elsasser has definite limits.

On the higher animal level the established mechanisms by which the microphysical indeterminacy is amplified to become macrophysically effective are represented by the neural mechanisms lodged in the cerebro-spinal system.

The body carries out voluntary movements by means of certain mechanisms set up in the nervous system and waiting only for the signal to start them; the brain is the point where the signal is given and also where the mechanism is operated. The Rollandic zone, where voluntary movement has been localized, is in fact comparable to a signal-box, from which the signalman shunts the coming train to its proper line. It is a sort of commutator, by which a given external stimulus can be put in communication with any motor disposition whatever.[10]

Or, in a wider phylogenetic perspective:

From the humblest Monera to the best-endowed insects, and up to the most intelligent vertebrates, the progress realized has been above all a progress of the nervous system, coupled at every stage with all the new constructions and complications of mechanism that this progress required. As we foreshadowed in the beginning of this work, the role of life is to insert some *indetermination* into matter. Indeterminate, i.e. unforeseeable, are the forms it creates in the course of its evolution. More and more indeterminate also, more and more free, is the activity to which these forms serve as the vehicle. A nervous system, with neurons placed end to end in such wise that, at the extremity of each, manifold ways open in which manifold questions present themselves, is a veritable *reservoir of indetermination.*[11]

The idea that an enormous amount of energy can be released by a tiny, almost infinitesimal quantity was not foreign to classical physics. The well known facts of explosion, of unstable equilibrium and of any triggering mechanism certainly belong within the purview of classical physics. Herbert Spencer had his own labels for facts of this kind: "the unstability of the homogeneous" and "multiplications of effects". Nor was it unknown that such phenomena are conspicuous in the organic realm where, as Schopenhauer observed, "effect is often not quantitatively equal to its cause". Needless to say that such phenomena contradict neither the law of conservation of energy nor the general idea of classical determinism. The amount of energy, no matter how spectacularly large, is not created *ex nihilo*, since it pre-existed in a potential form. Nor is the triggering agency, no matter how negligible, an undetermined event, comparable to the Lucretian *clinamen*. It is a definite physical occurrence with definite physical antecedents – at least as long as we remain on the ground of classical physics. From this point of view, the central impulse which steers the neural energy of the afferent stimulus into certain motor paths, no matter how small it remains and no matter how much it escapes our measurements, still remains a definite *physical* event which *must* have definite physical causes and thus be an inevitable outcome of the immediately preceding enormously complex state of the brain. The hair-trigger organization of the mammalian brain was well known to the last century's cerebral physiology, which, nevertheless, would not have called the nervous system "a reservoir of indetermination" except in a purely figurative sense.

Thus when William James suggested that the central activating impulse may be actually co-determined by a feeling of effort[12], it was possible to refute him in the name of classical determinism, which confined consciousness within the limbo of causal inefficacy. James himself was well aware of this difficulty, which stemmed from the fact that the *only* solution of the traditional mind-body problem compatible with classical physics was the *double-aspect theory* – called the theory of psychophysical parallelism on the continent – according to which mental events are merely passive epiphenomena or 'inner aspects' – of the rigorously determined neural processes. As Nietzsche said: "The will does not move anything; it does not explain anything; – it only accompanies the processes; it may be even absent."[13] Or in the words of another physiologically oriented psychologist of the *fin de siècle*:

We should not forget, however, that action is not produced because a concomitant psychical process is introduced. By no means. On the contrary, the material process which lies at the foundation of the action is completely in itself... the parallel psychical processes are useless and superfluous.[14]

Since these were the authoritative views of James' contemporaries, it is understandable that he himself conceded that "the feeling of effort *may* be an inert accompaniment and not the active element which it seems".[15] His option for indeterminism at that time was based almost exclusively on ethical grounds and certainly was altogether incompatible with the Laplacean universe in which genuine choice, contingency and freedom was nothing but a mere "*asylum ignorantiae*", a refuge of ignorance of all determining causes.

But to claim that the intellectual situation is the same today is nothing but a sheer dogmatism which is completely blind to all the profound and revolutionary changes which have occurred in physics in the last seventy years, and which, in particular, completely ignores the *widening* of the concept of causation which modern physics suggests. Today the classical concept of causality is being given up, or, more accurately, being replaced by a redefined concept purged of the inconsistencies and absurdities of the old Laplacean type. As we tried to point out, the widened concept of causation, advocated by Bergson before Whitehead, Reichenbach, Heisenberg, Bondi and Whitrow, affirms the reality of *a genuinely growing world* in which authentic novelties emerge – not from nothing, but from the past antecedents.

In such a world, the genuine efficacy of will ceases to be irrational since it is only a special instance of novelty. As Peirce prophetically observed a quarter of a century before Heisenberg:

On the other hand, by supposing the rigid necessity of causation to yield, I care not how little – be it by a strictly infinitesimal amount – we gain room to insert mind into our scheme, and to put it into the place where it is needed, into the position which, as the sole self-intelligible thing, it is entitled to occupy, that of the fountain of existence; and in doing so we resolve the problem of connection of soul and body.[16]

But this is more easily said than done. Peirce was undoubtedly right when he said that the physical universe, if compatible with a certain degree of indetermination, – no matter how small – ceases to be impervious to the causal efficacy of mind which remained a sheer irrationality in the universe of classical physics. It is well known that Descartes failed to reconcile psychophysical interaction with the rigorous determinism of his mechanistic universe. It is less well known that three centuries later Hans Driesch failed in a similar hopeless enterprise – to reconcile the action of his non-physical 'entelechy' with the determinism of Newtonian physics, to which he remained loyal through his whole life.[17] The present physics, anticipated in this respect by both Peirce and Bergson, removed two stumbling blocks which both Descartes and Driesch – as well as any dualist – faced: Laplacean determinism and the complete heterogeneity of the mental and the physical. As long as physical determinism remained complete and as long as the two realms – that of 'mind' and that of 'matter' – remained completely heterogeneous – any interaction between them remained both impossible and unintelligible. But within the panpsychistic perspective, which is so conspicuous in the passage from Peirce quoted above and which is so akin to the panpsychism of Bergson and Whitehead, there is no place for the sharp Cartesian dualism.

To conclude from this that the problem of the connection between the mental and the physical is 'solved' as Peirce did, is a bit premature. Some serious difficulties still remain of which Peirce was apparently unaware, but which became quite conspicuous considerably later, when Pascual Jordan – without knowing it – was trying to realize what may be called 'Peirce's program': a new formulation of the psychophysical problem within the framework of the contingentist physics.

Jordan's theory of organisms as "amplifiers of microphysical indetermination" stirred a lively discussion among the members of the Vienna circle. Moritz Schlick, Philipp Frank, Hans Reichenbach, Erwin Bünning, Edgar Zilsel, Otto Neurath and Hans Winterstein took part in it.[18] Their reaction was – with exception of Reichenbach – negative and even hostile; but the nature of the arguments and the tone of their polemic varied greatly from one person to another. Roughly speaking the objections fell into two different groups: those which were clearly irrelevant since they questioned and suspected the *motives* of Jordan's views instead of dealing with the views themselves; and those which critically analyzed Jordan's views. It would be unwise to deal extensively with objections of the first kind; all psychologizing and 'psychoanalyzing' of an opponent in discussion is, besides its irrelevancy, also a double-edged weapon which can be easily turned against those who use it. If Jordan was accused of doing an inveterate "metaphysics", (*veraltete Metaphysik* in Neurath's words), of being "prejudiced" in favor of the autonomy of life, of emotionally welcoming the alleged reconciliation of physics and freedom, he could have easily riposted by psychoanalyzing the unconscious or semiconscious attitudes of his opponents which showed themselves in the very language they used. It would be interesting to raise the question to what extent the psychological atmosphere of the imperial city of Vienna and of the old Habsburg Austro-Hungarian monarchy in general left lasting traumatic effects on the majority of members of the *Wiener Kreis*. The alliance of the reactionary dynastic regime with the hardly less reactionary church, with the resulting tight control of education, the morbid anti-semitism bordering on active persecution, and systematic suspicion of all anti-traditional ideas almost certainly contributed to this effect. It was in such an unhealthy atmosphere that the majority of the members grew up; it is then hardly surprising that words like 'metaphysics', 'freedom', and 'mind' aroused their suspicion and anger, that their very usage was regarded by them as an attempt to smuggle discredited supernaturalism back into philosophical discussions.

Instead of this, Jordan merely pointed out that the basic question was not the "emotional motives" of his views, but whether or not his views were correct. Hans Reichenbach also resolutely rejected any labelling of Jordan's views as "metaphysical" – metaphysical in the Viennese sense as confused, irresponsible speculations – and stated with admirable intellectual honesty and open-mindedness:

As philosophers, we must in principle admit the possibility that Jordan is right... And we must stress this possibility precisely against such biologists who are too easily inclined to reject any such suggestions from a physicist and to cling to the apriorist view of causality... Nothing is as dangerous as dogmatism; but, unfortunately, among many of them the dogmatic view still prevails that the quantum mechanical view of causality cannot affect in any way the concept of causality in biology.[19]

Far more worthy of attention are the objections which dealt with the *content* of Jordan's views. Among those, the critical analyses by Bünning and Schlick are most penetrating. Bünning tried to show that even the tiniest organisms are still macrophysical systems; even the monocellular organisms still have a diameter of several microns. The centrosome, with the diameter of a fraction of a micron, still consists of several thousands of huge molecules; and it is well known that such organic molecules consist of a large number of atoms. Even Bünning was aware that the zone of microphysical indeterminacy interacts with the realm of middle dimensions. He mentioned that no more than two alpha-particles can kill *bacillus prodigiosus*; that the gene consists of one giant molecule and that the question of whether change of the gene itself, i.e., *mutation*, is a strictly determined process must be left open. No wonder that Jordan in his rejoinder claimed that this part of Bünning's criticism strengthened rather than weakened his theory of organisms as amplifiers of microphysical contingency. Reichenbach himself regarded similar objections made by Zilsel as inconclusive.[20]

More serious was Bünning's objection that any amplification of microphysical contingency within an organism would seriously disturb its functioning and would thus lead to disease or even death. Moritz Schlick's penetrating analysis of Jordan's views was along the same lines. It argues that amplification of the microphysical chance-events – besides its enormous improbability – would result in a greater degree of contingency (*Zufall*) which is the *very opposite* of freedom. He points out that Jordan himself was aware that mere indetermination leaves the stability of organisms unexplained.[21]

These serious objections are far from fatal, though Jordan was unable to answer them. But he could have pointed out that the amplification process takes place by means of well established mechanisms whose functioning is adjusted to the normal functioning of the whole body and thus does not produce any destructive effect, as Bünning feared. This was, as we have seen, essentially the view of Bergson: that various ready-made motor mechanisms in the body can be triggered by an insignificant amount of energy which itself is of the order in which microphysical indeterminacy plays a role. Voluntary movements, despite their explosive character, do not threaten the well-being of organisms since the energy thus released is channeled by the established paths to the periphery of the body and to its surrounding.

But while this answers Bünning's objection concerning the allegedly destructive effect of amplification, it does not answer Schlick's question as to why the amplified microphysical contingency terminates almost always in an overt action tending to assure the survival of organism; in other words, why the resulting motor response generally has a *teleological* character. For if we concede that in an adult organism various ready-made motor responses, acquired by previous learning, are, so to speak, waiting to be triggered by an appropriate stimulus, the influence of a contingent factor in the critical area of the brain would bring about utter anarchy in the motor responses to various situations. The sight of approaching danger would with equal probability activate the reflex of running away from it as that of moving towards it. The objection becomes even more serious if we realize that in view of the 'hair-trigger organization of the brain', a teleological reaction would be a 'hit-or-miss' affair against which the odds would be overwhelmingly large. From a purely statistical point of view a correlation of the *appropriate* response with a given stimulus would be incomparably *less* probable than out-of-place or even a disastrous reaction. It would possess the extreme improbability of one single lucky hit contrasting with the enormous number, and consequently enormously greater probability, of all other possible unsuited combinations. This was pointed out by William James in his *Principles of Psychology* a long time ago, and although his objection was formulated against classical mechanism, it retains its full force against what David Bohm appropriately called 'indeterministic mechanism' as well. Whether we understand the term 'fortuitous' in the sense of the classical theory of gases, compatible with the Laplacean determinism, or in the modern quantum-mechanical sense, the difficulty remains. Thus Moritz Schlick and Philipp Frank were basically right when they pointed out that in Jordan's view man still would be "a purely physical system", amplifying a fortuitous microphysical indetermination; i.e. basically nothing but "*l'homme-machine*"[22], only, unlike the man-machine of Lamettrie, intrinsically unpredictable. In other words, neither vitalism nor any kind of interactionism gained anything by Jordan's theory.

To this Jordan retorted that it was not his intention "to save vitalism" as Edgar Zilsel charged; he insisted that from his point of view "autonomy of life" does not have any sense[23] since, contrary to the traditional dualistic vitalism, physical contingency is not confined to the organic realm, but is *omnipresent* in nature. But his view was not as unambiguous as he claimed; for other utterances of his clearly suggest that he did *not* regard the microphysical contingency *within* organisms on a par with the for-tuitous quantum-mechanical fluctuations. Otherwise why did he put so much stress on a close connection between such indetermination and the teleology of organisms? And – even more important – why did he stress the significance of *our feeling of freedom*, of *our consciousness of choice*, which he regarded as far more significant than the inveterate doctrines (*veralteten Doctrinen*) of classical determinism?[24] The only intelligible interpretation of his view which, while not explicitly present in his writings, is implied in them, is that he was proposing a modified version of the classical double-aspect theory. This theory regarded the awareness of choice which precedes the *fiat* of our will as an altogether illusory feeling, since in the classical scheme there are no such things as genuine possibilities, genuine alternatives; only *one* decision is possible from eternity and only *one* overt reaction at any given moment can take place. In Spinoza's graphical illustration, our feeling of freedom is as illusory as that which a falling stone would have, were it conscious; it, too, would probably regard its predetermined fall as a result of its own 'free will'. Or in the more recent words of Théodule Ribot, the feeling of effort, the experience of activity, is only "a subjective state corresponding to certain operations going on in the nerve centers and other parts of the organism, but resembling them as little as the sensations of sound and light resemble their objective cause".[25] It is as deceptive, as 'epiphe-nomenal' and, consequently, as causally inert as other conscious qualities; since it merely *accompanies* the incipient, but completely predetermined motor discharge that terminates in an overt reaction. Jordan apparently modified the classical version in one important aspect; the feeling of choice, according to him, is *not* illusory, since in the world of quantum physics there are genuine possibilities, truly open alternatives. This can mean only one thing – that psychological feeling of choice, of voluntary *fiat*, is an inner, subjective aspect of a microphysical indetermination occurring in a certain area of our brain and then amplified to an overt reaction. Briefly: while Jordan gave up the rigorous determinism of the classical double-aspect theory, he retained its basic claim about one-to-one correspondence of the physical and the mental.

It may be objected that thus formulated Jordan's theory makes an extremely improbable assumption. Since all the options of organisms are triggered by a statistically undetermined event, all teleological reactions are hit-and-miss affairs in the sense explained above. Jordan clearly does not explain why the initiating micro-event has – if not always, at least most frequently – the character of a spark thrown into an explosive material "at the desired moment, at the desired spot, in the desired direction" – to use Bergson's words. The improbability of such a solution was felt by Eddington when he wrote, not without humour:

I do not think that our decisions are precisely balanced on the conduct of certain key-atoms. Could we pick out one atom in Einstein's brain and say that if it had made the wrong quantum jump there would have been a corresponding flaw in the theory of relativity? Having regard to the physical influences and promiscuous collisions it is impossible to maintain this.[26]

Jordan apparently believed that not only the individual reaction of the organism, but the whole of phylogenetic evolution depends on the statistical improbabilities taking place within the key-atoms; hence, his emphasis on the microphysical character of mutations, mentioned above. But let us be fair to him. His idea that the evolution of all complex organic forms is a result of the cumulation of improbable microphysical events may appear grotesque; but is it really so different from the orthodox neo-Darwinian version of the evolutionary theory? For the latter claims that the development of the species resulted from fortuituous variations which were then sifted by natural selection. The only difference is that 'for-, tuituous' is understood by the neo-Darwinians in the classical sense compatible with Newtonian determinism while Jordan understands this term in the objectivist sense of Peirce's tychism or Boutroux's contingentism. It is rather amusing to see the mechanists of the Vienna circle accuse Jordan of failing to explain the teleology of the organism. They themselves proposed an equally non-teleological explanation.

But there is another decisive objection against Jordan's theory, even though it is less obvious. His key idea seems to be the claim that there is a one-to-one correlation between the introspective *fiat* of our will and a single microphysical event. A brief analysis will show the *impossibility* of this view. It is well known and was stressed repeatedly in this book that mental events or qualities have a certain *existential temporal minimum* and that they simply vanish *when* their temporal span is lowered below this minimum. We expressed this by saying that the mnenic span in the psychological realm is considerably larger than that of the physical events. It is unnecessary to repeat what was discussed before. Suffice it to say that the minimum psychological duration is of the order of $10^{-2}$ sec and as such it is enormously longer than the temporal minimum in the physical world. Even if we concede the non-existence of mathematical instants or zero-intervals, the estimated value of the chronon is of the order of $10^{-24}$ sec, i.e. $10^{22}$ shorter than the faintest pulsation of our conscious life. It is clear that no one-to-one correlation can exist between events of such an enormously different temporal span.*

The only way to save the one-to-one correlation between the physical and the mental would be to postulate that the conscious temporal *Gestalten* are only *apparently simple* and indivisible and that they, in reality, consist of an enormously large number of successive sub-sensations of a duration comparable to the duration of the corresponding physical events. But this claim was extensively discussed and dismissed in Part II, and it would be otiose to refute it again. The one-to-one correlation between the physical and the mental cannot be saved in an epistemologically tenable way without doing violence to the most obvious data of experience. The temporal indivisibility of the mental qualia is certainly such a datum. To dissect them artifically into some fictitious *'petites perceptions'* is clearly a heritage of the outdated associationistic 'mind-dust theory'. In the same way the related belief that the physical, i.e. instantaneous, present is *the only authentic present*, while the psychological present is merely 'specious', is a hangover of the nineteenth century physicalism and epiphenomenalism.

We must then agree with Zilsel that Jordan's attempt to "save vitalism" – if this is what he intended to do – failed for the reasons stated above; but we must add that if it failed, it was because it contained the classical physicalistic elements which made its failure inevitable. Since no such physicalistic elements are present in Bergson's thought, the similarity between Jordan's and Elsasser's views and those of Bergson is restricted to the two points, already mentioned: the belief in microphysical indeterminism, anticipated by Bergson as early as 1896, and the view of organisms as *the* amplifiers of physical micro-indetermination. Both Jordan, Elsasser and, still more recently, John R. Platt[28] approached the problem of freedom from almost an exclusively physical and physiological stand-

* This is what Edgar Zilsel saw clearly when he pointed out that, were Jordan's views correct, our stream of consciousness (*Bewustseinsablauf*) would go on at much quicker rate – I would add with *fantastically* much quicker rate.[27]

point. Yet, without a complementary introspective analysis of voluntary *fiat* as it was done in such classical studies as those of William James, Narciss Ach and Bergson himself, every attempted solution will remain necessarily incomplete.

Two basic difficulties of the Jordan-Elsasser view are absent in Bergson's "dualism" or "vitalism" – if we may be allowed to apply such lazy and simplifying *clichés* to the complex ideas of *Matter and Memory* and *Creative Evolution*. The first difficulty was created by the tendency of both Jordan and Elsasser to regard the organism as a sort of mechanical dice-shaker. An overt reaction, while being undetermined, is then teleological only *accidentally*, since nothing in their theory guarantees 'a happy outcome', that is, the correlation between a microphysical triggering event and an appropriate motor response. This difficulty is absent in Bergson's explanatory scheme, since the triggering microphysical event, instead of being an isolated and statistically improbable happening, is a *terminal phase* of the process by which the feeling of effort, voluntary *fiat*, mental option – the terminology itself is less relevant – becomes effective on the physical level. Such a view was an utter absurdity in the deterministic scheme of classical physics and a glaring inconsistency in the classical dualism; but it was not so in Bergson's conceptual scheme.

For matter in his scheme – like matter of contemporary physics – is no longer impervious to the causal efficacy of the mental qualia, since novelties exist both on the physical and the mental level, though in different degrees. In other words, the difference between the mental and the physical is far from being as sharp as in Cartesianism or neo-Cartesianism. As explained above, it is a difference of temporal span, with a correlated difference in degree of extension and degree of indeterminacy. Since this temporal span varies, the transition between the mental and the physical becomes intelligible and thus loses the thoroughly irrational character which it had in the classical dualism. If we realize this, we shall better understand the words by which Bergson concludes his essay 'L'effort intellectuel':

This operation, which is the very operation of life, consists in the gradual passage from the less realized to the more realized, from the inextensive to the extensive, from a reciprocal implication of parts to their juxtaposition. Intellectual effort is something of this kind. In analysing it, I have pressed as far as I could, on the simplest and at the same time the most abstract example, the growing materialization of the immaterial which is characteristic of vital activity.[29]

It would be beyond the scope of this book to analyze the examples to which Bergson refers in this passage. They range from the creation of new motor habits to a conscious effort to recall a reluctant recollection and, finally, to the highest instances of artistic and intellectual creation. Suffice it to say that the words 'the growing materialization of the immaterial' which has a deceptively mystical ring to our Cartesian ears, conditioned by the three centuries of the 'bifurcation of nature', have the precise meaning which we have lengthily analyzed in Chapters 4, 5 and 6 of Part III.

But this leads us to the second difficulty, present in Jordan's and, apparently, also in Elsasser's scheme. While Jordan avoids the fallacious exclusion of the very possibility of choice in the name of a non-existent rigorous determinism, he makes an epistemologically impossible assertion in correlating choice on the mental level with a single event on the physical – or rather microphysical – level. As already mentioned, Jordan, whether he realized it or not, was proposing a modified version of the double-aspect theory, adjusted to the principle of indeterminacy. Recall Zilsel's cogent objection against it. This objection might have been well formulated by Bergson himself for whom the absence of one-to-one correlation between the physical and the mental is the basis for his criticism of the double-aspect theory in *Matter and Memory*.

The central idea of this book may be characterized as a 'one-to-many correlation' between the mental and the physical. For what else is the fact, stressed by Bergson, that a simple, indivisible mental quality is *contemporary* with an exceedingly large number of successive events on the physical level? The term 'one-to-many correlation between the physical and the mental' is thus appropriate, provided we remain aware of the limitations which the usual connotations of the words 'one' and 'many' have when they are applied to any temporal process.

A more detailed analysis will shed some light on the mechanism or, more accurately, the *dynamism*, of the causal efficacy of mind as understood by Bergson. To make this analysis complete it would be necessary to take into account the cerebral *milieu* into which conscious initiative is inserted. In this respect, concrete analysis is very difficult and cannot be undertaken here. For every act is, as Bergson himself repeatedly stressed, a mixture of automatism and spontaneity, of the motor mechanisms and conscious initiative; in truth, from the time of his first book

Bergson stressed the *rare* character of free action.[30] We 'are acted upon' (*nous sommes agis*) far more frequently than we act (*nous agissons*) in the authentic sense. We must refer again to the books *Matière et mémoire* and *L'energie spirituelle* in which the relation of consciousness to cerebral events is discussed concretely and in detail. But even if we consider the physical efficacy of the mental in general, that is, without taking into account the concrete cerebral *milieu* into which it is placed, we reach interesting conclusions.

As pointed out above, the act of will, considered introspectively, has a certain temporal, existential *minimum*, a certain shortest duration which still enormously surpasses the duration of the elementary physical events. "The growing materialization of the immaterial" in this case cannot mean anything except that the durational tension, which made a single temporal *Gestalt* of the mental quality of volition, is *loosened* in the sense analyzed in detail in Chapter 4 of Part III. In other words, the indivisible psychological present is disintegrated into a large number of successive and extremely short 'sub-presents'. When the relaxation of this tension is sufficiently quickened, the resulting 'pulsations' will be of the same mnemic span as the elementary events constituting the physical reality. There is then, on the physical level the sudden appearance of a great number of contingent events whose effect is added precisely because of their nearly simultaneous appearance. This seems to be the meaning of the following passage in Bergson's Huxley lecture:

Placed at the confluence of consciousness and matter, sensation condenses, into the duration which belongs to us and characterizes our consciousness, immense periods of what we can call by analogy the duration of things. Must we not think, then, that if our perception contracts material events in this way it is in order that our action may dominate them? Supposing the necessity inherent in matter be such that at each of its moments it can be forced, but only within extremely restricted limits, how in such case must a consciousness proceed, if it would insert a free action into the material world, let the action to be no more than releasing a spring or directing a movement? Would it not have to adopt precisely this method? Should we not expect between its duration and the duration of things a difference of tension such that innumerable instants of the material world could be held within one single instant of the conscious life, *so that the desired action (l'action voulue), accomplished by consciousness in one of its moments, could be distributed over an enormous number of the moments of matter and so sum up within it the indetermination almost infinitesimal which each of them admits?* In other words, is not the tension of the duration of conscious being a measure of its power of acting, of the quantity of free creative activity it can introduce into the world? I hold that it is....[31] (Italics added.)

To forestall some pedantic linguistic objections, let us add that the terms 'consciousness' and 'things' are used by Bergson for the purpose of abbreviation. He neither separates consciousness in the old substantialistic sense from concrete psychological changes nor is he unaware that substantial 'things' do not exist on the microphysical level. All this is quite obvious to anybody who recalls insistence, on Bergson's part, on the reality of "change without substratum" in both physics and psychology.

The passage is similar to that which appeared two decades later in *La Pensée et le Mouvant* and to another passage in *Matter and Memory* fifteen years yearlier – another indication of the continuity of Bergson's thought.[32] The durational tension of the perceptual qualities is necessary for our macroscopic and macrochronic, that is, *deterministic*, perspective of matter. Without it "our action would be dissolved" or, more accurately, it would lose its macroscopic efficacy; it would be on a par with other short-lived, random, microphysical events. Paradoxically, the deterministic perspective in perception is necessary for action *transcending* determinism. But the paradox is only apparent; the understanding of it will help us to see what is common and what is different in a microphysical chance-event and an action stemming from conscious choice.

An element of novelty is present on both the physical and mental level; but on the former it is 'infinitesimal', more accurately negligible, for practical purposes. In the Bergsonian terminology, each physical event is an *almost* identical repetition of its immediate ancestor; its novelty barely transcends its own past.[33] Furthermore, the effect of this novelty is cancelled in the over-all interaction of the elementary chance-events so that on the macroscopic level it can be safely disregarded. This mutual cancelling of the elementary chance-activities is the basis of the stability and regularity of matter as it appears to us in the macrochronic perspective. The reason for this is not difficult to see; for, as Ralph S. Lillie observed, when chance events are repeated a sufficient number of times, they distribute themselves equally in all possible directions; hence an impulse in one direction is sooner or later compensated for or reversed by an impulse in the opposite direction. "Accordingly, chance activities cancel one another out and cannot by themselves lead to progressive differentiation, evolution, or complex organization."[34] Bergson expressed the same idea in his own panpsychistic language: when in the concluding part of *Matter and Memory* he wrote that "nature may be regarded as a neutralized and consequently latent consciousness, consciousness of which the eventual manifestations hold each other in check, and annul each other precisely at the moment at which they might appear."[35] Only very exceptionally do the elementary chance-events happen to act in the same sense so that a cumulative effect results. Such additive interference of the elementary events may be the cause of the radioactive explosions in which the extremely small probability of the alpha-particles escaping from the nucleus is realized. (A macroscopic analogy – imperfect, because it is deterministic – of this phenomenon is an accidental additive interference of elementary wavelets inside of a glass of water which results in the temporary creation of a high crest of the liquid spilling over the rim of the vessel.)

Now the conscious biological reaction *outwardly* resembles such a phenomenon in the sense that its immediate antecedent consists in a very complex and improbable correlation of the elementary contingencies. It is their correlation, that is, their nearly simultaneous occurrence which assures their combined, interferentially reinforced effect instead of their normal cancellation. This is in agreement with Elsasser's definition of life as "the presence of correlation among the multitude of physically indeterminate events".[36] But this correlation is *not* an accidental coincidence, as Jordan, and apparently also Elsasser, thought. It is an outward manifestation of the total temporal *Gestalt*, introspectively perceived as voluntary *fiat*; in truth, the correlated physical events were generated by a sudden temporal deconcentration ('dilution', 'distension') of this mental quality. Only in this way can Driesch's concept of *Ganzheitskausalität* become intelligible; it was not so in his neo-Cartesian dualistic and deterministic scheme. The microphysical effect thus produced is then amplified by activating the energy stored in the motor mechanism of the cerebro-spinal system with the resulting overt reaction taking place 'in the desired moment, at the desired spot, in the desired direction'. In this way the generally teleological character of conscious reactions, which remains so puzzling in the Jordan-Elsasser theory, is explained. Bergson's view is similar to the view expressed considerably later by the biologist Ralph S. Lillie:

According to the present view, some element of indeterminacy – in the sense of present determination, or internal determination, or what may be called "spontaneity" – is always present in a natural event, but to a degree which varies greatly in the different instances. Within the vital organization this internal indetermination is apparently able to express itself in a way which is not possible in nonliving systems... This condition reaches its maximum in the more complex types of biological causation such as human activity, although it is there limited in its possibilities by physical and physiological constants already established, as well as by social and other environmental conditions.[37]

It is also immediately clear why this view is compatible with the fact that organic life does and always will appear to the eyes, to the microscope – and unfortunately even to the intellect of an experimental biologist – only in its physico-chemical, that is, its deterministic aspect. Every process taking place in the organism and in its nervous system appears to an external observer as an extremely complicated network of causal chains and only by their complexity do they differ from other physico-chemical causal chains in inorganic nature. The fact that some of these causal chains are activated by a conscious initiative whose immediate, triggering, physical effect is too tiny to be detected, since it is on the scale of quantum mechanical indeterminacies, naturally escapes the attention of the majority of biologists and psychologists.[38] Hence the prevailing tendency among them to regard the difference between a reflex action and conscious decision as one of degree; the latter, according to them, is nothing but a more complex form of the former – a mere extension of the external stimulus into the motor paths, modified by its detour through the cortical engrams.

Bergson's view is considerably different from that ridiculed by Eddington and by the positivistic critics of Jordan according to which the overt reactions would depend on a single microphysical contingency somewhere in the neural tissue. The simile of a single switch activating a complex motor mechanism is obviously grossly inaccurate; it would be far more appropriate to speak of a whole *keyboard of switches* activated by a single decision in the sense of the previously exposed one-to-many correlation.

In this sense Bergson's view is different from that of Ralph S. Lillie's correlation between the physical and the mental[39] while it is very close to the view of Eddington who wrote in 1928 that "in the physical part of the brain directly affected by the mental decision" the crowd of atoms has "a configuration which the secondary laws of physics would set aside as too improbable"[40] and the interdependence of behavior "which is not present in inorganic matter". If we abstract from the traditional and crudely visual connotations of words like 'atom' and 'configuration' which

Eddington used for abbreviative purposes only, it is clear that the "improbable configuration of atoms affected by mental decision" is nothing
but another term for the additive interference of the microphysical events
which were generated by the deconcentration of a single mental quality.

We must also not forget that the term 'improbable' is justified only
if we confine ourselves to the exclusively physical, so to speak 'external',
point of view. If a physicist or a physiologist were able to observe under
the microscope the conscious reaction of the organism originated in the
motor zone of the brain – which, as Bohr convincingly pointed out, is and
will be impossible – he would be overwhelmed by the marvellously precise
cooperation of the enormous number of physical elements corresponding
to the introspectively simple mental quality of decision. Either this
astonishment overwhelms him entirely and then he slips into anthro-
morphism in imagining "statically ready-made particles juxtaposed to
one another, and, also statically, an external cause which plasters upon
them a skilfully contrived organization".[41] Or he will remain on the terrain
of physics and will then speak of "realized improbability" or "chance
coincidence". While the first view is naively anthropomorphic – the
*entelechy* (whether of Aristotle or that of Hans Driesch) always bears a
similarity to the planning activity of *homo faber* – the latter errs in
regarding the central microphysical events triggering the outgoing reaction
as *externally* related; from such a point of view their correlation is *always*
nothing but a miraculous chance-coincidence and the resulting purpose-
fulness is *always*, as Taine observed, "a happy accident". Bergson, as we
know, rejected naive finalism as much as mechanism; finalism for him
is nothing but "inverted mechanism".[42] But he would have been equally
opposed to what Bohm called "indeterministic mechanism". While not
denying the mutual externality of the triggering contingent events, he
regarded them as *the final phase of the process* of "the growing materialization of the immaterial" referred to above. The correlation and cooperation of the contingent events betray their common origin which their
discontinuity belies. In a similar way the letters on a piece of paper – from
the physical point of view the mere irregular and scattered black spots –
only by their mutual correlations – by their 'order' or 'arrangement' –
betray their origin in the indivisible inspiration which threw them on
the paper.

If the self-appointed guardians of rationality are shocked by Bergson's dualism and interactionism, it is because, first, they confuse the rationality of mechanism with rationality in general; second, they fail to understand that Bergson's dualism is basically different from the traditional dualism of Descartes and Driesch – "para-mechanistic" as Gilbert Ryle called it – and it is thus free of their difficulties and absurdities inherent in any sharp 'bifurcation of nature'; that this bifurcation is avoided in Bergson's view which, instead of opposing the fictitious absolute indeterminacy of 'free will' to the rigid necessity of the Laplacean type, proposes the theory of different degrees of freedom and – correlatively – of *different degrees of causality* in the sense hinted at much more recently by Norbert Wiener.[43] Only in this context can John Dewey's profound remark that microphysical indetermination is a *necessary*, though not *sufficient* condition of freedom acquire an intelligible meaning.[44] In other words: the psychophysical interaction which remained an utter irrationality in the classical deterministic framework, ceased to be so in the genuinely growing and temporalistic universe whose most important features were foreseen by Bergson prior to its discovery by contemporary physics.

## NOTES

[1] L. Cuénot, *Invention et finalité en biologie*, Paris, 1941.

[2] N. Bohr, *Die Atomtheorie und Naturbeschreibung*, Berlin, 1931, p. 89.

[3] P. Jordan, 'Quantenmechanik und Grundprobleme der Biologie und Psychologie', *Naturwissenschaften* XX (1932) 815–821; 'Quantenphysikalische Bermerkungen zur Biologie und Psychologie', *Erkenntnis* IV (1934) 215–252.

[4] W. Elsasser, 'Quantum Mechanics, Amplifying Processes and Living Matter', *Philosophy of Science* XXVIII (1951) 300–325; 'A Reformulation of Bergson's Theory of Memory', *Philosophy of Science* XX (1953) 7–21.

[5] 'Life and Consciousness' in *M.E.*, p. 14.

[6] M.E., *ibid*.

[7] *C.E.*, p. 127.

[8] W. Elsasser, the first article quoted above, pp. 306–307.

[9] *C.E.*, pp. 127–128; cf. also a similar passage in *M.E.*, p. 35: "I quite agree that, if the will is capable of creating energy, the quantity created may be so small that it would not affect sensibly our instruments of measurement. Yet its effect might be enormous, like that of the spark which explodes in a powder magazine ... It will then require an almost negligible action, such as the slight pressure on the hair-trigger of a pistol, in order to liberate at the required moment, in the direction chosen, as great an amount as is wanted of accumulated energy. The glycogen lodged in the muscles is, in fact, a real explosive; by it voluntary movement is accomplished: to make and utilize explosives of this kind seems to be the unvarying and essential preoccupation of life, from its first apparition in protoplasmic masses, deformable at will, to its complete expansion in organisms capable of free action."

10 *M.E.*, p. 43.

11 *C.E.*, pp. 139–140.

12 W. James, *The Principles of Psychology*, I, Ch. V ('The Automaton Theory'), esp. p. 140.

13 F. Nietzsche, *Götzendämmerung*, *Werke*, VIII, Leipzig, 1895, p. 96.

14 Th. Ziehen, *Introduction to Physiological Psychology* (London, 1892), p. 273–274.

15 James, *op. cit.*, I, p. 452. James added significantly: "No measurements are yet performed (it is safe to say that none ever will be performed) which can show that it contributes energy to the result." – On James' views on this problem cf. my article 'James' Early Criticism of the Automaton Theory', *Journal of the History of Ideas* XV, pp. 260–279.

16 C. S. Peirce, 'Doctrine of Necessity Examined', in *Collected Papers of C. S. Peirce*, (ed. by C. Harthsorne and P. Weiss), VI, pp. 42–43.

17 Cf. Driesch's book *Relativitätstheorie und Weltanschauung*, Leipzig, 1930.

18 *Erkenntnis* V 1935, pp. 178–184; H. Winterstein, 'Der mikrophysikalische Vitalismus', *Erkenntnis* VII (1937–38), 81–91.

19 H. Reichenbach, 'Metaphysik bei Jordan?' *op. cit.*, pp. 178–179.

20 E. Bünning, 'Sind die Organismen die mikrophysikalische Systeme?', *Erkenntnis* V (1935) 337–347; Reichenbach, *loc. cit.*, p. 179; Jordan's rejoinder in *Erkenntnis* V, pp. 348–353.

21 M. Schlick, 'Ergänzende Bermerkungen über P. Jordan's Versuch einer quantenmechanischer Deutung der Lebenserscheinungen, *Erkenntnis* V, pp. 181–183.

22 P. Frank, 'Jordan und radikale Positivismus', *ibid.*, p. 184.

23 Cf. Note 20.

24 Cf. *Erkenntnis* IV, p. 243.

25 Th. Ribot, *The Diseases of Will* (transl. from the 8th French edition by M. M. Snell), Chicago, 1896, p. 96.

26 *The Nature of the Physical World*, p. 313.

27 E. Zilsel, 'P. Jordan's Versuch den Vitalismus zu rechnen', *Erkenntnis* V, p. 63.

28 John R. Platt, 'Amplification Aspects of Biological Response and Mental Activity', *American Scientist* **44** (1956) 180–197.

29 *M.E.*, p. 7.

30 *T.F.W.*, p. 167.

31 *M.E.*, pp. 16–17.

32 *M.M.*, pp. 218–219; *C.M.*, pp. 68–69.

33 *M.M.*, p. 220: "If matter does not remember the past, it is because it repeats the past unceasingly, because, subject to necessity, it unfolds a series of moments of which each is the equivalent of the preceding moment and may be deduced from it; thus its past is truly given in the present." In the passage already quoted (p. 244–245) Bergson foresaw that the physical necessity is not absolute, but only very approximate and as such hides to us the elementary contingency of the physical events.

34 Ralph S. Lillie, *General Biology and Philosophy of Organism*, University of Chicago Press, 1945, p. 197.

35 *M.M.*, p. 245.

36 W. Elsasser, *Philosophy of Science* XX (1953) 8.

37 Lillie, *op. cit.*, p. 107; cf. also his article 'Biological Causation' in *Philosophy of Science* VII (1940) 336.

38 This deceptively deterministic aspect of life which hides to us the quantitatively insignificant, but crucially important creative innovations was stressed by the out-

standing American biologist Sewall Wright in the concluding part of his article 'Gene and Organism', *The American Scientist* **87** (1953) 16–17.

[39] Ralph S. Lillie (*op. cit.*, pp. 81–82, 130) seems to share Leibniz's view about the completely extra-spatial character of the mental and the vital activity which leads him logically to the "single key-atom" theory which Elsasser (*Philosophy of Science* **XVIII**, p. 300) rejects.

[40] A. S. Eddington, *The Nature of the Physical World*, pp. 312–314. A similar view is that of Hermann Weyl in *Was ist Materie?*, Berlin, 1924, p. 84.

[41] *C.E.*, p. 272.

[42] *C.E.*, pp. 45–50.

[43] N. Wiener, *I am a Mathematician*, M.I.T. Press, Cambridge, Mass., 1964, p. 323.

[44] John Dewey, *The Quest for Certainty*, Hilton & Balch Co., New York, 1929, pp. 289–250.

APPENDIX III

## BERGSON'S THOUGHTS ON ENTROPY
## AND COSMOGONY

In his very complimentary report submitted to the French Academy of
Sciences on January 23, 1909 on Émile Meyerson's book *L'Identité et
Réalité*, Bergson briefly outlined the main theses of its author. First, that
the essential feature of every explanation is *identification* in time and
space which, pushed to its extreme consequences, leads to the elimination
of both succession and diversity; hence the latent Eleatism of human
intellect. Second, that while reality yields to a considerable degree to such
identifying explanations, it nevertheless resists a complete elimination of
succession and diversity; third, that this resistance shows itself in the
discovery of the law of entropy (or 'the principle of Carnot', as it is called
in French) which, contrary to the identifying tendencies of the classical
explanations, discloses the intrinsic irreversibility of time.[1]

It was only natural that Bergson welcomed Meyerson's epistemological
theses which were so much in agreement with his own criticism of the
static and monistic tendencies, characterizing the classical form of human
intellect.. Nor was it surprising that he welcomed Meyerson's interpretation of the significance of the law of entropy. Only one year before the
publication of *L'Identité et réalité*, he expressed the same view in his
*L'Evolution créatrice* i.e. that the irreversibility of time found its expression in the second principle of thermodynamics: "it is the most metaphysical of the laws of physics since it points out without interposed symbols,
without artificial devices of measurements, the direction in which the
world is going."[2] To him the irreversibility of physical time was a mere
consequence of the irreversibility of becoming in general. For, since, according to him, becoming constitutes the very nature of *every* reality, it
cannot be absent even in the physical world, contrary to the trends of classical physics and contrary even to his original view stated in his first book.

This metaphysical context of his own doctrine accounts for what differentiates Bergson's view of the second law of thermodynamics from

that of many physicists, who *defined* the irreversibility or – as they call it – 'direction' of physical time by the increase of entropy. Now it is clear that for several reasons Bergson could not share such view. In the first place, the gradual increase of entropy was for him merely *one manifestation* of the 'unidirectionality' of time or irreversibility of becoming in general, not anything *identical* with it or something by which becoming could be exhaustibly characterized. There are other kinds of irreversibility which attracted Bergson's attention before: the irreversibility of psychological duration in his first two books and the irreversibility of organic evolution in *Creative Evolution*. As pointed out in this book, he believed that an attentive analysis of psychological duration will disclose the general structure of *every* duration. The exclusive identification of irreversibility with the increase of entropy would confine duration to the physical realm only, contrary to his basic claim that irreversible becoming is an *all pervasive* feature of reality.

Second, the identification of 'the direction of time' with the increase of entropy has been – and is still being – carried out within the larger framework of the relational theory of time. Now it is true that we can speak meaningfully of the relational theory of time in Bergson. But we must define it carefully in order to differentiate it clearly from the relational theory of the physicists speculating about the significance of the second law of thermodynamics. For there are two very different kinds of the relational theory: one, the tradition of which goes back to Aristotle and Heraclitus; the second, whose roots reach back to Lucretius and ancient atomism. Bergson clearly belongs to the Heraclitus-Aristotle tradition. While both these theories agree that "time is nothing in itself", that is, that it is inseparable from concrete changes and events, they differ substantially on the nature of these events. According to Heraclitus, Aristotle and Bergson, time is constituted by qualitative, irreversible changes: "one cannot step twice into the same river." According to atomism, time is correlated with the changing configurations of the particles which themselves are *beyond change*, that is, immutable, indestructible and uncreatable. This view re-emerged in the modern corpuscular-kinetic model of nature in an improved and more exact form. It still retained the basic claim of Epicurus that time is nothing but an 'accident of accidents', that is, the function of the changing configurations of the unchanging particles. (It is true that there were some important dissenters who, like

Gassendi and Newton, insisted on the independence of time from its physical content; but since they correlated time with 'the divine duration', they were very close to the first type of the relational theory of time. In truth, it is doubtful that any theory of completely absolute, *altogether empty* time can be even meaningfully formulated.) Now since nothing in principle prevents the possibility that any configuration will eventually recur, the past situation can return, the 'direction of time' may be reversed and time itself, whether on the cosmic or a local scale, be cyclical. If the giant configuration of all the elements constituting the universe and defining its 'state at a certain moment', will recur, the cosmic time itself would be cyclical as the Stoics in antiquity and Nietzsche in the last century believed; if such a situation will occur only on a local scale, there would be no cosmic time at all since it would be resolved into the multiplicity of local 'time segments', running in opposite directions and each of them changing its direction after sufficiently long intervals of time. In other words, the irreversibility of becoming would cease to be the universal feature and would become contingent and local – local both in the spatial and temporal sense. "Change *must* be reducible to an arrangement and rearrangement of parts; the irreversibility of time *must* be an appearance relative to our ignorance; the impossibility of turning back *must* be only the inability of man to put things back to their original place." [3] Those are the very words by which Bergson characterized the corpuscular-kinetic model of the universe; and since he rejected this model, he *ipso facto* rejected the relational theory of time implied in it.

It was then only consistent for him to try to express the meaning of the second law of thermodynamics in a *language free of corpuscular-kinetic terms*. If the universe is an *intrinsically irreversible* process, then every physical law must, in principle at least, be expressible in the terms of *events* or *changes* without any surreptitious reference to some immutable particles which merely change their positions without changing themselves:

It [i.e. the second law of thermodynamics] tells us that changes that are visible and heterogeneous will be more and more diluted into changes that are invisible and homogeneous, and that the instability to which we owe the richness and variety of the changes taking place in our solar system will gradually give way to the relative stability of elementary vibrations continually and perpetually repeated... From this point of view, a world like our solar system is seen to be ever exhausting something of the mutability it contains. In the beginning it had the maximum of possible utilization of energy; this mutability has gone on diminishing unceasingly. [4]

By the word "mutability" (*mutabilité* in French) Bergson translates the term *Umwandelbarkeit*, literally 'transformability' (i.e. of energy) which Boltzmann used in that part of his book to which Bergson in this context refers. In Boltzmann's words, the difference between the entropy and its maximum value, which is the 'driving force' (*das Treibende*) of all natural processes, is continually diminishing; thus despite the constancy of the total energy its mutability (*Umwandelbarkeit*) will become smaller and smaller, the development of nature becoming "more and more languid" (*das Naturgeschehen immer matter*).[5]

Bergson then considers the question which had been raised by a number of physicists and cosmogonists: what was the origin of the initial improbable state of the 'maximum mutability' (i.e. the minimum entropy) within the solar system? If we assume that its source was in some region of space *outside* our solar system, we merely postpone the solution since the same question may be raised about that external region. The difficulty may be avoided by assuming the infinite number of the worlds "capable of passing the mutability to each other"; thus "the sum of mutability contained in the universe is infinite" and therefore there is no reason to seek its origin and its end. This idea of the universe everlastingly regenerating itself can be traced to some Presocratics, and in its modern form it appeared among those astronomers and physicists who wanted to avoid the inevitability of the 'heat death' of the universe. Thus Arrhenius, whose book appeared in the same year as *Creative Evolution*, believed that the cold portions of distant nebulae are warmed up by the impact of the fine particles ejected by the radiation of the stars; in this way 'the entropy clock' is re-wound at least on the local scale. Boltzmann, without being as specific as Arrhenius, also postulated the possibility of the entropy clock 'running backwards' in very remote parts of the universe.[6] To the objection of Poincaré that a local re-winding of the entropy clock would merely delay the thermal death of the universe,[7] Arrhenius could retort that this would not be true if the universe were infinite so that the radiation, instead of being dissipated in the surrounding void, would circulate from one system to another, compensating the lowering of the temperature in one system by increasing it in another. The assumption of the infinity of the universe was thus essential, if the final levelling of temperatures could be avoided. The infinity of the universe is impossible without the reality of infinite space. This Bergson rejected since he rejected any reification of space. Such reification would mean, according to him, "an absolute externality of all the parts of matter in relation to one another" which was incompatible with his opposition to the doctrine of external relations and, more specifically, with his view that geometrical space is a mere fictitious instantaneous cut in the four-dimensional 'extensive becoming.'[8]

Bergson then considered the second attempt to explain the original state of maximum 'mutability': "Again it might be supposed that the general instability has arisen from a general state of stability; that the period in which we are now, and in which the utilizable energy is diminishing, has been preceded by a period in which the mutability was increasing, and that the alternation of increase and diminution succeeded each other forever." In other words, that the entropy of the systems is alternatingly increasing and decreasing. The same idea is implied in the hypothesis previously considered; for although it stresses the *simultaneous* existence of the systems with the opposite entropy gradients, it assumes that by their interaction the situation will be reversed: the entropy originally increasing in one system will tend to decrease while the opposite will be true for the other system. Thus the only essential difference between these two hypotheses, which Bergson considered, is that the first one explicitly assumed the infinity of the universe while the second does not. An example of the second was Rankine's model of the universe which assumed not only the finiteness of the cosmic mass, but also the finite extent of the world-aether; the boundaries of the aether were then acting as 'reflecting walls' which prevented the luminous vibrations from escaping into the surrounding void and thus made 'the reconcentration of energy' possible.[9]

In rejecting the second hypothesis, Bergson erroneously enlisted the support of Boltzmann. It is true that Boltzmann spoke about an extremely small probability of the reversion of the entropy slope, and Bergson rather loosely concluded that "the mathematical improbability of it passes all imagination and practically amounts to absolute impossibility". He failed to realize that, as Reichenbach stressed more recently, by the very nature of probability sequences "every combination of attributes that has a non-zero probability must occur with a non-zero frequency."[10] In other words, with the limitless duration at our disposal the reversion of the entropy slope on the local and even on the cosmic scale eventually *must* happen. This, indeed, was Boltzmann's view since only in this way is his dismissal of the unique cosmic time justifiable: "For the universe both directions of time are indistinguishable in the same way as in space no 'up' and 'down' exists.[11] This, needless to say, is completely contrary to Bergson's view and it is rather strange to see him to claim the support of the thinker who so consistently adhered to the corpuscular-kinetic model of the universe.[12] This is even stranger since the rejection of such a model, as we have seen, is one of the cornerstones of Bergson's theory of matter. The only explanation of it is that Bergson overlooked the broader mechanistic context of Boltzmann's view which logically implied the reversibility of time, and, by identifying 'extreme improbability' with impossibility, he involuntarily 'bergsonized' Boltzmann's view. While he was entirely consistent in rejecting the reversibility of time, he needlessly burdened his argument by his claim of Boltzmann's alleged agreement.

In truth, Bergson in the very next sentence, following the quotation from Boltzmann, states clearly that the solution to the problem of the initial improbable state of the maximum energy *cannot* be solved in the terms of the classical, corpuscular-kinetic model:

In reality, the problem remains insoluble as long as we keep on the ground of physics, for the physicist is obliged to attach energy to extended particles, and, even if he regards the particles as reservoirs of energy, he remains in space: he would belie his role if he sought the origin of these energies in an extra-spatial process. It is there, however, that it must be sought.[13]

Does this mean that the initial improbable state of the universe can be explained only in metaphysical, perhaps theological way? Does Bergson advocate the creation of the world in time? It apparently sounds so and it was understood so both by Bergson's disciple Jacques Chevalier as well as by Bergson's opponent René Berthelot. But Bergson's view was more ambiguous than both Chevalier and Berthelot believed.[14] An attentive analysis of the texts will show it clearly.

It is true that the term 'extra-spatial' is to Bergson synonymous with 'ideal' or 'psychological'; thus it was easy for Chevalier to interpret this term in this particular context as "ideal on the cosmic scale", that is, in the sense of the instantaneous divine creation. What confirmed Chevalier in this interpretation was not only the wishful thinking of a loyal Roman Catholic who for years was exerting on Bergson a subtle, continuous and almost successful pressure to join Chevalier's own church; there were some other, less personal reasons. There is no question that Bergson had a very pronounced sympathy toward philosophical *finitism* which rejected the concept of actual infinity as self-contradictory. This view was upheld by Herbart and Eugen Dühring in Germany and by Renouvier's neo-criticism in France. It is true that Bergson never quotes Renouvier; but he highly praised one of his followers, François Evellin whose discussion of Zeno's paradoxes he regarded as "decisive."[15] Now Evellin, like Renou-vier, held the view that Kant's antinomies are solvable, in other words, that they are no antinomies at all, since only the *proofs of the theses* are cogent, while those of the antitheses are spurious.[16] This means that in the first antinomy only the thesis must be accepted, the antithesis rejected: the world, according to Renouvier and Evellin, is finite both in its spatial and temporal extent. The elapsed eternity of past events is logically as impossible as any other actually infinite aggregate. Since Bergson, as we have seen, rejected the infinitude of space, his rejection of the infinity of the past would be only consistent.

Furthermore, if we trust the information supplied relatively recently by Chevalier, Bergson rejected as early as in 1901 *both* spatial and tem-poral infinity of the universe in linking the thesis in the first antinomy with the thesis in the *third* antinomy:

If there is the origin of all, if the duration is finite, there is freedom at the beginning of everything and, consequently, in the things themselves. If, on the contrary, there is no origin of the whole, if the duration is infinite and eternal, if there is no absolute begin-ning, there cannot be freedom within the series itself.[17]

The influence of Renouvier, who never tired of stressing the correlation of infinitism and necessitarianism, seems to be quite probable here; in particular, the term 'absolute beginning' (*le commencement absolu*) was the very term Renouvier repeatedly used. Bergson's rejection of necessi-tarianism thus naturally led him in the direction of finitism and this prob-ably was the reason why he regarded the initial maximum state of entropy as 'zero time', the absolutely first, initial moment, antecedent to every other event, the beginning of the whole cosmic duration.

But Bergson's rejection of the bottomless past or beginningless eternity was more hesitant and more ambiguous than Chevalier wanted us to believe. First, in speaking of the origin of the initial state of the universe he referred not to a single extra-spatial act, but to 'an extra-spatial *process*'. To be aware of this is more than a mere philological pedantry. For the word 'process' (*processus* in the original) suggests a continuous ac-tion rather than an instantaneous act. Furthermore, in the very next paragraph in which Bergson tries to clarify the meaning of that 'extra-spatial process', we find the sentence containing Bergson's philosophy of matter in its maximum conciseness: *"Extension,* we said, appears only as a *tension* which interrupts itself."[18] Needless to repeat our extensive commentaries made about it in Part III of this book. It has been made sufficiently clear that this process, which Bergson calls "the gradual passage from the inextensive to the extensive, from a reciprocal implication of parts to their juxtaposition", is not an instantaneous act since it goes on *all the time*. It constitutes what the traditional metaphysics called "action of mind on matter" and, more generally, it is present wherever the rhythm of duration is 'quickened' in the sense explained in Part III.

This interpretation is strengthened if we take into account another passage from the same section of *Creative Evolution* dealing with 'the ideal genesis of matter.' It precedes only by a few pages the discussion of the law of entropy:

The mystery that spreads over the existence of the universe comes in great part from this, that we want the genesis of it to have been accomplished at one stroke or the whole matter to be eternal. Whether we speak of creation or posit an uncreated matter, it is the totality of the universe that we are considering at once. At the root of this habit of mind lies the prejudice... the idea, common to materialists and to their opponents, that there is no really acting duration, and that the absolute – matter or mind – can have no place in concrete time... From which it follows that everything is given once for all, and that it is necessary to posit from all eternity either material multiplicity itself, or the act creating this multiplicity, given in bloc in the divine essence. Once this prejudice is eradicated, the idea of creation becomes more clear, for it is merged in that of growth. But it is no longer then of the universe in its totality that we must speak.[19]

We can hardly have a more unambiguous rejection of both instantaneous creation and of the eternity of the universe. This apparently implies that Bergson did not regard the thesis and the antithesis of the first Kant's antinomy as logically exhaustive. But is it meaningful to say that the universe is neither eternal nor created in time (or, possibly 'with time' – *cum tempore*, as St. Augustine said)? And then how to reconcile this with Bergson's above mentioned leanings toward finitism?

In this context two important clarifications are in place: one which considers the purely logical and linguistic aspect of this problem; the other which is based on modern cosmogony. As we shall see, they are not unrelated.

First, taking into account all Bergson's utterances relevant to this problem, it is clear that he, indeed, *did accept* the finitistic thesis. But what he *did question* was the *linguistic formulation* of the thesis, in particular the use of the term 'world' which, according to him, illegitimately suggested the connotation of 'the universe in its totality.' The assumption of 'the universe in its totality' was, in his view, the postulate common to both the creationistic and eternalist thesis: in both views, "everything is given once for all', whether it is given in the eternal multiplicity of matter or "in the act creating this multiplicity, given in bloc in the divine essence." In other words, the true dilemma is: "the cosmic past is either finite or infinite", which must not be confused with the false dilemma "the universe is either eternal or was created *at once*."

This is an important distinction which classical cosmogony entirely ignored. Kant's first antinomy was certainly understood as the opposition between creationism and eternalism. Kant himself understood it in this way, and so was it understood by the majority of the nineteenth century physicists and philosophers, whether they favored the thesis or the anti-thesis. The finiteness of the past, suggested by the generalization of the law of entropy, was interpreted as an argument for the creation. This is quite often true even now; in this respect the situation has hardly changed since Stewart-Tait's book *The Unseen Universe* to Sir James Jeans' *The Universe around Us*, to E. T. Whittaker's *The Beginning and the End of the World* or Milne's *Modern Cosmology and the Christian Idea of God*. On the other side, it was the fear of 'supernaturalistic creationism' that motivated the opposition to the generalization of the second law of thermo-dynamics among scientists and philosophers, such as Herbert Spencer, Ernest Häckel, Friedrich Nietzsche and Svante Arrhenius. The case of Eugen Dühring who found the finiteness of the cosmic past compatible with his uncompromising atheism, was an exception, – perhaps the only one.

This situation ceased to be so clear-cut in comtemporary cosmology and cosmogony and nothing indicates their revolutionary character more than the way in which they undermined the very conceptual foundations on which the first Kantian antinomy was erected. As far as space is concerned, the general theory of relativity together with Riemannian geometry brought about a very important distinction: the terms 'infinite' and 'limitless' which were synonymous in classical Euclidian cosmology, are *not* so any longer. Kant, in this respect following Euclid and Newton, used these terms interchangeably as his formulation of the first antinomy clearly shows; the possibility that the universe could be both limitless and finite clearly did not occur to him.

Modern cosmology did nothing analogous at first to the part of the antinomy dealing with time. Time in Eistein's original cosmological model – the so called 'cylindrical' one – is as infinite and without beginning as the Newtonian time. The modification which modern cosmogony brought to the formulation of the first antinomy came later and was of a different kind: it challenged the traditional identification of finitism and creationism described above. But this challenge came only with Lemaître's theory of the expanding universe.

Professor Gonseth in his introduction to Lemaître's book *The Primeval Atom* wrote that its author succeeded in overcoming Kant's first antinomy.[20] It would probably have been more accurate to say that Lemaître showed that the antinomy was formulated in a misleading, inadequate way. Neither the thesis nor antithesis denied the infinity of both time and space. They only disagreed whether *the physical universe* – not space nor time – are limited or without limits; in other words, whether the universe began *in* time (and has its limits *in* space) or not. This is evident in the very wording of the thesis: "The world has a beginning *in* time..."[21] Thus what the first thesis denied was the elapsed eternity of the *material* universe, but not the beginningless eternity of the physically empty duration prior to 'the date of creation.' This was especially clear in the thought of Gassendi and Newton. Thus the so-called finitistic thesis in Kant's first antinomy was in truth *crypto-infinitistic* since it silently assumed the infinity of both space and time. This assumption was based on another basic distinction characterizing classical physics and cosmology: that between the container-like space and time and their physical content. This was the very essence of Newton's absolutism: both space and time because of their independence from their physical content are unlimited even if the physical reality is neither infinite nor eternal. It would hardly be fair to judge Kant too severely for his inability to formulate the first thesis in other terms than those which the Newtonian physics provided – the only ones available at that time. It is hardly necessary to state again that this distinction between the physical 'content' and its spatio-temporal 'container' has been wiped out and in this way the conditions have been created for a completely new reformulation of the first antinomy. From now on the question is not any longer whether *the world* is infinite *in* space and time, but whether *space and time themselves are such.* This re-formulation is a direct consequence of the merging of matter-energy and its changes with the dynamic structure of time-space. We have seen how this reformulation affected the wording and the solution of that part of the antinomy which deals with space. Now let us see how it affects the time-part of the same antinomy.

Einstein's 'cylindrical' model of the universe was of a short duration. It still retained two classical features; it claimed that on the cosmic scale both space and time could be regarded as separable and thus in a sense *absolute;* time, in addition, was infinite in both directions. Space, being Riemannian, possesses a constant radius of curvature since the fluctuations of its curvature on the local scale which account for the phenomena of gravitation can be disregarded when the universe at large is considered. Thus, in Whitrow's words, "relativity is reduced to a local phenomenon."[22] Einstein's model was thus static; its short life was due first to the mathematical proof of its instability, but, primarily, to the empirical discovery of the recession of the nebulae and Hubble's law. This led to its replacement by various dynamical models of the expanding universe.

Lemaître's model was one of these. It rather boldly linked the recession of the galaxies with the second law of thermodynamics, reinterpreted in the terms of quantum theory. While the total amount of energy remains the same, it is continually being fragmented into an ever increasing number of quanta, each of which possesses lesser and lesser energy and, consequently (according to Planck's relation $E = hf$) a longer and longer wave length. The dissipation of energy consists precisely in this continual fragmentation. "If we go back in the course of time we must find fewer and fewer quanta, until we find all the energy of the universe packed in a few or even in a unique quantum."[23] In this way Lemaître reached his conclusion about the temporal beginning of the universe: the world history began by a 'super-radioactive explosion' of the original single quantum and the subsequent development of the universe is a continuation of this process of fragmentation into the increasing number of less energetic quanta. Since, according to Einstein's equation $E = mc^2$, to every energy corresponds a certain mass, Lemaître's original superphoton can be equally well called "the primeval atom" – the term which the title of Lemaître's book made fairly familiar.

This was not the only consequence of the relativistic physics in this cosmogony. Another one was that the expansion of space and the explosion of the primeval superquantum had to be regarded as *two aspects of one and the same process*. This was an obvious consequence of the relativistic fusion of time-space with its physical content. The explosion of the primeval atom did not take place *in* some pre-existing container-like space, nor was it an event which took place *in time*, but space and time themselves came into being *with* this initial, privileged event. This was an obvious consequence of the relational theory of time-space, that is, its inseparability from its physical content. To ask what was *before* the 'zero time' is as meaningless as to ask what is 'behind' Lemaître's elliptical space. Such question would be meaningful in the absolutist Newtonian scheme which explicitly admitted the reality of physically empty space and time existing *prior* to the physical universe; but it is clearly devoid of meaning in *any* relativistic cosmology.

The main reason why we dwelt on Lemaître's cosmogony is that of all contemporary cosmogonies it comes closest to that sketched by Bergson in those few pages of *Creative Evolution* to which we referred. In the first place, it is *finitistic*: like that of Bergson, it rejects the infinity of both space and time. At the same time, it is not creationistic in the traditional sense: according to both Bergson and Lemaître, the universe was created neither instantaneously nor *ex nihilo* since it has come into being by a *gradual process*. Although this process had a definite beginning, it did not come out of some antecedent physical and temporal void such as that postulated by Newton. Hardly anyone else criticized the concepts of nothingness and of temporal void ('le temps homogène') more severely than Bergson. Bergson stressed the indeterminacy of the future; so did Lemaître when he wrote:

Clearly the initial quantum could not conceal in itself the whole course of evolution; but, according to the principle of indeterminacy, that is not necessary. Our world is now understood to be a world *where something really happens*; the whole story of the world need not have been written down in the first quantum like a song on the disc of a phonograph. (*Italics added.*)[24]

The basic difference between the classical relativistic cosmology and that of Lemaître is contained in this passage. The former was deterministic since it was believed that the microphysical indeterminacy can be safely disregarded on the megacosmic level; the latter by its very nature tried to bridge the gap between the megacosmic and microcosmic levels. For such a gap has not existed in the past; the universe originated from the primeval superquantum which was subject to the principle of indeterminacy.

Thus the italicized words *a world where something really happens* have more than a mere Bergsonian ring. The dynamic aspect of the physical reality is as prominent in Lemaître as in Bergson to the extent that even *space itself is incorporated into the process*. The dynamization of space, suggested by a proper interpretation of the Einstein's time-space, is further accentuated in Lemaître's cosmogony; not only space in virtue of its expansion has no rigid immutable structure, but it is itself of derivative character, generated by the first cosmic event. Within the original undivided superquantum prior to its explosion, there was no spatial diversity at all; it began to appear only with the incipient fragmentation of the primeval atom. In Bergson's thought spatiality was also of derivative nature. By defining matter as "the tendency toward homogeneity and juxtaposition", Bergson tried in *Matter and Memory* to dim out a too sharp distinction which he drew between duration and extension in the first book. This was conceptually possible only by incorporating spatiality into duration, that is, by denying the reality of geometrical durationless space. Such space, as explained in Part III, is a mere ideal limit toward which physical reality *tends* without ever attaining it completely: "in conferring on matter the properties of pure space, we are transporting ourselves to the terminal point of the movement of which matter simply indicates the direction."[25] Similarly, in Lemaître's model, the Euclidian space is a mere ideal limit of the cosmic evolution, never entirely attained: although the radius of curvature of the elliptical space is perpetually increasing, it will never become infinitely large. In other words, the continually diminishing curvature of space will never entirely disappear as, in a two dimensional analogy, the surface of an ever increasing sphere will never become altogether flat. The march of physical reality toward a complete, though never entirely attained homogeneity of Euclidian space seems to be a feature common both to *The Primeval Atom* and *Creative Evolution*.

This does not mean that Bergson anticipated the theory of the expanding space in any specific way. Such merit belongs to another Frenchman – A. Calinon who envisioned the possibility of a variable space constant as early as in 1889.[26] But in 1907 when *Creative Evolution* was published, the general relativity theory still was not in existence; in truth, even when it was formulated, Bergson failed to be interested in it and to see the convergence of some of its conclusions with his own.[27] Thus we should not be deceived by some of Bergson's metaphors as, for instance, when he wrote about "the center from which worlds shoot out like rockets in a fire-works display".[28] The similarity with Lemaître's image of the exploding atom is indeed, striking, making the sarcasms of Berthelot and Bertrand Russell far less effective than they appeared before the First World War.[29] Yet, since Bergson clearly did not have in mind any non-Euclidian space, this similarity is largely coincidental. But *not entirely* coincidental: two general philosophical themes – the finiteness of space and time and, in particular, the incorporation of space into the cosmogonic process – are, without doubt common to both Bergson and Lemaître. It was his emphasis on process and his general distrust of the classical mechanistic models that led Bergson's thought in the *general direction* of Lemaître's cosmogony.

Furthermore, there was the *third* common philosophical theme which was really only a specific form of the seond one: the general tendency of matter toward homogeneity. As mentioned above, Bergson correlated this tendency with the increase of entropy, interpreted in a language free of the corpuscular-kinetic terms. Such interpretation was then not exceptional; it was upheld by Mach, Duhem and, in general, by the energeticists. Mach in his *Die Principien der Wärmelehre*, written a decade before *Creative Evolution*, called the corpuscular interpretation of the entropy increase as an increase of molecular disorder 'highly artificial' (*recht künstlich*)[30] and welcomed F. Wald's hint that the roots of the entropy law lie far deeper. Such views, though prophetic in a long run, nevertheless belonged to what now appears as an *exaggerated reaction against mechanism*, exaggerated because it failed to recognize the fruitfulness of mechanistic models on the molecular level. Bergson, in welcoming this premature reaction against mechanism, failed to realize that the successes of the kinetic theory of heat are perfectly compatible with his philosophy and, in particular, with his biological epistemology. For the molecular level still lies on the boundaries of the 'realm of the middle dimensions' where the inadequacy of the corpuscular-kinetic models is not yet visible. The real inadequacy of these models showed itself clearly *only on the submolecular and subatomic level.*

What appeared particularly doubtful in 1907 was Bergson's equation of the increase of entropy with the tendency toward homogeneity in nature, that is, with a gradual elimination of all physical differences. René Berthelot, writing a few years later, reminded Bergson that the increase of entropy does not eliminate the differences in the chemical constitution of bodies.[31] Berthelot failed to realize that the immutability of chemical elements ceased to be a dogma after the discovery of radioactivity; but he could hardly have anticipated that the subsequent development of physics would make questionable the very concept of immutable particle even on the intranuclear level. (In truth, his attitude toward modern physics remained exceedingly negative as late as in 1934, when the inadequacy of the classical models had become undeniable. In fairness to him, it must be said that this was true of some other traditional rationalists such as Brunschwicg in France, Driesch in Germany, Blanshard in the U.S.[32]) What above all was not anticipated in 1912 was a *far more general interpretation of the second law of thermodynamics* which regards the gradual elimination of thermal differences as a mere special case of the passage from less probable to more probable states in nature. Although the probability considerations played a very important role in classical thermodynamics, the assumption of the permanency of the basic material particles was retained and Boltzmann himself regarded it as indispensable. In this respect even the electron theory of matter was regarded as a new and improved version of atomism and its successes were hailed as a decisive refutation of the doubts expressed by Stallo, Mach, Ostwald and Duhem which were inspired more by epistemological reasons than by observations.

The situation is obviously different now and it would be otiose to repeat what had been said about the crisis of the concept of substantial particle. In this way the conditions are created for a new interpretation of the second law of thermodynamics which is free of the corpuscular language in a way similar to that sketched in Bergson's passage quoted above. As mentioned there, within the framework of classical physics the process of gradual dissipation of energy could be characterized as 'homogeneization' only in a restricted sense: it would be merely the tendency to an equalization of temperatures. Even if we assume that all celestial and macroscopic bodies were eventually transformed into a molecular or atomic dust with the temperature approaching absolute zero, one fundamental distinction would be preserved: that between space and matter, between the 'full' and the 'empty'. For this distinction was the very basis of the classical concept of matter – and matter itself, no matter how far pulverized, would not disappear; the basic particles were supposed to persist everlastingly even in the 'thermal death' of the universe. In other words, the law of conservation of matter was perfectly compatible with the law of entropy. Today, after the coming of the relativity theory and the theory of quanta, the situation is significantly different. First, the special theory of relativity overcame the classical distinction between mass and energy: to every energy belongs a certain mass and vice versa. But then does not the law of dissipation of energy apply also to the *internal energy* which is contained in every material particle, including the electrons and nucleons? An affirmative answer would be only consistent and its plausibility is increased when the same question is approached from the standpoint of quantum theory. As mentioned above, in the language of the quantum theory the process of the dissipation of energy means that there is a gradual fragmentation of energy into smaller and smaller quanta. In Jeans's words, the increase of entropy is a special case of the tendency of every radiation to *lengthen its wave length*.[33] But according to wave mechanics, every *energy*, including the internal energy of the material particles, is of a *vibratory* kind; the difference between the electromagnetic radiation and matter itself is that the former is characterized by a smaller frequency and correspondingly longer wave length. If the law of dissipation of energy means a lengthening of the wave length of *every* kind of energy, it must apply to matter itself; which means a gradual transformation of matter into radiation.[34]

This was suggested by Sir James Jeans even prior to the discovery of the dematerialization of positive electrons. It is true that the opposite process – the materialization of photons – was discovered at the same time; but it is far less frequent. Thus in the words of Jacques Merleau-Ponty, "Einstein's famous equation $E = mc^2$ conceals an essential dissymmetry."[35] Within the framework of the generalized law of entropy this is understandable: the electrons and nucleons represent *more organized* and, consequently, *less probable* structures than the photons and thus the transformation of matter into the field is far more probable and, therefore, more frequent than the opposite process. This continuous transformation of matter into radiation – 'into field', as Weyl would say – must be considered jointly with the expansion of space: the waves into which 'the waves of matter' (*les ondes matérielles* of de Broglie) are lengthened are spreading into the ever expanding elliptical space. In truth, from the standpoint of the general relativity theory, the preposition 'into' is misleading since it is tinged by the outdated classical distinction between a container-like space and its physical content. It is far more appropriate to say that the expansion of space and the dilution of its energetic content go on together, both being two probably complementary aspects of one and the same process since they both result in a growing homogeneization of the physical universe.

There is no place here to dwell on empirical difficulties of Lemaître's theory. The very fact that they exist explains why other cosmogonic models have been proposed since then. They fall into the three main categories: (a) *other evolutionary models*, according to which the universe either began or at least went through a major reorganization in the past; (b) *oscillating models* which assume that there are the alternating phases of the expansion and contraction. It is clear that the boundaries between these models are far from definite since it is always possible to assume that prior to the so-called 'zero-time', assumed by Lemaître, Jeans, Milne and Whittaker, there was the period of contraction preceded by another expansion … and so forth *in infinitum*. Thus the oscillating models assume the infinity of the past; in this respect they agree with the *third* group (c) the *stationary or steady state theories*, proposed by Bondi, Gold and Hoyle. This group is closest to the eternalistic tradition of classical physics and classical philosophy; the so called 'perfect cosmological principle' is nothing but another formulation of the principle of 'the unity of nature in space and time', proclaimed by Bruno and Spinoza at the beginning of the classical period and since then continually reaffirmed. (In truth, nature is one even according to the finitistic models of the universe since there is literally nothing 'beyond' the elliptical space or 'before' zero time.)

Will there be any decisive *empirical* test that would settle forever the dispute between finitism and infinitism? While the situation does not appear altogether hopeless concerning space, no decisive test is conceivable which would bring a definitive solution as far as time is concerned. It is true that the steady state theory in claiming that the universe had always the same appearance as it has now, implies consequences which are in principle falsifiable. But even if we assume that the observation of very distant galaxies and of their distribution would decisively falsify the steady state theory, the finiteness of the past would still remain a mere hypothesis. A scientist or a philosopher resolutely committed to infinitism, can always postulate that prior to the so-called 'zero-time' there were the alternating cycles of contraction and expansion without any beginning. It is obvious that such a theory cannot be falsified; this was the reason why Gamow, whose views superficially appear to be finitistic, dismissed any question about the pre-expansion state of the universe as meaningless – as meaningless as the question "what was God doing before the creation of the world."[36] But Gamow apparently conceded that such a stage *had existed*; in other words, his model, appearances notwithstanding, is eternalistic.

With the impossibility of any conclusive empirical test the only way to approach this problem – that is, if it should be approached at all – is on a metaphysical or logical level. Since Bergson's leaning toward finitism was, as we had said, very probably influenced by the arguments of the French thinkers Renouvier and Evellin, it will be worth while to re-examine these arguments and confront them with those of their opponents.

The finitistic thesis, following Aristotle in this respect, holds that the only legitimate concept of infinity is that of an *open* or *potential* kind; in this sense, time is potentially infinite since none of its moments is the last one; there is always a future beyond each provisionally last moment, and in the inexhaustibility of this future consists the infinity of time. On the other hand, 'actual infinity' is a contradiction in terms since it joins two incompatible features – completeness and inexhaustibility. Any infinite aggregate is both a whole and a negation of the whole which is impossible. Renouvier pointed out that Boethius' famous definition of eternity exhibits this contradiction in its very wording: *Interminabilis vitae tota simul et perfecta possessio*.[37] Now 'interminabilis' is clearly incompatible with 'tota simul'; nothing can be both at once ('simul'), that is complete, and at the same time incomplete. (In truth, Boethius uses even a stronger term: 'interminabilis' means *incapable* of completion.) From this point of view the only meaningful eternity is that of potential, *futuristic* kind; *static* eternity, such as the elapsed eternity of the past moments, is impossible. For this reason, in the first antinomy of Kant it is the *thesis* which is correct and the antithesis which is wrong and the finitists certainly did accept the concluding words of Kant's argument for the thesis: "An infinite aggregate of actual things cannot therefore be viewed as a given whole, nor consequently as simultaneously given." [38]

As mentioned above, modern finitists understand the first thesis in a sense far more radical than Kant did: not only the universe, but *time itself* is finite since it is coextensive with the finite duration of the universe. In adopting the relational theory of time (and space), the post-Kantian finitists effectively answered the objection which Kant raised in his 'proof' of the first antithesis: that the first thesis implies the existence of empty time prior to the beginning of the world. This, as mentioned already, was true of Kant's own formulation of the thesis which was phrased in a Newtonian language; but it is certainly not true of the thesis once it is formulated in the terms of the relational theory. The cogency of the answer which the relational theory gives to Kant's objection was recently recognized by C. D. Broad:

> Then to say that the world had a beginning is simply to say that there was a certain event which was followed by others but was not preceded by any other event. To say that this event would "have been preceded by empty time" would come to this. It would amount to saying that it is logically possible that there should have been events which preceded the event which was *in fact* the first event. On this relational view of time the question: "Why did the world begin when it did, and not at some earlier or later moment?" would reduce to the question: "Why did the particular event, which in fact had no predecessors, not have predecessors?" [39]

Such question is based obviously on a *petitio principii*; the assumption of beginningless time is present in its very wording.

Among those who gave serious attention to the French neocriticist rejection of actual infinity was Louis Couturat whose book *De l'infini mathématique* appeared in the same year as *Matière et mémoire* (1896). Since this book was written *after* the epoch making work of George Cantor which he takes into account, it is instructive to survey the essence of his defense of actual infinity, especially since contemporary infinitists hardly improve upon it.

Couturat presented the controversy in the form of a fictitious dialogue between the critics and the defenders of actual infinity. It is only natural that he identified himself with the defenders, even though he tried to present the finitistic views fairly and accurately. The main finitistic argument against the existence of infinite aggregates was the impossibility of their enumeration; but such an objection remains valid only as long as the successive genesis enters into their definition. Hence the necessity of a modified definition from which any reference to successive construction is eliminated. This is the meaning of Couturat's statement when, in the true spirit of Cantor, he says that infinite aggregates exist *prior to, and independent of* the enumeration of their members. Yet, Couturat is not very consistent in this approach to the problem. Only two pages after the passage in which he insisted on the independence of infinite wholes from enumeration, we come across the following curious lines:

When you say that an infinite aggregate can never be completely enumerated, such impossibility is neither intrinsic nor logical, but practical and material: *it is merely a question of time*, (Italics added.) [40]

Does it mean that an infinite aggregate *can* be enumerated, i.e. its last ∞th term *can* be reached? It is true that Couturat insists that this cannot be done in a finite time; in the sentence immediately following the passage quoted above, he adds: 'Give me an infinite time and I take upon myself to enumerate an infinite aggregate." Couturat apparently does not realize that to say: "I shall complete this operation after an infinitely long time" is equivalent to: "I shall *never* complete this operation." The last sentence clearly shows that the illusion of actual infinity arises from the psychological fusion of two incompatible meanings: the first, suggesting the completion of the operation ('I shall complete...') and the second, suggesting its incompleteness ('never').

The same contradiction comes up in another passage which has a direct bearing on the problem of the infinity of the past. Only one page after the quoted passage Couturat castigated Jean Bernoulli for 'imprudently conceding' that in any infinite sequence there is the ∞th term "in the same sense that in the collection of ten terms there is the tenth term." [41] This view, according to Couturat, is based on a false analogy: from the fact that there is an infinite *cardinal* number of the terms in the sequence, it does not follow that there is the last, ∞th (*infinitième*) *ordinal* number in it. It is not difficult to see that this passage directly contradicts his claim on a previous page that an enumeration of the infinite aggregate is possible, which means – if it means anything at all – that the last ordinal number *can* be reached. Couturat would undoubtedly stress that it can be reached only 'after an infinite interval of time' which means 'never', – the conclusion with which the finitists would heartily agree. But why then speak of 'completion' at all?

Couturat's criticism of Bernoulli was much more recently restated by Adolf Grünbaum; he insists that "it is the very essence of a progression not to have a last term and not to be completable in that ordinal sense!"[42] Neither Couturat nor Grünbaum nor any other infinitist realized that this argument, if accepted, is fatal to the doctrine of the beginningless past. For, if the totality of past events is an actually infinite whole, it should not have any attainable last term. Yet the same theory holds that this last, infinitieth and supposedly unattainable term is actually *attained* in the present moment in which I am living and, furthermore, that *it has been attained* in any previously present moment of the universe! Grünbaum would probably retort by insisting that 'now' has a mere subjective status, like becoming, from which it is inseparable; the objective physical world is devoid of either of them. It would be otiose to repeat my criticism of the neo-Eleatic metaphysics lurking behind such views.

The self-contradictory character of elapsed eternity is the main reason why the finitists – some rephrasing apart – accept Kant's proof of the thesis in the first antinomy. How badly Kant's proof can be misunderstood even by an outstanding mind is shown by Bertrand Russell. He accused Kant of committing 'an elementary blunder'; but it is clear from the following comment that it was Russell who completely missed the essential part of Kant's argument:

Kant, in his first antinomy, seems to hold that it is harder for the past to be infinite than for the future to be so, on the ground that the past is now completed, and that nothing infinite can be completed. It is very difficult to see how he can have imagined that there was any sense in this remark; but it seems most probable that he was thinking of the infinite as 'unended.' It is odd that he did not see that the future too has one end at the present, and is precisely on a level with the past.[43]

A strange passage indeed, whose purported meaning is revealed only in the last sentence: Russell clearly believes in a complete symmetry between the past and the future and in a complete subjectivity or irrelevance of the 'direction of time'; only thus he could call the present the 'end' of the future! In this way he missed the essential part of Kant's argument – the distinction between the *potential* infinity of the future contrasting with the *actual* infinity of the past, since for him, both are ontologically on the same level: in truth, according to him they are both equally *actual* in the Eleatic sense:

His [Kant's] regarding the two as different in this respect illustrates just that kind of slavery to time which, as we agreed in speaking of Parmenides, the true philosopher must learn to leave behind him.[44]

What is characteristic of Russell is that his thinking, despite its highly abstract character, is hopelessly tinged by visual elements. When he thinks of interval of time, he unwittingly *sees* in his imagination a *geometrical* interval, that is, a straight segment bounded by two points; and since their order is spatial and completely devoid of succession, they both can be called 'ends'. But it would obviously be not only odd, but plainly incorrect to call the year 1872 the 'end' of Russell's life. Yet, this is what Russell really says when he calls the present 'the end of the future.' It is always the same old confusion of succession with its static geometrical symbol which Bergson never tired of criticizing.

The same geometrical symbolism underlies another prejudice which has a direct bearing on the controversy between finitism and infinitism. It is clear that if time can be symbolized by a straight Euclidian line, it can be extended in either direction; from this the inconceivability of the finiteness of the past follows. Kant's first antithesis is precisely based on the inconceivability of the 'zero-time' or of The First Moment; for – Kant argues – if the universe began, then there must have been empty time *before* its beginning. (In truth, this is what Gassendi and Newton did believe, even though their original temporal void was only *physically*, not metaphysically empty, being filled by the divine everlastingness.) Augustine's rejoinder that the universe began *with* time, not *in* time, convinces only a few; yet, it remains unanswerable since within the framework of the relational theory of time (which is inseparable from finitism) it is meaningless to posit the temporal void *prior* to time. Yet, our unconsciously Newtonian view of time as a linear receptacle, unlimited in either direction and existing prior to any concrete events stubbornly resists rational argumentation. Even C. D. Broad's lucid restatement of the meaninglessness of the objections against the relational finitistic theory of time was strangely misunderstood as an attempt 'to block the way of possible inquiry,'[45] to wit, the inquiry into what was *before* the zero-time. Such misunderstanding of Broad's argument is based on a *petitio principii*; time is simply *assumed* to be infinite in the direction of the past and its limitation in this direction appears as impossible as any arbitrary boundary of Euclidian line.

But probably the strongest motive behind the reluctance to admit the possibility of the finite past is the apprehension that it will be interpreted in a theological sense. This was in a sense natural; for only too often has 'the beginning of the world' been interpreted as the decisive argument for its creation, especially in the English speaking world. We mentioned the names of B. Stewart, G. P. Tait, James Jeans, E. T. Whittaker and E. A. Milne. It is significant that pope Pius XII welcomed the evolutionary cosmology for the same reason: he regarded it as conclusive, scientific evidence for the theistic creation.[46] On the other hand, it is only fair to see the other side of the coin. The antitheological prejudice was and is equally strong. It often lay behind the reluctance of the cosmogonists to extrapolate the validity of the law of entropy to the whole universe and it was certainly conspicuous in the thought of Häckel and Nietzsche; the universe *must* be eternal, – if we do not want to capitulate to theologians.

Under such conditions the discussion clearly ceases to be dispassionate and fruitful. Fortunately, we now can hear an increasing number of voices insisting that this cosmological controversy should be separated from theological and religious issues. Jacques Merleau-Ponty pointed out that Eddington was always opposed to the theological interpretation of the initial state of the universe and that, more surprisingly, the same attitude was eventually adopted by Lemaître. More recently, Ronald W. Hepburn denied that the finitistic cosmogony is theistic in its strict interpretation; the most it is entitled to assert is that the world 'just started to be'.[47] Of course, finitism in cosmogony is *compatible* with theism, even though it does not *imply* it. But according to the old tradition going as far back as Origen, theism is equally compatible with the eternity of the universe – a clear indication that the question of the finiteness of the past can be dissociated from theology. This is virtually conceded even by Milne; while he equated 'zero-moment' with the act of creation, he stressed that this act itself is 'for ever invisible',[48] transcendent to our experience, since it is the event preceding all other events and in this sense 'prior to physics'. The last expression is that of Lemaître who said that "physics begins with the multiplicity, and no physics is possible in the absolute simplicity of the primordial being (*dans l'absolue simplicité de l'être initial*). Now time is a multiple being." While Lemaître interpreted this 'absolute simplicity of the primordial being' in an Aristotelian-Thomistic sense, he, more cautious than Milne, regarded it as his own private interpretation and conceded to the materialist the freedom 'to deny any transcendent Being'.[49]

It is significant that, as already mentioned, the disbelief in the eternity of the past did not convert Eugen Dühring to theism and that Herbert Spencer labelled the doctrine of 'the universe creating itself' as pantheism.[50]

All this indicated that the doctrine of the finite past should be considered on its own merit and in comparison to its rival theory. Its opponents must state their theory in a more satisfactory language; they must avoid the contradiction which they commit in saying that the infinite ordinal number – the last, $\infty$th term – is unattainable by its own nature and at the same time that it is being attained in every actual moment of the universe. They must realize that finitism is based on *the relational theory of time* which makes it invulnerable to any objection based on the absurdity of empty time, supposedly preceding the zero-moment. They must face the fact that certain questions *are* unanswerable because they are meaningless: it is probably meaningless to ask what is 'inside' the electron, especially if we interpret it, as Riemann and Weyl suggested, as a 'hole' in space; it is equally meaningless to ask what was 'before' time since the relation of 'before' is a temporal relation and cannot be applied outside of time. I fail to see any difference between this question and that which naively asks what is 'behind' the spherical or elliptical space.

From the standpoint of the biological theory of knowledge the alleged logical necessity of actual infinity is a combined effect of *macroscopic conditioning* and of *the illusion of spatialization:* since physical space on the human level appears approximately very Euclidian, the second axiom of Euclid – the possibility of extending a straight line in either direction – is accepted by our imagination and extrapolated beyond any limit; and since time is customarily symbolized by an Euclidian line, the infinity of time appears to us as inevitable as the infinity of space. By extending such line from the 'now'-point 'backwards' indefinitely, we obtain the beginningless eternity; by extending it 'forwards', another illusion – very much discussed in this book – arises, the illusion of the ready made, prefabricated future, whose apparent 'becoming' is, in Grünbaum's words, merely 'coming into our awareness'. If we realize the limited empirical adequacy of the Euclidian geometry, and, in particular, if we become aware of the thoroughly misleading character of all spatial symbols when they are applied to time, both illusions are dissipated. It is true that 'the line being extended' conveys, despite its inherent limitations, the *potential* infinity or, rather *indefiniteness* of the future; for the successive character of our operation of extending mirrors, so to speak, the successive character of reality itself; the fallacy of the pre-existing future arises only if we substitute *the already drawn line* for the line in the process of being extended. But a backward extending of the line from the present to the past runs, so to speak, in the sense 'opposite' to that in which the present came into being and in this sense, it is doubly inadequate. In truth, considered psychologically, even this operation of 'backward extending' like any other mental operation is forward-going, 'future-oriented', and the 'infinity' of its future, is only potential, like the infinity of any other process. The idea of the elapsed eternity of time is thus nothing other than the potential infinity of our mental operation fictitiously projected into the past.

Whether Bergson's commitment to finitism – which has been implicit rather than explicit – was related to his theism is uncertain. It is not impossible, especially if the neo-criticism of Renouvier and Evellin had not been without influence on him; in their thought finitism and theism were associated very closely. On the other hand we must not forget that Bergson used the theistic language for the first time only in *Creative Evolution* when he spoke of God 'as a center from which worlds spring' provided that this center is regarded not as a thing, but as a continuity of springing (*jaillissement*). This language with its Plotinian overtones and by the usage of the present tense suggest the idea of continuous creation rather than a single act in the past. One has the same impression when reading Bergson's letter to Father Joseph de Tonquedec dated February 20, 1915; while he explicitly rejected monism and pantheism[51], his language again suggested a continuous divine creation and there was no reference to the primordial creative act of which he spoke in his conversation with Jacques Chevalier more than a decade before. Certainly, the final and more explicit stage of Bergson's theism in his *Two Sources of Morality and Religion* was hardly determined by his leanings toward finitism in his cosmogony. It is characteristic that in his brief restatement of the central ideas of *Creative Evolution* in 1932, Bergson failed even to mention entropy.[52]

As far as the future of the universe and of evolution is concerned, Bergson's views do not seem to be unambiguous; but this is certainly due more to one central feature of Bergson's philosophy than to its own intrinsic ambiguity. This central idea is the open, indeterminate character of the future: one might say that the ambiguity of the future is the least ambiguous idea of Bergson's thought. His rejection of both mechanism and finalism in *Creative Evolution* as well as his rejection of historical determinism in *Two Sources* were inspired by this conviction: that the future outcome of the cosmic history, including the history of the earth, of its life and of mankind itself is not preordained either in the mechanistic or in the theological sense.

In the light of this central idea some apparent discrepancies in what may be called 'Bergson's eschatology' disappear. In the lyrical passage at the end of the third chapter of *Creative Evolution* Bergson in highly poetic terms describes life, 'vital impetus' as overcoming every obstacle, 'perhaps even death'.[53] This has always been regarded as a sure indication of Bergson's optimism and as such it was welcomed, among others, by Bernard Shaw; it certainly stands in a stark contrast to the current twentieth century pessimism as it found its most eloquent expression in Russell's gloomy vision of the hostile universe and of the ultimate doom of life in his *Free Man's Worship*. (This was written in 1903 when Russell still believed that man, "his origin, his growth, his hopes and fears, his loves and beliefs, are but the outcome of accidental collocation of atoms"[54]; although he later, under the pressure of new facts, gave up his original Lucretian view, he retained his basically pessimistic outlook until his death.) Bergson himself in the passage, referred to above, admitted that his own doctrine 'does not only facilitate speculation; it gives also more power to act and to live."

Yet, to call Bergson an 'optimist' is certainly an oversimplification. There are some lines in the very same chapter of *Creative Evolution* with a distinctly pessimistic overtone. In characterizing the *ectropic* (anti-entropic) nature of life, Bergson writes as follows:

It [i.e. life] has not the power to reverse the direction of physical changes, such as the principle of Carnot determines it. It does, however, behave absolutely as a force would have behaved which, left to itself, would work in the inverse direction. Incapable of *stopping* the course of material changes downwards, it succeeds in *retarding* it. The evolution of life really continues, as we have shown, an initial impulsion ...[which] has determined the development of the chlorophyllian function in the plant and of the sensory-motor system in the animal, brings life to more and more efficient acts by the fabrication and use of more and more powerful explosives. Now, what do these explosives represent if not a storing-up of the solar energy, the degradation of which energy is thus provisionally suspended on some of the points where it was being poured forth? The usable energy which the explosivie contains will be expended, of course, at the moment of explosion; but it would have been expended sooner if an organism had not happened to be there to arrest its dissipation, in order to retain and to save it up.[55]

Does not this then mean that the dissipation of energy will ultimately take place and will prevail over the 'ectropism' of life?[56] And does not this contradict a biological overcoming of death, envisioned by Bergson? The only answer to this question, though not explicitly given is implicitly present in Bergson's philosophy, which certainly transcends the simplistic dichotomy 'optimism versus pessimism'. The first term usually suggests the inevitability of some metaphysical 'happy end', whether it is the inevitable victory of God, of Good or of Life. When he wrote that '*perhaps* even death' will be overcome, the word 'perhaps' is crucial; it is one of the *possibilities* in the open future of mankind and of organic life, – not any necessity. In this respect Bergson's vision of 'creative evolution' differs essentially from the 'inevitable progress' of Spencer's mechanistic evolutionism and classical 'laissez-faire' liberalism. Bergson repeatedly warned against such illusions. He pointed out that in the evolutionary process "failure seems the rule, success exceptional and always imperfect". Even the exceptional success of evolution – the human species – does not have its future guaranteed and its progress assured; the last words of *Two Sources of Morality and Religion*, quoted in the last chapter of Part III, betray a great concern in this respect. The transition from the closed to the open society on which the future of mankind clearly depends, is *not* a historical necessity; there is still a terrifying possibility that mankind, split and fragmented into closed societies antagonistic to each other, may eventually destroy itself completely.

But it is obvious that any fatalistic pessimism is equally incompatible with the philosophy of creative evolution. The extinction of human race as well as that of organic life on the earth is a *possibility* only, not a predetermined and unavoidable event. Furthermore, life is, according to Bergson, probably not confined to this planet and may exist in other solar systems "under forms of which we have no idea, in physical conditions to which it seems to us, from the point of view of our physiology, to be absolutely opposed".[57] Under such conditions it is conceivable that the ultimate dissipation of energy may be not only delayed, but delayed indefinitely. This would be in conformity with the statistical character of the law of entropy which was recognized even by classical physics; it would be rather odd if this statistical, i.e. non-necessitarian, character were not recognized by modern physics which regards *all* macroscopic laws as merely statistical. It would have been equally odd for Bergson, opposing historical inevitability on either the planetary or cosmic scale, to ignore it. Furthermore, he explicitly raised the question whether we have the right "to extend to the entire universe considerations drawn from the present state of our solar system".[58] It is true that his questioning of the universal applicability of the law of dissipation of energy was based on considerations completely different from those which inspired Boltzmann. Boltzmann's doubts were inspired by his kinetic-corpuscular view of nature and by the reversibility of all mechanical processes. The only alternative to the 'thermal death' of the universe was for him the everlasting cyclical process; for Bergson, it is an unlimited, irreversible evolution with possibilities of local regresses and even disasters, but with no necessity of final and ultimate death of life.

In Bergson's *broader* metaphysical perspective the ultimate death of life is utterly inconceivable. Even if all its organic forms were destroyed, the physical events would remain – and let us remember, that these events are still *proto-mental*, i.e. *proto-vital* in their character. Furthermore, the destruction of all organisms would mean only the destruction of all *manifestations* of life and not of the principle of life itself; this distinction is essential for understanding Bergson's modified dualism. This point had been discussed in some chapters of Part III and in Appendix II; it would, certainly deserve a far more extensive discussion, at least as extensive as Bergson's theory of matter. But such discussion is clearly beyond the scope of this book.

## NOTES

[1] H. Bergson, *Écrits et Paroles*, II, Presses Universitaires de France, Paris, 1959, pp. 311–312. Meyerson dealt with the second law of thermodynamics in Chapter VIII of *L'Identité et réalité*; later (1921) in *De l'explication dans les sciences*, 1, pp. 198–210.
[2] *C.E.*, p. 265.
[3] *C.E.*, p. 21.
[4] *C.E.*, pp. 265–266.
[5] *Vorlesungen über die Gastheorie*, Leipzig 1898, p. 256. In the English translation of Boltzmann's book (*Lectures on the Gas Theory*, transl. by Stephen C. Bush, Univ. of California, Berkeley and Los Angeles, 1964, p. 445), 'das Treibende' is translated rather misleadingly as 'goal'.
[6] Svante Arrhenius, *Worlds in the Making* (transl. by H. Borns), Harper & Bros, New York, 1908, p. 209; L. Boltzmann, *op. cit.*, pp. 256–258.
[7] H. Poincaré, *Leçons sur les hypothèses cosmogoniques*, Paris 1911, Ch. X, esp. p. 254.

[8] *C.E.*, p. 266.

[9] W. Rankine, 'On the Reconcentration of the Mechanical Energy in the Universe', *Phil. Mag.* **IV** (1852) 358–360.

[10] *C.E.*, p. 267; Boltzmann, *op. cit.*, p. 254; H. Reichenbach, *The Direction of Time*, Univ. of California Press, Los Angeles, 1956, p. 111.

[11] Boltzmann, *op. cit.*, p. 257.

[12] 'Über die Unentbehrlichkeit der Atomistik in der Naturwissenschaft', in Boltzmann's *Populäre Schriften*, Leipzig 1905, pp. 141–157.

[13] *C.E.*, p. 267.

[14] Jacques Chevalier, *Entretiens avec Bergson*, Paris 1959, pp. 5–6; R. Berthelot, *Un romantisme utilitaire*, II, Paris 1913, p. 241.

[15] *C.E.*, p. 338.

[16] F. Evellin, *L'Infini et quantité*, Paris 1880, pp. 203–208.

[17] J. Chevalier, *op. cit.*, pp. 5–6. The term 'absolute beginning' as well as the insistence on the correlation between determinism and infinitism can be found in nearly all Renouvier's writings, including one of his last books *Les dilemmes de la métaphysique pure*, 2nd Ed., Paris 1927, esp. in Ch. IV and VI.

[18] *C.E.*, p. 267.

[19] *C.E.*, p. 262–263.

[20] G. Lemaître, *The Primeval Atom. An Essay on Cosmology* (transl. by Betty, H. and Serge A. Korff, Van Nostrand, New York, 1950, pp. 13–15.

[21] I. Kant, *Critique of Pure Reason* (transl. by Norman Kemp Smith), Macmillan, London, 1953, p. 396.

[22] G. J. Whitrow, *The Structure and Evolution of the Universe. An Introduction to Cosmology*, Harper & Bros, New York, 1959, pp. 103–104. Bergson could have used this reinstatement of absolute time as an argument for his temporalistic interpretation of space-time in his *Duration and Simultaneity*, but his lack of interest in general relativity prevented him from doing so. Cf. my article 'Bergson's Theory of Matter and Modern Physics', in *Bergson and the Evolution of Physics*, (ed. and transl. by P. A. Y. Gunter), Univ. of Tennessee Press, Knoxville, 1969, p. 315 note.

[23] Lemaître, *op. cit.*, pp. 17f.

[24] Lemaître, *ibid.* pp. 18–19.

[25] *C.E.*, p. 223.

[26] A. Calinon, 'Les espaces géométriques', *Revue philosophique* **27** (1889) 588–595.

[27] See note 22.

[28] *C.E.*, p. 271.

[29] Berthelot ridicules the 'pyrotechnics' of Bergson's metaphors in *Creative Evolution* (*op. cit.*, II, p. 63). Cf. Russell's mockeries in his article 'The Philosophy of Bergson', *The Monist* **22** (1912) 333.

[30] E. Mach, *Die Principien der Wärmelehre*, 4th Ed., Leipzig 1923, p. 364.

[31] R. Berthelot, *Un romantisme utilitaire*, II, p. 244.

[32] L. Brunschwicg, *La physique du XXe siècle et la philosophie*, Paris 1936; H. Driesch, *Relativitätstheorie und Weltanschauung*, Leipzig 1930; B. Blanshard, *The Nature of Thought*, Allen & Unwin, London, 1939, pp. 392–393. In the meeting of Société française de philosophie, 17 November, 1934 René Berthelot insisted that the impossibility to establish absolute simultaneity of distant events is merely technical and provisional, not intrinsic; the microphysical indeterminacy is, according to him, of the same kind. (*Bulletin de la Société française de philosophie*, 34e année No. 5 (1934) 172–183.)

[33] Sir James Jeans, *The Universe Around Us*, 4th ed., The Cambridge University Press, 1944, p. 278.

[34] *Op. cit.*, p. 281.

[35] Jacques Merleau Ponty, *Cosmologie du XXe siècle. Etude épistemologique et historique des théories de la cosmologie contemporaine*, Paris 1965, p. 353.

[36] G. Gamow, 'Modern Cosmology', in *Scientific American* **190** (1954) 63.

[37] Ch. Renouvier, *Les dilemmes de la métaphysique pure*, Paris 1927, p. 106.

[38] Kant, *op. cit.*, p. 398.

[39] C. D. Broad, 'Kant's Mathematical Antinomies', *Proceedings of the Aristotelian Society* **55** (1954–55), 7.

[40] Louis Couturat, *De l'infini mathématique*, Paris 1896, p. 462.

[41] Couturat, *ibid.*, p. 463.

[42] A. Grünbaum, *Modern Science and Zeno's Paradoxes*, Wesleyan University Press, Middletown, 1967, pp. 124–5.

[43] B. Russell, *Our Knowledge of the External World*, Allen & Unwin, London, 1914, pp. 157, 179–180.

[44] Russell, *ibid.*, p. 180.

[45] M. Munitz, *The Mystery of Existence*, Appleton-Century-Croft, New York, 1965, p. 140.

[46] E. L. Mascall, *Christian Theology and Natural Science*, Longmans, Green & Co. London, 1956, pp. 149–153. Pius XII referred to E. T. Whittaker's book *Space and Spirit. Theories of the Universe and the Arguments for the Existence of God*, H. Regnery Co., Hinsdale, 1948.

[47] Ronald Hepburn, 'Creation', in *The Encyclopedia of Philosophy*, Macmillan, New York, 1967, p. 255.

[48] E. A. Milne, *Modern Cosmology and the Christian Idea of God*, Clarendon Press, Oxford, 1952, pp. 76–77. On Milne, cf. R. S. Cohen, 'E. A. Milne's Theory of Relativity, A Critical Study', *The Review of Metaphysics* **3** (1949–50) 385–405.

[49] Quoted by Merleau Ponty, *op. cit.*, p. 333; 344.

[50] H. Spencer, *First Principles*, 4th ed., Appleton, New York, 1896, p. 33. On Dühring, cf. H. Bois, 'Le finitisme de Dühring', *L'année philosophique XX* (1910) 93–124.

[51] H. Bergson, *Écrits et Paroles*, II, p. 365. On the other hand, in his previous letter to the same person May 12, 1908 Bergson stressed that his criticism of the concept of nothingness does not mean that the universe is eternal (*ibid.*, p. 296). Since, however, even Chevalier conceded (*loc. cit.*, p. 5) that Bergson in 1901 did not express himself dogmatically about the problem of the origin of the universe, the most appropriate characterization of his views would be as "cautious leaning toward finitism".

[52] *Les deux sources de la morale et de la religion*, Paris 1932, Ch. II, pp. 116–122.

[53] *C.E.*, p. 295.

[54] *Mysticism and Logic*, Longmans, Green & Co., 1919, p. 47.

[55] *C.E.*, p. 268.

[56] The term 'ectropism' was coined by the physicist Felix Auerbach in his book *Ektropism oder die physikalische Theorie des Lebens*, Jena 1910.

[57] *C.E.*, p. 279.

[58] *C.E.*, p. 269, note.

# ADDITIONAL SELECTED BIBLIOGRAPHY

## PART I

Biological approach to epistemology is also present in the following writings: W. James, *The Principles of Psychology*, vol. II, Ch. XXVIII 'Necessary Truths and the Effects of Experience'; A. Fouillé, 'Les origines de notre structure intellectuelle et cérebrale', *Revue philosophique* **XXXII** (1892) 433–66, 570–602; Georg Simmel, 'Über eine Beziehung der Selectionslehre zur Erkenntnistheorie', *Archiv für systematische Philosophie* **L** (1895) 34–45; O. Wiener, *Die Erweiterung unserer Sinne*, Leipzig 1900; J. M. Baldwin, *Darwin and Humanities*, Allen and Unwin, London 1910, esp. Ch. IV 'Darwinism and Logic'; Henri Piéron, *La sensation, guide de vie*, Paris 1945; Louis Rougier, *Traité de la connaissance*, Paris 1955, esp. Ch. XXV 'La nouvelle théorie de la connaissance'; Donald T. Campbell, 'Evolutionary Epistemology', to be published in *The Philosophy of Karl R. Popper* in The Library of Living Philosophers (ed. by Paul A. Schilpp), The Open Court Publishing Company, La Salle, Illinois. This article contains a very extensive bibliography.

## PART II

The bibliography of the books and articles dealing with Bergson's thought is enormous; that contained in Rose-Marie Mossé-Bastide's *Bergson éducateur* (Paris 1955) extends until 1952; that given by G. Mourélos, *Bergson et les niveaux de réalité* (Paris 1964) until 1961. Among the older books the significance of A. Bazaillas' *La vie personnelle. Etude sur quelques illusions de la perception intérieure* (Paris 1904) is still recognized, par example by V. Jankélevitch, *Henri Bergson* (Paris 1959) p. 40; for it disposed of the still current myth of 'amorphous continuity' of Bergsonian *durée*. This alleged amorphousness was the target of both Lovejoy and Ushenko and in France of Gaston Bachelard in his *La dialectique de la durée* (Paris 1936). Another older book significant within the context of this study is Frank Grandjean, *La raison et la vue* (Paris 1920) dealing even more systematically than Bergson with the distorting influence of visual and spatial imagery. Léon Husson's *L'intellectualisme de Bergson* (Paris 1947) is even by its own title opposed to the persistent misrepresentation of Bergson as an irrationalist. Two more recent studies on Bergson's style are: E. Bréhier, 'Images plotiniennes, images bergsoniennes', in *Les Etudes bergsoniennes* **II** (1949) 107–28; and L. Adolphe, *La dialectique des images chez Bergson*, Paris 1951. On the relation of Bergson to Whitehead, cf. P. Devaux, 'Le bergsonisme de Whitehead', in *Revue internationale de philosophie* **56–57** (1961) 217–36; F. Cesselin, *La philosophie organique de Whitehead* (Paris 1950). Concerning the perception of time and epistemological problems involved in it, *The Problem of Time* (University of California Publications in Philosophy, vol. 18, 1935) remains valuable while P. Fraisse's *The Psychology of Time* (Eyre and Spottiswoode, Ltd., London, 1964) combines both a psychological and philosophical approach and contains a very extensive bibliography. Cf. also *The Voices of Time* (ed. by J. T. Fraser) Brazillach, New York 1966, Part I and II, in particular the articles of C. Benjamin, F. Kümmel, W. Dürr, J. Piaget, M.-L. Franz and J. Cohen. On the difficult problem of the status of the past see P. Weiss, 'The Past: its Nature and Reality', *The Review of Metaphysics*, V (1952–3) 507–22; Ch. Hartshorne, 'The Immortality of the Past: Critique of Prevalent Misconceptions' (the *Review of Metaphysics* **VII** (1954–5) 92–112) and Weiss' rejoinder, 'The Past: Some recent Discussions', *ibid.*, pp. 299–306. Bergson's view that rigorous determinism eliminates time and change entirely was shared among American philosophers by F. J. E. Woodbridge in his *Nature and Mind* (Columbia University Press, 1937, p. 52).

PART III

The main reason why Bergson's philosophy of physics was either ignored or completely misunderstood was – besides the influence of classical physics – the fact that the development of his thought was not taken into account. The difference between the first phase in which the physical world was still regarded as timeless and the second phase when becoming was reinstated even in the realm of matter was simply ignored. Only thus could S. Radhakrishnan claim that "Bergson's account of matter is riddled with inconsistencies and contradictions" (*The Mind* **26** (1917) 329).

A similar view was upheld as late as in 1941 by Dominique Parodi ('La durée et la matière chez Bergson', *Revue de métaphysique et de morale* **48**). Only after the Second World War the studies dealing more attentively with this aspect of Bergson's thought began to appear: V. Mathieu, 'Il tempo ritrovato. Bergson e Einstein', *Filosofia* **IV** (1953) and 'Scienza e metafisica in Bergson', *Giornale di metafisica* **XIV** (1959); L. Adolphe, *L'univers bergsonien* (Paris 1955) and 'Bergson et la science d'aujour-d'hui', *Etudes philosophiques* **19** (1959); F. Heidsieck, *Henri Bergson et la notion d'espace* (Paris 1957) dealing with the topic which previously had been discussed only by Wilko Emmens (*Das Raumproblem bei H. Bergson*, Leiden 1931); finally, the papers and discussions at *Congrès Bergson* at Paris, 1959 by M. Ambacher, M. Čapek, O. Costa de Beau-regard, I. Dambska, A. Kremer-Marietti M. Matchinski, A. Mercier and A. Metz. The three studies dealing with Bergson's criticism of Einstein by G. Pflug, J. F. Busch and W. Berteval together with the lively dis-cussion between Bergson and A. Metz were recently translated and in-cluded in the book *Bergson and the Evolution of Physics* (ed. by P. A. Y. Gunter), University of Tennessee Press, Knoxville 1969. This collection contains also besides three essays referred to in this book (that of L. de Broglie, of R. Blanché and my own), also the translations of two articles of O. Costa de Beauregard and one of S. Watanabe as well as the dis-cussion by Vere C. Chapell and D. Sipfle of Bergson's treatment of Zeno's paradoxes.

Appendix II. Cf. Henry Margenau's recent article 'Quantum Mecha-nics, Free Will and Determinism', *Journal of Philosophy* **64** (1967) 714–25; and my article, 'The Main Difficulties of the Identity Theory', *Scientia* **104** (1969) 1–17.

Appendix III. It is worth mentioning that A. N. Whitehead came closest to the Bergsonian polarity of ectropism-entropy in the concluding pages of *The Function of Reason* (Princeton University Press, 1929). Cf. also Introductory Summary of the same book.

# EXTRACT FROM THE LETTER TO M. ČAPEK

... Il était impossible de mieux comprendre l'essentiel de mes vues sur la durée et la matière. En particulier, vous avez admirablement montré comment, dans quel sens et dans quelle mesure, la conception de la matière que j'ai de plus en plus precisée dans mes ouvrages successifs anticipait sur les conclusions de la physique d'aujourd'hui. Ce point n'avait guere été aperçu, pour la raison très simple que mes vues sur la question, emises à une époque ou l'on considérait comme évident que les élements ultimes de la matière doivent être conçus à l'image du tout, déroutèrent les lecteurs, et furent le plus souvent laissées de côté comme étant la partie incompréhensible de mon œuvre. Ils jugèrent d'ailleurs, probablement, que c'en était une partie accessoire. Aucun (sauf peut-être, dans une certaine mesure, le profond mathématicien et philosophe Whitehead) ne s'est pas aperçu comme vous qu'il y avait là pour moi quelque chose d'essentiel, qui se rattachait étroitement à la théorie de la durée, et qui était en même temps *dans la direction* où la physique s'engagerait tôt ou tard...

Laissez-moi vous adresser, Monsieur, avec mes compliments et mes remerciements, l'assurance de mes sentiments bien sympathiques.

H. BERGSON

# INDEX OF NAMES AND SUBJECTS

60–1; to Reichenbach, 65–7; to Piaget, 67–71; on the logic of solid bodies, 56, 69, 72–4, 336; his intuition, 59, 86–91; on the inadequacy of mechanistic models, 62, 273–4, on the fallacy of spatialization, 85, 129, 136–7, 251; criticizes psychological atomism, 92–7; links succession with novelty, 99–101, 104, 107–9, 111–12, 116–17; on the structure of psychological duration, 116–121, 124–31; denies durationless instants, 134–7, 139; his relation to Weyl, 140; rejects the atomistic theory of time, 142, 144–5; on the unity and multiplicity of duration, 147–150; his relation to Brouwer, 150, 180, 183–5; on the status of the past, 152–165; commented on by Royce, 165–7; by Ingarden, 167–72; by Croce, 172–3; rejects nominalism, 174–5; his philosophy of mathematics, 176–185; on duration in the physical world, 189–193, 195; on different temporal spans, 196, 198–206, 214–19, 223, 249–50, 289–90, 294, 360; on psychological extension, 208–12; on degrees of spatiality and their relation to differences in temporal span, 218–222; denies instantaneous space, 224–5, 234–5; his view on the relativity theory, 236–256; his agreement with Einstein, 236–8; his correct insights, 239–44; his errors and inconsistencies, 244–52; on matter as 'extensive becoming', 212, 231, 235, 255, 326; criticizes corpuscular-kinetic models, 257, 259–62, 268–71; on the substantiality of change, 273–76; ignores contemporary independence, 278–9, 282; evaluation by Louis de Broglie, 292, 294–6; upholds microphysical indeterminacy, 283–291, 299–301; similarity to Whitehead's and Bohm's views, 303–12; on primary and secondary qualities, 313–14; his use of auditory metaphors, 316–18, 326, 328–9; rejects Heraclitus, 329; his world contrasted with the world of Laplace, 331–3; his relation to Russell, 335–345; on the relation of microphysical indeterminacy to life and freedom,

347–8, 350–1, 356, 358–65; his thoughts on entropy and cosmogony, 368–75, 379–82; his finitism, 373–5, 385, 392; on the future of life, 392–3
Bergsonism, literary, IX–X, 58
Berigard, Claude 93
Berkeley, G. 84
Bernoulli, J. 387–8
Berthelot, Marcellin, 9, 11, 14
Berthelot, René, criticizes Poincaré, 22, 26, 28; criticizes Bergson, IX, XI–XII, 180–1, 185, 192, 194, 202, 204–5, 237, 273, 277, 279, 283, 287–8, 291, 373; on the difference between Nietzsche and Bergson, 80; his negative attitude toward modern physics, 231, 382, 396
Bessel, F. W. XI
Beth, E. W. 163, 183, 186
Bifurcation of nature, 311, 359
Birtwistle, G. 267
Black, M. 186
Blanché, R. 179, 185, 327
Blanshard, B. 217, 299, 301, 382, 396
Boethius 385
Bohm, D. XII, 175, 246, 249, 255–6, 297–8, 301, 308–9, 312, 328, 330, 354, 364
Bohr, N. 266, 268, 296, 346, 365
Bois, H. 397
Boltzmann, L. 266, 371–3, 382, 395–6
Borel, E. 350, 384
Born, M. 240, 242, 255, 296
Boscovich, R. 51
Boutroux, E., his contingentism, 25–5, 113–14, 117, 299, 356; anticipates microphysical indeterminacy, 286–7, 290; his relation to Bergson, XII, 291; opposed to bifurcation of nature, 310–11
Bouvier, R. 28
Bradley, F. H. 111, 159
Bridgman, P. W. 58, 63, 276–7
Broad, C. D., on the reality of the past, 154, 161; on Kant's first antinomy, 386, 389, 391
Broglie, de, Louis X, XII, 49, 54, 328, 383; discovers the undulatory nature of matter, 47, 267; on Bergson's anticipations, 288, 291–6; returns to de-

# SYNTHESE LIBRARY

Monographs on Epistemology, Logic, Methodology,
Philosophy of Science, Sociology of Science and of Knowledge, and the
Mathematical Methods of Social and Behavioral Sciences

*Editors:*

DONALD DAVIDSON (Rockefeller University and Princeton University)
JAAKKO HINTIKKA (Academy of Finland and Stanford University)
GABRIEL NUCHELMANS (University of Leyden)
WESLEY C. SALMON (Indiana University)

‡JAAKKO HINTIKKA, *Models for Modalities. Selected Essays.* 1969, IX + 220 pp.
Dfl. 34,—

‡D. DAVIDSON and J. HINTIKKA (eds.), *Words and Objections: Essays on the Work of W. V. Quine.* 1969, VIII + 366 pp. Dfl. 48,—

‡J. W. DAVIS, D. J. HOCKNEY and W. K. WILSON (eds.), *Philosophical Logic.* 1969, VIII + 277 pp. Dfl. 45,—

‡ROBERT S. COHEN and MARX W. WARTOFSKY (eds.), *Boston Studies in the Philosophy of Science.* Volume V: *Proceedings of the Boston Colloquium for the Philosophy of Science 1966/1968.* 1969, VIII + 482 pp. Dfl. 58,—

‡ROBERT S. COHEN and MARX W. WARTOFSKY (eds.), *Boston Studies in the Philosophy of Science*, Volume IV: *Proceedings of the Boston Colloquium for the Philosophy of Science 1966/1968.* 1969, VIII + 537 pp. Dfl. 69,—

‡NICHOLAS RESCHER, *Topics in Philosophical Logic.* 1968, XIV + 347 pp. Dfl. 62,—

‡GÜNTHER PATZIG, *Aristotle's Theory of the Syllogism. A Logical-Philological Study of Book A of the Prior Analytics.* 1968, XVII + 215 pp. Dfl. 45,—

‡C. D. BROAD, *Induction, Probability, and Causation. Selected Papers.* 1968, XI + 296 pp.
Dfl. 48,—

‡ROBERT S. COHEN and MARX W. WARTOFSKY (eds.), *Boston Studies in the Philosophy of Science.* Volume III: *Proceedings of the Boston Colloquium for the Philosophy of Science 1964/1966.* 1967, XLIX + 489 pp. Dfl. 65,—

‡GUIDO KÜNG, *Ontology and the Logistic Analysis of Language. An Enquiry into the Contemporary Views on Universals.* 1967, XI + 210 pp. Dfl. 38,—

*EVERT W. BETH and JEAN PIAGET, *Mathematical Epistemology and Psychology.* 1966, XXII + 326 pp. Dfl. 58,—

*EVERT W. BETH, *Mathematical Thought. An Introduction to the Philosophy of Mathematics.* 1965, XII + 208 pp. Dfl. 32,—

‡PAUL LORENZEN, *Formal Logic.* 1965, VIII + 123 pp. Dfl. 22,—

‡GEORGES GURVITCH, *The Spectrum of Social Time.* 1964, XXVI + 152 pp. Dfl. 20,—

‡A. A. ZINOV'EV, *Philosophical Problems of Many-Valued Logic.* 1963, XIV + 155 pp.
Dfl. 28,—

‡MARX W. WARTOFSKY (ed.), *Boston Studies in the Philosophy of Science.* Volume I: *Proceedings of the Boston Colloquium for the Philosophy of Science, 1961–1962.* 1963, VIII + 212 pp. Dfl. 22,50

‡B. H. KAZEMIER and D. VUYSJE (eds.), *Logic and Language. Studies dedicated to Professor Rudolf Carnap on the Occasion of his Seventieth Birthday.* 1962, VI + 246 pp. Dfl. 32,50

*EVERT W. BETH, *Formal Methods. An Introduction to Symbolic Logic and to the Study of Effective Operations in Arithmetic and Logic.* 1962, XIV + 170 pp. Dfl. 30,—

*HANS FREUDENTHAL (ed.), *The Concept and the Role of the Model in Mathematics and Natural and Social Sciences. Proceedings of a Colloquium held at Utrecht, The Netherlands, January 1960.* 1961, VI + 194 pp. Dfl. 30,—

‡P. L. R. GUIRAUD, *Problèmes et méthodes de la statistique linguistique.* 1960, VI + 146 pp. Dfl. 22,50

*J. M. BOCHEŃSKI, *A Precis of Mathematical Logic.* 1959, X + 100 pp. Dfl. 20,—

# SYNTHESE HISTORICAL LIBRARY

Texts and Studies
in the History of Logic and Philosophy

*Editors:*

N. KRETZMANN (Cornell University)
G. NUCHELMANS (University of Leyden)
L. M. DE RIJK (University of Leyden)

‡KARL WOLF and PAUL WEINGARTNER (eds.), *Ernst Mally: Logische Schriften.* 1971,
X + 340 pp.                                                                    Dfl. 80,—

‡LEROY E. LOEMKER (ed.), *Gottfried Wilhelm Leibnitz: Philosophical Papers and Letters.*
A Selection Translated and Edited, with an Introduction. 1969, XII + 736 pp.
                                                                              Dfl. 125,—

‡M. T. BEONIO-BROCCHIERI FUMAGALLI, *The Logic of Abelard.* Translated from the
Italian. 1969, IX + 101 pp.                                                    Dfl. 25,—

Sole Distributors in the U.S.A. and Canada:

*GORDON & BREACH, INC., 150 Fifth Avenue, New York, N.Y. 10011
‡HUMANITIES PRESS, INC., 303 Park Avenue South, New York, N.Y. 10010